HOLT SCIENCE & TECHNOLOGY

Introduction to Matter

HOLT, RINEHART AND WINSTON

A Harcourt Classroom Education Company

Austin • New York • Orlando • Atlanta • San Francisco • Boston • Dallas • Toronto • London

Acknowledgments

Chapter Writers

Christie Borgford, Ph.D.
Professor of Chemistry
University of Alabama
Birmingham, Alabama

Andrew Champagne
Former Physics Teacher
Ashland High School
Ashland, Massachusetts

Mapi Cuevas, Ph.D.
Professor of Chemistry
Santa Fe Community College
Gainesville, Florida

Leila Dumas
Former Physics Teacher
LBJ Science Academy
Austin, Texas

William G. Lamb, Ph.D.
Science Teacher and Dept. Chair
Oregon Episcopal School
Portland, Oregon

Sally Ann Vonderbrink, Ph.D.
Chemistry Teacher
St. Xavier High School
Cincinnati, Ohio

Lab Writers

Phillip G. Bunce
Former Physics Teacher
Bowie High School
Austin, Texas

Kenneth E. Creese
Science Teacher
White Mountain Junior High
School
Rock Springs, Wyoming

William G. Lamb, Ph.D.
Science Teacher and Dept. Chair
Oregon Episcopal School
Portland, Oregon

Alyson Mike
Science Teacher
East Valley Middle School
East Helena, Montana

Joseph W. Price
Science Teacher and Dept. Chair
H. M. Browne Junior High
School
Washington, D.C.

Denice Lee Sandefur
Science Teacher and Dept. Chair
Nucla High School
Nucla, Colorado

John Spadafino
Mathematics and Physics Teacher
Hackensack High School
Hackensack, New Jersey

Walter Woolbaugh
Science Teacher
Manhattan Junior High School
Manhattan, Montana

Academic Reviewers

Paul R. Berman, Ph.D.
Professor of Physics
University of Michigan
Ann Arbor, Michigan

Russell M. Brengelman, Ph.D.
Professor of Physics
Morehead State University
Morehead, Kentucky

John A. Brockhaus, Ph.D.
*Director, Mapping, Charting and
Geodesy Program*
Department of Geography and
Environmental Engineering
United States Military Academy
West Point, New York

Walter Bron, Ph.D.
Professor of Physics
University of California
Irvine, California

Andrew J. Davis, Ph.D.
Manager, ACE Science Center
Department of Physics
California Institute of
Technology
Pasadena, California

Peter E. Demmin, Ed.D.
*Former Science Teacher and
Department Chair*
Amherst Central High School
Amherst, New York

Roger Falcone, Ph.D.
*Professor of Physics and
Department Chair*
University of California
Berkeley, California

Cassandra A. Fraser, Ph.D.
Assistant Professor of Chemistry
University of Virginia
Charlottesville, Virginia

L. John Gagliardi, Ph.D.
*Associate Professor of Physics and
Department Chair*
Rutgers University
Camden, New Jersey

Gabriele F. Giuliani, Ph.D.
Professor of Physics
Purdue University
West Lafayette, Indiana

Roy W. Hann, Jr., Ph.D.
Professor of Civil Engineering
Texas A&M University
College Station, Texas

John L. Hubisz, Ph.D.
Professor of Physics
North Carolina State University
Raleigh, North Carolina

Samuel P. Kounaves, Ph.D.
Professor of Chemistry
Tufts University
Medford, Massachusetts

Karol Lang, Ph.D.
Associate Professor of Physics
The University of Texas
Austin, Texas

Gloria Langer, Ph.D.
Professor of Physics
University of Colorado
Boulder, Colorado

Phillip LaRoe
Professor
Helena College of Technology
Helena, Montana

Joseph A. McClure, Ph.D.
Associate Professor of Physics
Georgetown University
Washington, D.C.

LaMoine L. Motz, Ph.D.
Coordinator of Science Education
Department of Learning
Services
Oakland County Schools
Waterford, Michigan

R. Thomas Myers, Ph.D.
Professor of Chemistry, Emeritus
Kent State University
Kent, Ohio

Hillary Clement Olson, Ph.D.
Research Associate
Institute for Geophysics
The University of Texas
Austin, Texas

David P. Richardson, Ph.D.
Professor of Chemistry
Thompson Chemical
Laboratory
Williams College
Williamstown, Massachusetts

John Rigden, Ph.D.
Director of Special Projects
American Institute of Physics
Colchester, Vermont

Acknowledgments (cont.)

Peter Sheridan, Ph.D.
Professor of Chemistry
Colgate University
Hamilton, New York

Vederaman Sriraman, Ph.D.
Associate Professor of Technology
Southwest Texas State University
San Marcos, Texas

Jack B. Swift, Ph.D.
Professor of Physics
The University of Texas
Austin, Texas

Atiq Syed, Ph.D.
Master Instructor of Mathematics and Science
Texas State Technical College
Harlingen, Texas

Leonard Taylor, Ph.D.
Professor Emeritus
Department of Electrical Engineering
University of Maryland
College Park, Maryland

Virginia L. Trimble, Ph.D.
Professor of Physics and Astronomy
University of California
Irvine, California

Martin VanDyke, Ph.D.
Professor of Chemistry, Emeritus
Front Range Community College
Westminster, Colorado

Gabriela Waschewsky, Ph.D.
Science and Math Teacher
Emery High School
Emeryville, California

Safety Reviewer

Jack A. Gerlovich, Ph.D.
Associate Professor
School of Education
Drake University
Des Moines, Iowa

Teacher Reviewers

Barry L. Bishop
Science Teacher and Dept. Chair
San Rafael Junior High School
Ferron, Utah

Paul Boyle
Science Teacher
Perry Heights Middle School
Evansville, Indiana

Kenneth Creese
Science Teacher
White Mountain Junior High School
Rock Springs, Wyoming

Vicky Farland
Science Teacher and Dept. Chair
Centennial Middle School
Yuma, Arizona

Rebecca Ferguson
Science Teacher
North Ridge Middle School
North Richland Hills, Texas

Laura Fleet
Science Teacher
Alice B. Landrum Middle School
Ponte Vedra Beach, Florida

Jennifer Ford
Science Teacher and Dept. Chair
North Ridge Middle School
North Richland Hills, Texas

Susan Gorman
Science Teacher
North Ridge Middle School
North Richland Hills, Texas

C. John Graves
Science Teacher
Monforton Middle School
Bozeman, Montana

Dennis Hanson
Science Teacher and Dept. Chair
Big Bear Middle School
Big Bear Lake, California

David A. Harris
Science Teacher and Dept. Chair
The Thacher School
Ojai, California

Norman E. Holcomb
Science Teacher
Marion Local Schools
Maria Stein, Ohio

Kenneth J. Horn
Science Teacher and Dept. Chair
Fallston Middle School
Fallston, Maryland

Tracy Jahn
Science Teacher
Berkshire Junior-Senior High School
Canaan, New York

Kerry A. Johnson
Science Teacher
Isbell Middle School
Santa Paula, California

Drew E. Kirian
Science Teacher
Solon Middle School
Solon, Ohio

Harriet Knops
Science Teacher and Dept. Chair
Rolling Hills Middle School
El Dorado, California

Scott Mandel, Ph.D.
Director and Educational Consultant
Teachers Helping Teachers
Los Angeles, California

Thomas Manerchia
Former Science Teacher
Archmere Academy
Claymont, Delaware

Edith McAlanis
Science Teacher and Dept. Chair
Socorro Middle School
El Paso, Texas

Kevin McCurdy, Ph.D.
Science Teacher
Elmwood Junior High School
Rogers, Arkansas

Alyson Mike
Science Teacher
East Valley Middle School
East Helena, Montana

Donna Norwood
Science Teacher and Dept. Chair
Monroe Middle School
Charlotte, North Carolina

Joseph W. Price
Science Teacher and Dept. Chair
H. M. Browne Junior High School
Washington, D.C.

Terry J. Rakes
Science Teacher
Elmwood Junior High School
Rogers, Arkansas

Beth Richards
Science Teacher
North Middle School
Crystal Lake, Illinois

Elizabeth J. Rustad
Science Teacher
Crane Middle School
Yuma, Arizona

Rodney A. Sandefur
Science Teacher
Naturita Middle School
Naturita, Colorado

Helen Schiller
Science Teacher
Northwood Middle School
Taylors, South Carolina

Bert J. Sherwood
Science Teacher
Socorro Middle School
El Paso, Texas

Patricia McFarlane Soto
Science Teacher and Dept. Chair
G. W. Carver Middle School
Miami, Florida

David M. Sparks
Science Teacher
Redwater Junior High School
Redwater, Texas

Larry Tackett
Science Teacher and Dept. Chair
Andrew Jackson Middle School
Cross Lanes, West Virginia

Elsie N. Waynes
Science Teacher and Dept. Chair
R. H. Terrell Junior High School
Washington, D.C.

Sharon L. Woolf
Science Teacher
Langston Hughes Middle School
Reston, Virginia

Alexis S. Wright
Middle School Science Coordinator
Rye Country Day School
Rye, New York

Lee Yassinski
Science Teacher
Sun Valley Middle School
Sun Valley, California

John Zambo
Science Teacher
Elizabeth Ustach Middle School
Modesto, California

K Introduction to Matter

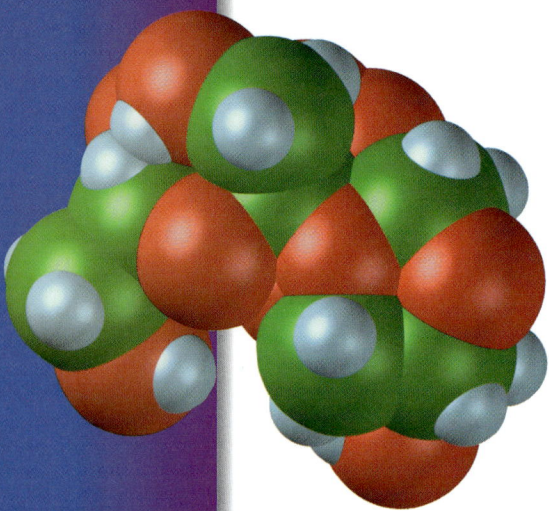

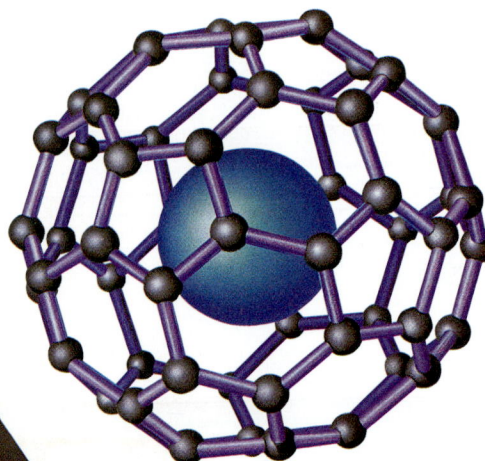

Skills Development

Process Skills

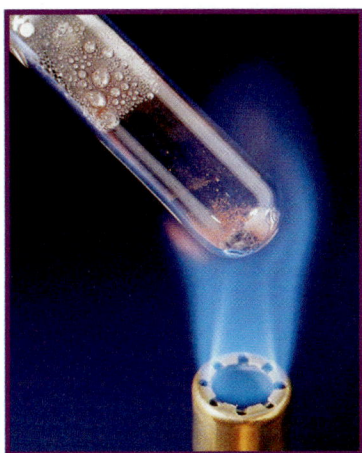

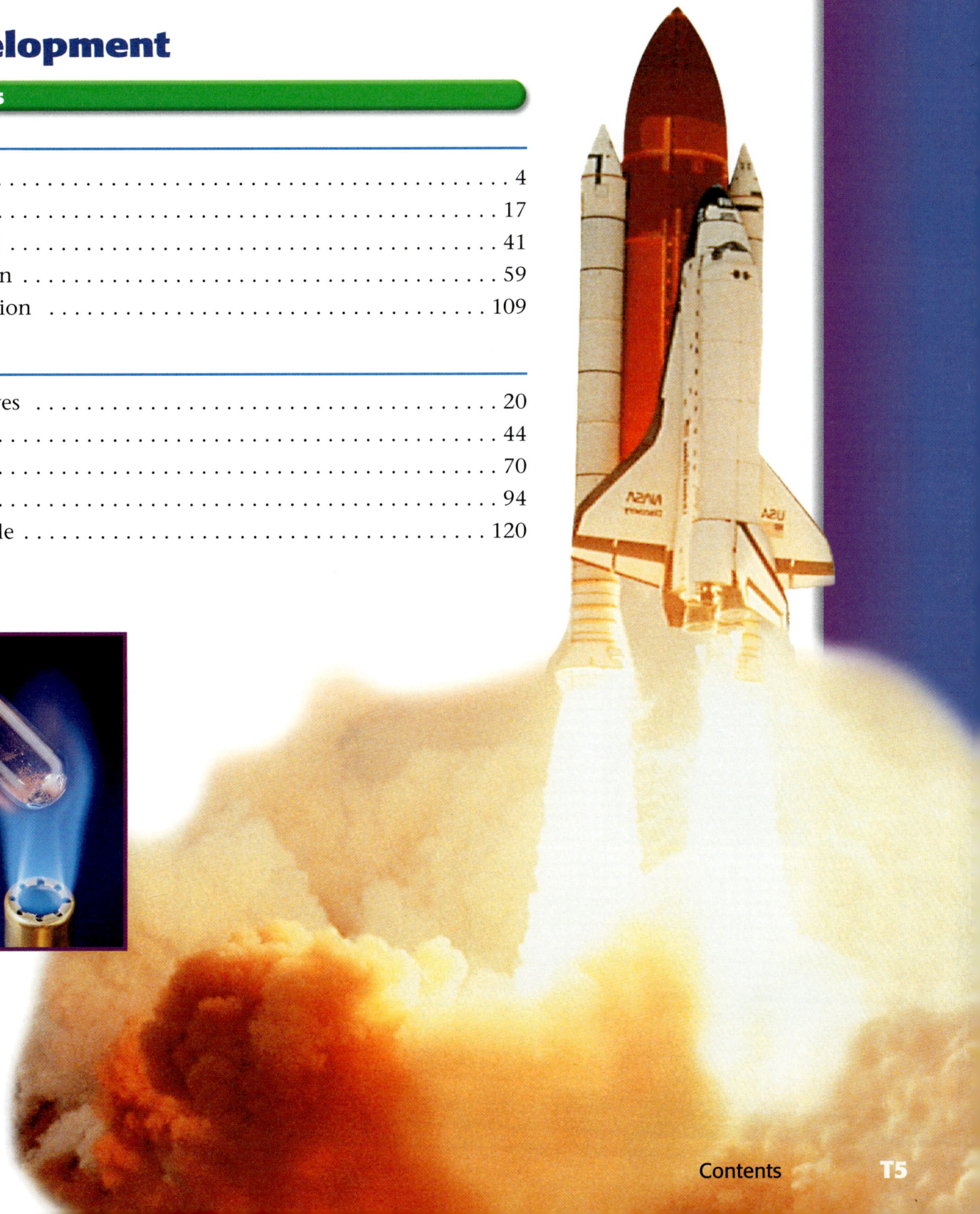

Skills Development *(continued)*

Research and Critical Thinking Skills

Program Scope and Sequence

Selecting the right books for your course is easy. Just review the topics presented in each book to determine the best match to your district curriculum.

A MICROORGANISMS, FUNGI, AND PLANTS

B ANIMALS

CHAPTER 1

A — It's Alive!! Or, Is It?
- ❑ Characteristics of living things
- ❑ Homeostasis
- ❑ Heredity and DNA
- ❑ Producers, consumers, and decomposers
- ❑ Biomolecules

B — Animals and Behavior
- ❑ Characteristics of animals
- ❑ Classification of animals
- ❑ Animal behavior
- ❑ Hibernation and estivation
- ❑ The biological clock
- ❑ Animal communication
- ❑ Living in groups

CHAPTER 2

A — Bacteria and Viruses
- ❑ Binary fission
- ❑ Characteristics of bacteria
- ❑ Nitrogen-fixing bacteria
- ❑ Antibiotics
- ❑ Pathogenic bacteria
- ❑ Characteristics of viruses
- ❑ Lytic cycle

B — Invertebrates
- ❑ General characteristics of invertebrates
- ❑ Types of symmetry
- ❑ Characteristics of sponges, cnidarians, arthropods, and echinoderms
- ❑ Flatworms versus roundworms
- ❑ Types of circulatory systems

CHAPTER 3

A — Protists and Fungi
- ❑ Characteristics of protists
- ❑ Types of algae
- ❑ Types of protozoa
- ❑ Protist reproduction
- ❑ Characteristics of fungi and lichens

B — Fishes, Amphibians, and Reptiles
- ❑ Characteristics of vertebrates
- ❑ Structure and kinds of fishes
- ❑ Development of lungs
- ❑ Structure and kinds of amphibians and reptiles
- ❑ Function of the amniotic egg

CHAPTER 4

A — Introduction to Plants
- ❑ Characteristics of plants and seeds
- ❑ Reproduction and classification
- ❑ Angiosperms versus gymnosperms
- ❑ Monocots versus dicots
- ❑ Structure and functions of roots, stems, leaves, and flowers

B — Birds and Mammals
- ❑ Structure and kinds of birds
- ❑ Types of feathers
- ❑ Adaptations for flight
- ❑ Structure and kinds of mammals
- ❑ Function of the placenta

CHAPTER 5

A — Plant Processes
- ❑ Pollination and fertilization
- ❑ Dormancy
- ❑ Photosynthesis
- ❑ Plant tropisms
- ❑ Seasonal responses of plants

CHAPTER 6

CHAPTER 7

Life Science

C — CELLS, HEREDITY, & CLASSIFICATION

Cells: The Basic Units of Life
- ❏ Cells, tissues, and organs
- ❏ Populations, communities, and ecosystems
- ❏ Cell theory
- ❏ Surface-to-volume ratio
- ❏ Prokaryotic versus eukaryotic cells
- ❏ Cell organelles

The Cell in Action
- ❏ Diffusion and osmosis
- ❏ Passive versus active transport
- ❏ Endocytosis versus exocytosis
- ❏ Photosynthesis
- ❏ Cellular respiration and fermentation
- ❏ Cell cycle

Heredity
- ❏ Dominant versus recessive traits
- ❏ Genes and alleles
- ❏ Genotype, phenotype, the Punnett square and probability
- ❏ Meiosis
- ❏ Determination of sex

Genes and Gene Technology
- ❏ Structure of DNA
- ❏ Protein synthesis
- ❏ Mutations
- ❏ Heredity disorders and genetic counseling

The Evolution of Living Things
- ❏ Adaptations and species
- ❏ Evidence for evolution
- ❏ Darwin's work and natural selection
- ❏ Formation of new species

The History of Life on Earth
- ❏ Geologic time scale and extinctions
- ❏ Plate tectonics
- ❏ Human evolution

Classification
- ❏ Levels of classification
- ❏ Cladistic diagrams
- ❏ Dichotomous keys
- ❏ Characteristics of the six kingdoms

D — HUMAN BODY SYSTEMS & HEALTH

Body Organization and Structure
- ❏ Homeostasis
- ❏ Types of tissue
- ❏ Organ systems
- ❏ Structure and function of the skeletal system, muscular system, and integumentary system

Circulation and Respiration
- ❏ Structure and function of the cardiovascular system, lymphatic system, and respiratory system
- ❏ Respiratory disorders

The Digestive and Urinary Systems
- ❏ Structure and function of the digestive system
- ❏ Structure and function of the urinary system

Communication and Control
- ❏ Structure and function of the nervous system and endocrine system
- ❏ The senses
- ❏ Structure and function of the eye and ear

Reproduction and Development
- ❏ Asexual versus sexual reproduction
- ❏ Internal versus external fertilization
- ❏ Structure and function of the human male and female reproductive systems
- ❏ Fertilization, placental development, and embryo growth
- ❏ Stages of human life

Body Defenses and Disease
- ❏ Types of diseases
- ❏ Vaccines and immunity
- ❏ Structure and function of the immune system
- ❏ Autoimmune diseases, cancer, and AIDS

Staying Healthy
- ❏ Nutrition and reading food labels
- ❏ Alcohol and drug effects on the body
- ❏ Hygiene, exercise, and first aid

E — ENVIRONMENTAL SCIENCE

Interactions of Living Things
- ❏ Biotic versus abiotic parts of the environment
- ❏ Producers, consumers, and decomposers
- ❏ Food chains and food webs
- ❏ Factors limiting population growth
- ❏ Predator-prey relationships
- ❏ Symbiosis and coevolution

Cycles in Nature
- ❏ Water cycle
- ❏ Carbon cycle
- ❏ Nitrogen cycle
- ❏ Ecological succession

The Earth's Ecosystems
- ❏ Kinds of land and water biomes
- ❏ Marine ecosystems
- ❏ Freshwater ecosystems

Environmental Problems and Solutions
- ❏ Types of pollutants
- ❏ Types of resources
- ❏ Conservation practices
- ❏ Species protection

Energy Resources
- ❏ Types of resources
- ❏ Energy resources and pollution
- ❏ Alternative energy resources

Scope and Sequence (continued)

	F INSIDE THE RESTLESS EARTH	**G** EARTH'S CHANGING SURFACE
CHAPTER 1	**Minerals of the Earth's Crust** ❏ Mineral composition and structure ❏ Types of minerals ❏ Mineral identification ❏ Mineral formation and mining	**Maps as Models of the Earth** ❏ Structure of a map ❏ Cardinal directions ❏ Latitude, longitude, and the equator ❏ Magnetic declination and true north ❏ Types of projections ❏ Aerial photographs ❏ Remote sensing ❏ Topographic maps
CHAPTER 2	**Rocks: Mineral Mixtures** ❏ Rock cycle and types of rocks ❏ Rock classification ❏ Characteristics of igneous, sedimentary, and metamorphic rocks	**Weathering and Soil Formation** ❏ Types of weathering ❏ Factors affecting the rate of weathering ❏ Composition of soil ❏ Soil conservation and erosion prevention
CHAPTER 3	**The Rock and Fossil Record** ❏ Uniformitarianism versus catastrophism ❏ Superposition ❏ The geologic column and unconformities ❏ Absolute dating and radiometric dating ❏ Characteristics and types of fossils ❏ Geologic time scale	**Agents of Erosion and Deposition** ❏ Shoreline erosion and deposition ❏ Wind erosion and deposition ❏ Erosion and deposition by ice ❏ Gravity's effect on erosion and deposition
CHAPTER 4	**Plate Tectonics** ❏ Structure of the Earth ❏ Continental drifts and sea floor spreading ❏ Plate tectonics theory ❏ Types of boundaries ❏ Types of crust deformities	
CHAPTER 5	**Earthquakes** ❏ Seismology ❏ Features of earthquakes ❏ P and S waves ❏ Gap hypothesis ❏ Earthquake safety	
CHAPTER 6	**Volcanoes** ❏ Types of volcanoes and eruptions ❏ Types of lava and pyroclastic material ❏ Craters versus calderas ❏ Sites and conditions for volcano formation ❏ Predicting eruptions	

Earth Science

Scope and Sequence *(continued)*

	K INTRODUCTION TO MATTER	**L** INTERACTIONS OF MATTER
CHAPTER 1	**The Properties of Matter** ❏ Definition of matter ❏ Mass and weight ❏ Physical and chemical properties ❏ Physical and chemical change ❏ Density	**Chemical Bonding** ❏ Types of chemical bonds ❏ Valence electrons ❏ Ions versus molecules ❏ Crystal lattice
CHAPTER 2	**States of Matter** ❏ States of matter and their properties ❏ Boyle's and Charles's laws ❏ Changes of state	**Chemical Reactions** ❏ Writing chemical formulas and equations ❏ Law of conservation of mass ❏ Types of reactions ❏ Endothermic versus exothermic reactions ❏ Law of conservation of energy ❏ Activation energy ❏ Catalysts and inhibitors
CHAPTER 3	**Elements, Compounds, and Mixtures** ❏ Elements and compounds ❏ Metals, nonmetals, and metalloids (semiconductors) ❏ Properties of mixtures ❏ Properties of solutions, suspensions, and colloids	**Chemical Compounds** ❏ Ionic versus covalent compounds ❏ Acids, bases, and salts ❏ pH ❏ Organic compounds ❏ Biomolecules
CHAPTER 4	**Introduction to Atoms** ❏ Atomic theory ❏ Atomic model and structure ❏ Isotopes ❏ Atomic mass and mass number	**Atomic Energy** ❏ Properties of radioactive substances ❏ Types of decay ❏ Half-life ❏ Fission, fusion, and chain reactions
CHAPTER 5	**The Periodic Table** ❏ Structure of the periodic table ❏ Periodic law ❏ Properties of alkali metals, alkaline-earth metals, halogens, and noble gases	
CHAPTER 6		

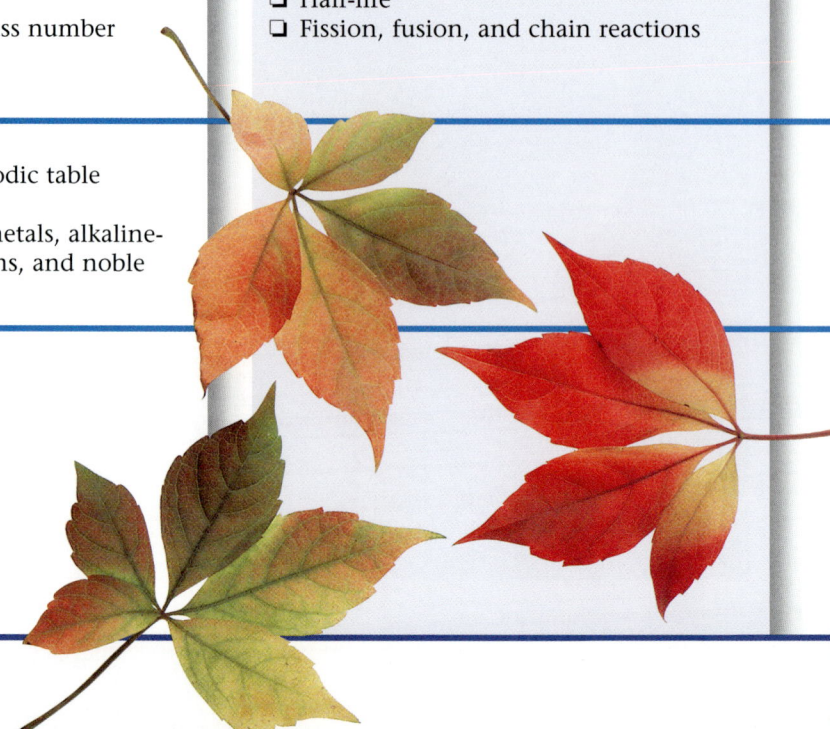

Physical Science

M — FORCES, MOTION, AND ENERGY

Matter in Motion
- ❏ Speed, velocity, and acceleration
- ❏ Measuring force
- ❏ Friction
- ❏ Mass versus weight

Forces in Motion
- ❏ Terminal velocity and free fall
- ❏ Projectile motion
- ❏ Inertia
- ❏ Momentum

Forces in Fluids
- ❏ Properties in fluids
- ❏ Atmospheric pressure
- ❏ Density
- ❏ Pascal's principle
- ❏ Buoyant force
- ❏ Archimedes' principle
- ❏ Bernoulli's principle

Work and Machines
- ❏ Measuring work
- ❏ Measuring power
- ❏ Types of machines
- ❏ Mechanical advantage
- ❏ Mechanical efficiency

Energy and Energy Resources
- ❏ Forms of energy
- ❏ Energy conversions
- ❏ Law of conservation of energy
- ❏ Energy resources

Heat and Heat Technology
- ❏ Heat versus temperature
- ❏ Thermal expansion
- ❏ Absolute zero
- ❏ Conduction, convection, radiation
- ❏ Conductors versus insulators
- ❏ Specific heat capacity
- ❏ Changes of state
- ❏ Heat engines
- ❏ Thermal pollution

N — ELECTRICITY AND MAGNETISM

Introduction to Electricity
- ❏ Law of electric charges
- ❏ Conduction versus induction
- ❏ Static electricity
- ❏ Potential difference
- ❏ Cells, batteries, and photocells
- ❏ Thermocouples
- ❏ Voltage, current, and resistance
- ❏ Electric power
- ❏ Types of circuits

Electromagnetism
- ❏ Properties of magnets
- ❏ Magnetic force
- ❏ Electromagnetism
- ❏ Solenoids and electric motors
- ❏ Electromagnetic induction
- ❏ Generators and transformers

Electronic Technology
- ❏ Properties of semiconductors
- ❏ Integrated circuits
- ❏ Diodes and transistors
- ❏ Analog versus digital signals
- ❏ Microprocessors
- ❏ Features of computers

O — SOUND AND LIGHT

The Energy of Waves
- ❏ Properties of waves
- ❏ Types of waves
- ❏ Reflection and refraction
- ❏ Diffraction and interference
- ❏ Standing waves and resonance

The Nature of Sound
- ❏ Properties of sound waves
- ❏ Structure of the human ear
- ❏ Pitch and the Doppler effect
- ❏ Infrasonic versus ultrasonic sound
- ❏ Sound reflection and echolocation
- ❏ Sound barrier
- ❏ Interference, resonance, diffraction, and standing waves
- ❏ Sound quality of instruments

The Nature of Light
- ❏ Electromagnetic waves
- ❏ Electromagnetic spectrum
- ❏ Law of reflection
- ❏ Absorption and scattering
- ❏ Reflection and refraction
- ❏ Diffraction and interference

Light and Our World
- ❏ Luminosity
- ❏ Types of lighting
- ❏ Types of mirrors and lenses
- ❏ Focal point
- ❏ Structure of the human eye
- ❏ Lasers and holograms

HOLT SCIENCE & TECHNOLOGY

Components Listing

Effective planning starts with all the resources you need in an easy-to-use package for each short course.

Directed Reading Worksheets Help students develop and practice fundamental reading comprehension skills and provide a comprehensive review tool for students to use when studying for an exam.

Study Guide Vocabulary & Notes Worksheets and Chapter Review Worksheets are reproductions of the Chapter Highlights and Chapter Review sections that follow each chapter in the textbook.

Science Puzzlers, Twisters & Teasers Use vocabulary and concepts from each chapter of the Pupil's Editions as elements of rebuses, anagrams, logic puzzles, daffy definitions, riddle poems, word jumbles, and other types of puzzles.

Reinforcement and Vocabulary Review Worksheets Approach a chapter topic from a different angle with an emphasis on different learning modalities to help students that are frustrated by traditional methods.

Critical Thinking & Problem Solving Worksheets Develop the following skills: distinguishing fact from opinion, predicting consequences, analyzing information, and drawing conclusions. Problem Solving Worksheets develop a step-by-step process of problem analysis including gathering information, asking critical questions, identifying alternatives, and making comparisons.

Math Skills for Science Worksheets Each activity gives a brief introduction to a relevant math skill, a step-by-step explanation of the math process, one or more example problems, and a variety of practice problems.

Science Skills Worksheets Help your students focus specifically on skills such as measuring, graphing, using logic, understanding statistics, organizing research papers, and critical thinking options.

LAB ACTIVITIES

ALL LABS ARE CLASSROOM TESTED & APPROVED

Datasheets for Labs These worksheets are the labs found in the *Holt Science & Technology* textbook. Charts, tables, and graphs are included to make data collection and analysis easier, and space is provided to write observations and conclusions.

Whiz-Bang Demonstrations Discovery or Making Models experiences label each demo as one in which students discover an answer or use a scientific model.

Calculator-Based Labs Give students the opportunity to use graphing-calculator probes and sensors to collect data using a TI graphing calculator, Vernier sensors, and a TI CBL 2™ or Vernier Lab Pro interface.

EcoLabs and Field Activities Focus on educational outdoor projects, such as wildlife observation, nature surveys, or natural history.

Inquiry Labs Use the scientific method to help students find their own path in solving a real-world problem.

Long-Term Projects and Research Ideas Provide students with the opportunity to go beyond library and Internet resources to explore science topics.

ASSESSMENT

Chapter Tests Each four-page chapter test consists of a variety of item types including Multiple Choice, Using Vocabulary, Short Answer, Critical Thinking, Math in Science, Interpreting Graphics, and Concept Mapping.

Performance-Based Assessments Evaluate students' abilities to solve problems using the tools, equipment, and techniques of science. Rubrics included for each assessment make it easy to evaluate student performance.

TEACHER RESOURCES

Lesson Plans Integrate all of the great resources in the *Holt Science & Technology* program into your daily teaching. Each lesson plan includes a correlation of the lesson activities to the National Science Education Standards.

Teaching Transparencies Each transparency is correlated to a particular lesson in the Chapter Organizer.

Concept Mapping Transparencies, Worksheets, and Answer Key

Give students an opportunity to complete their own concept maps to study the concepts within each chapter and form logical connections. Student worksheets contain a blank concept map with linking phrases and a list of terms to be used by the student to complete the map.

TECHNOLOGY RESOURCES

One-Stop Planner CD-ROM

Finding the right resources is easy with the One-Stop Planner CD-ROM. You can view and print any resource with just the click of a mouse. Customize the suggested lesson plans to match your daily or weekly calendar and your district's requirements. Powerful test generator software allows you to create customized assessments using a databank of items.

The One-Stop Planner for each level includes the following:

- All materials from the Teaching Resources
- Bellringer Transparency Masters
- Block Scheduling Tools
- Standards Correlations
- Lab Inventory Checklist
- Safety Information
- Science Fair Guide
- Parent Involvement Tools
- Spanish Audio Scripts
- Spanish Glossary
- Assessment Item Listing
- Assessment Checklists and Rubrics
- Test Generator

sciLINKS

sciLINKS numbers throughout the text take you and your students to some of the best on-line resources available. Sites are constantly reviewed and updated by the National Science Teachers Association. Special "teacher only" sites are available to you once you register with the service.

go.hrw.com

To access Holt, Rinehart and Winston Web resources, use the home page codes for each level found on page 1 of the Pupil's Editions. The codes shown on the Chapter Organizers for each chapter in the Annotated Teacher's Edition take you to chapter-specific resources.

Smithsonian Institution

Find lesson plans, activities, interviews, virtual exhibits, and just general information on a wide variety of topics relevant to middle school science.

CNNfyi.com

Find the latest in late-breaking science news for students. Featured news stories are supported with lesson plans and activities.

CNN Presents Science in the News Video Library

Bring relevant science news stories into the classroom. Each video comes with a Teacher's Guide and set of Critical Thinking Worksheets that develop listening and media analysis skills. Tapes in the series include:

- Eye on the Environment
- Multicultural Connections
- Scientists in Action
- Science, Technology & Society

Guided Reading Audio CD Program

Students can listen to a direct read of each chapter and follow along in the text. Use the program as a content bridge for struggling readers and students for whom English is not their native language.

Interactive Explorations CD-ROM

Turn a computer into a virtual laboratory. Students act as lab assistants helping Dr. Crystal Labcoat solve real-world problems. Activities develop students' inquiry, analysis, and decision-making skills.

Interactive Science Encyclopedia CD-ROM

Give your students access to more than 3,000 cross-referenced scientific definitions, in-depth articles, science fair project ideas, activities, and more.

ADDITIONAL COMPONENTS

Holt Anthology of Science Fiction

Science Fiction features in the Pupil's Edition preview the stories found in the anthology. Each story begins with a Reading Prep guide and closes with Think About It questions.

Professional Reference for Teachers

Articles written by leading educators help you learn more about the National Science Education Standards, block scheduling, classroom management techniques, and more. A bibliography of professional references is included.

Holt Science Posters

Seven wall posters highlight interesting topics, such as the Physics of Sports, or useful reference material, such as the Scientific Method.

Holt Science Skills Workshop: Reading in the Content Area

Use a variety of in-depth skills exercises to help students learn to read science materials strategically.

> **Key**
>
> These materials are blackline masters.
>
> All titles shown in green are found in the *Teaching Resources* booklets for each course.

Science & Math Skills Worksheets

The *Holt Science and Technology* program helps you meet the needs of a wide variety of students, regardless of their skill level. The following pages provide examples of the worksheets available to improve your students' science and math skills, whether they already have a strong science and math background or are weak in these areas. Samples of assessment checklists and rubrics are also provided.

In addition to the skills worksheets represented here, *Holt Science and Technology* provides a variety of worksheets that are correlated directly with each chapter of the program. Representations of these worksheets are found at the beginning of each chapter in this Annotated Teacher's Edition. Specific worksheets related to each chapter are listed in the Chapter Organizer. Worksheets and transparencies are found in the softcover *Teaching Resources* for each course.

Many worksheets are also available on the HRW Web site. The address is **go.hrw.com.**

Science Skills Worksheets: Thinking Skills

BEING FLEXIBLE

USING YOUR SENSES

THINKING OBJECTIVELY

UNDERSTANDING BIAS

USING LOGIC

BOOSTING YOUR MEMORY

IMPROVING YOUR STUDY HABITS

READING A SCIENCE TEXTBOOK

Science Skills Worksheets: Experimenting Skills

SAFETY RULES!

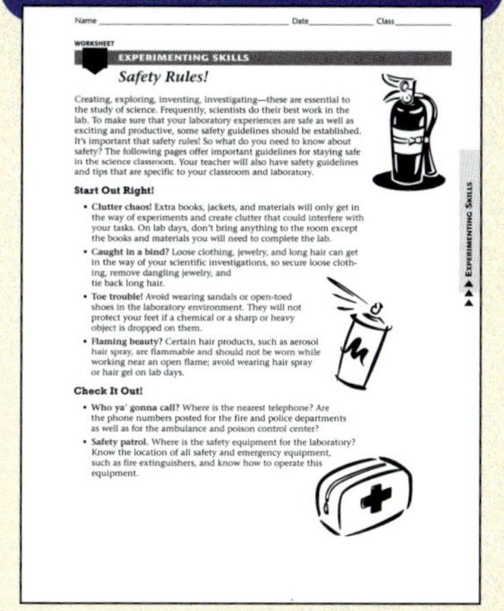

Safety Rules!

Creating, exploring, inventing, investigating—these are essential to the study of science. Frequently, scientists do their best work in the lab. To make sure that your laboratory experiences are safe as well as exciting and productive, some safety guidelines should be established. It's important that safety rules! So what do you need to know about safety? The following pages offer important guidelines for staying safe in the science classroom. Your teacher will have safety guidelines and tips that are specific to your classroom and laboratory.

Start Out Right!

- **Clutter chaos!** Extra books, jackets, and materials will only get in the way of experiments and create clutter that could interfere with your tasks. On lab days, don't bring anything to the room except the books and materials you will need to complete the lab.
- **Caught in a bind?** Loose clothing, jewelry, and long hair can get in the way of your scientific investigations, so secure loose clothing, remove dangling jewelry, and tie back long hair.
- **Toe trouble!** Avoid wearing sandals or open-toed shoes in the laboratory environment. They will not protect your feet if a chemical or a sharp or heavy object is dropped on them.
- **Flaming beauty?** Certain hair products, such as aerosol hair spray, are flammable and should not be worn while whether your arm is near an open flame; avoid wearing hair spray or hair gel on lab days.

Check It Out!

- **Who ya' gonna call?** Where is the nearest telephone? Are the phone numbers posted for the fire and police departments as well as for the ambulance and poison control center?
- **Safety patrol.** Where is the safety equipment for the laboratory? Know the location of all safety and emergency equipment, such as fire extinguishers, and know how to operate this equipment.

DOING A LAB WRITE-UP

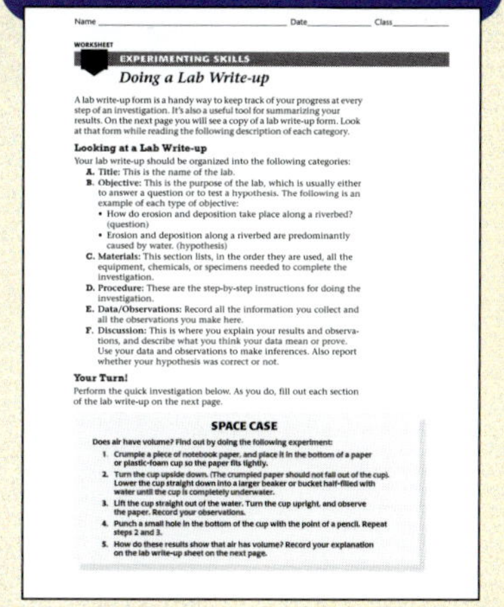

Doing a Lab Write-up

A lab write-up form is a handy way to keep track of your progress at every step of an investigation. It's also a useful tool for summarizing your results. On the next page you will see a copy of a lab write-up form. Look at that form while reading the following description of each category.

Looking at a Lab Write-up

Your lab write-up should be organized into the following categories:
- **A. Title:** This is the name of the lab.
- **B. Objective:** This is the purpose of the lab, which is usually either to answer a question or to test a hypothesis. The following is an example of each type of objective:
  - How do erosion and deposition take place along a riverbed?
  - Erosion and deposition along a riverbed are predominantly caused by water. (hypothesis)
- **C. Materials:** This section lists, in the order they are used, all the equipment, chemicals, or specimens needed to complete the investigation.
- **D. Procedure:** These are the step-by-step instructions for doing the investigation.
- **E. Data/Observations:** Record all the information you collect and all the observations you make here.
- **F. Discussion:** This is where you explain your results and observations, and describe what you think your data mean or prove. Use your data and observations to make inferences. Also report whether your hypothesis was correct or not.

Your Turn!

Perform the quick investigation below. As you do, fill out each section of the lab write-up on the next page.

SPACE CASE

Does air have volume? Find out by doing the following experiment
1. Crumple a piece of notebook paper, and place it in the bottom of a paper or plastic-foam cup so the paper fits tightly.
2. Turn the cup upside down. (The crumpled paper should not fall out of the cup.) Lower the cup straight down into a larger beaker or bucket half-filled with water until the cup is completely underwater.
3. Lift the cup straight out of the water. Turn the cup upright, and observe the paper. Record your observations.
4. Punch a small hole in the bottom of the cup with the point of a pencil. Repeat step 2 and 3.
5. How do these results show that air has volume? Record your explanation on the lab write-up sheet on the next page.

UNDERSTANDING VARIABLES

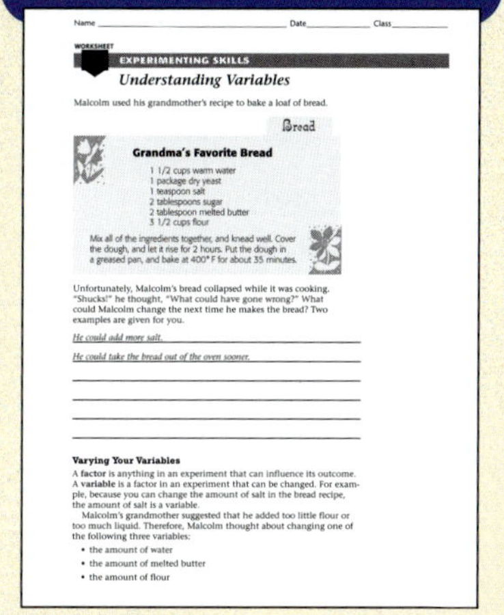

Understanding Variables

Malcolm used his grandmother's recipe to bake a loaf of bread.

Grandma's Favorite Bread

1 1/2 cups warm water
1 package dry yeast
1 weapoon salt
2 tablespoons sugar
2 tablespoon melted butter
3 1/2 cups flour

Mix all of the ingredients together, and knead well. Cover the dough, and let it rise for 2 hours. Put the dough in a greased pan, and bake at 400°F for about 35 minutes.

Unfortunately, Malcolm's bread collapsed while it was cooking. "Shucks!" he thought, "What could have gone wrong?" What could Malcolm change the next time he makes the bread? Two examples are given for you.

He could add more salt.

He could take the bread out of the oven sooner.

Varying Your Variables

A factor is anything in an experiment that can influence its outcome. A **variable** is a factor in an experiment that can be changed. For example, because you can change the amount of salt in the bread recipe, the amount of salt is a variable.

Malcolm's grandmother suggested that he added too little flour or too much liquid. Therefore, Malcolm thought about changing one of these three variables:
- the amount of water
- the amount of melted butter
- the amount of flour

WORKING WITH HYPOTHESES

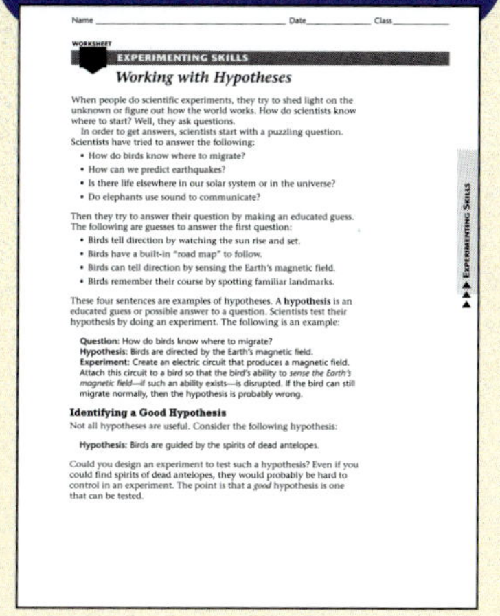

Working with Hypotheses

When people do scientific experiments, they try to shed light on the unknown or figure out how the world works. How do scientists know where to start? Well, they ask questions. To get answers, scientists start with a puzzling question. Scientists have tried to answer the following:
- How do birds know where to migrate?
- Can we predict earthquakes?
- Is there life elsewhere in our solar system or in the universe?
- Do elephants use sound to communicate?

Then they try to answer their question by making an educated guess. The following are guesses to answer the first question:
- Birds tell direction by watching the sun rise and set.
- Birds have a built-in "road map" to follow.
- Birds can tell direction by sensing the Earth's magnetic field.
- Birds remember their course by spotting familiar landmarks.

These four sentences are examples of hypotheses. A **hypothesis** is an educated guess or possible answer to a question. Scientists test their hypothesis by doing an experiment. The following is an example:

Question: How do birds know where to migrate?
Hypothesis: Birds are directed by the Earth's magnetic field.
Experiment: Create an electric circuit that produces a magnetic field. Attach this circuit to a bird so that the bird's ability to sense the Earth's magnetic field—if such an ability exists—is disrupted. If the bird can still migrate normally, then the hypothesis is probably wrong.

Identifying a Good Hypothesis

Not all hypotheses are useful. Consider the following hypothesis:

Hypothesis: Birds are guided by the spirits of dead antelopes.

Could you design an experiment to test such a hypothesis? Even if you could find spirits of dead antelopes, they would probably be hard to control in an experiment. The point is that a good hypothesis is one that can be tested.

DESIGNING AN EXPERIMENT

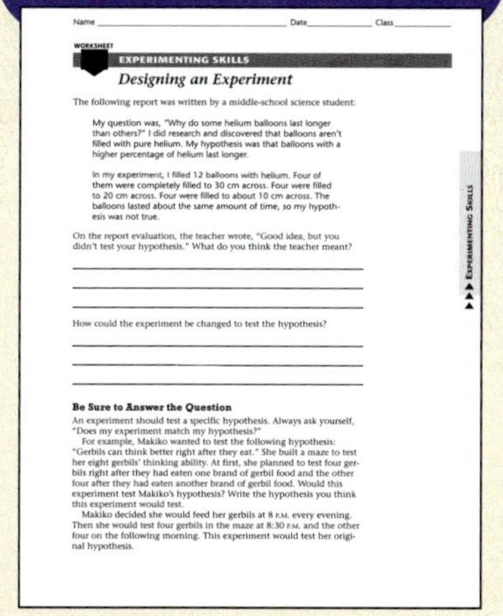

Designing an Experiment

The following report was written by a middle-school science student:

My question was, "Why do some helium balloons last longer than others?" I did research and discovered that balloons aren't all the same. Some balloons have a higher percentage of helium last longer.

In my experiment, I filled 12 balloons with helium. Four of them were completely filled to 30 cm across. Four were filled to 20 cm across. Four were filled to about 10 cm across. The balloons lasted about the same amount of time, so my hypothesis was not true.

On the report evaluation, the teacher wrote, "Good idea, but you didn't test your hypothesis." What do you think the teacher meant?

How could the experiment be changed to test the hypothesis?

Be Sure to Answer the Question

An experiment should test a specific hypothesis. Always ask yourself, "Does my experiment match my hypothesis?"

For example, Makiko wanted to test the following hypothesis: "Gerbils can think better right after they eat." She built a maze to test her eight gerbils' thinking ability. At first, she planned to test four gerbils right after they had eaten one brand of gerbil food and the other four after they had eaten another brand of gerbil food. Would this experiment test Makiko's hypothesis? Write the hypothesis you think this experiment would test.

Makiko decided she would feed her gerbils at 8 P.M. every evening. Then she would test four gerbils in the maze at 8:30 P.M. and the other four on the following morning. This experiment would test her original hypothesis.

USING THE INTERNATIONAL SYSTEM OF UNITS (SI)

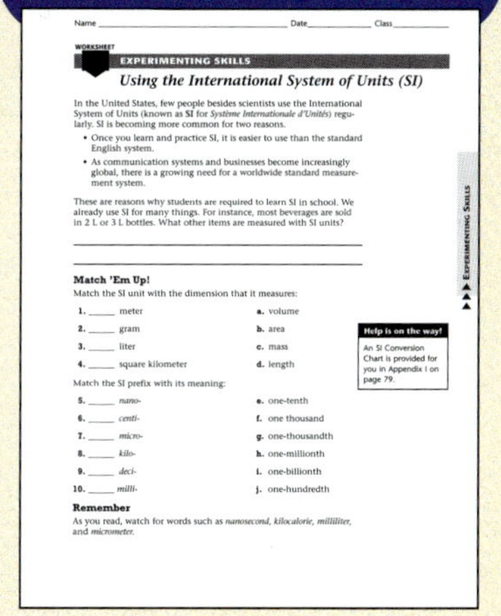

Using the International System of Units (SI)

In the United States, few people besides scientists use the International System of Units (known as SI for *Système Internationale d'Unités*) regularly. SI is becoming more common for two reasons:
- Once you learn and practice SI, it is easier to use than the standard English system.
- As communication systems and businesses become increasingly global, there is a growing need for a worldwide standard measurement system.

There are reasons why students are required to learn SI in school. We already use SI for many things. For instance, most beverages are sold in 2 L or 3 L bottles. What other items are measured with SI units?

Match 'Em Up!

Match the SI unit with the dimension that it measures:
1. _____ meter
2. _____ gram
3. _____ liter
4. _____ square kilometer

a. volume
b. area
c. mass
d. length

Match the SI prefix with its meaning:
5. _____ nano-
6. _____ centi-
7. _____ micro-
8. _____ milli-
9. _____ deci-
10. _____ kilo-

e. one-tenth
f. one thousand
g. one-thousandth
h. one-millionth
i. one-billionth
j. one-hundredth

Remember

As you read, watch for words such as *nanosecond*, *kilocalorie*, *milliliter*, and *micrometer*.

Help is on the way!
An SI Conversion Chart is provided for you in Appendix I on page 79.

MEASURING

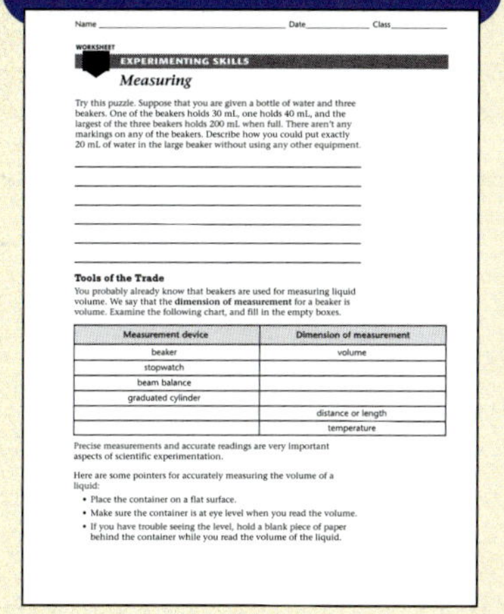

Measuring

Try this puzzle. Suppose that you are given a bottle of water and three beakers. One of the beakers holds 30 mL, one holds 40 mL, and the largest of the three beakers holds 200 mL when full. There aren't any markings on any of the beakers. Describe how you could put exactly 20 mL of water in the large beaker without using any other equipment.

Tools of the Trade

You probably already know that beakers are used for measuring liquid volume. We say that the **dimension of measurement** for a beaker is volume. Examine the following chart, and fill in the empty boxes.

Measurement device	Dimension of measurement
beaker	volume
stopwatch	
beam balance	
graduated cylinder	
	distance or length
	temperature

Precise measurements and accurate readings are very important aspects of scientific experimentation.

Here are some pointers for accurately measuring the volume of a liquid:
- Place the container on a flat surface.
- Make sure the container is at eye level when you read the volume.
- If you have trouble seeing the level, hold a blank piece of paper behind the container while you read the volume of the liquid.

Science Skills Worksheets: Researching Skills

CHOOSING YOUR TOPIC

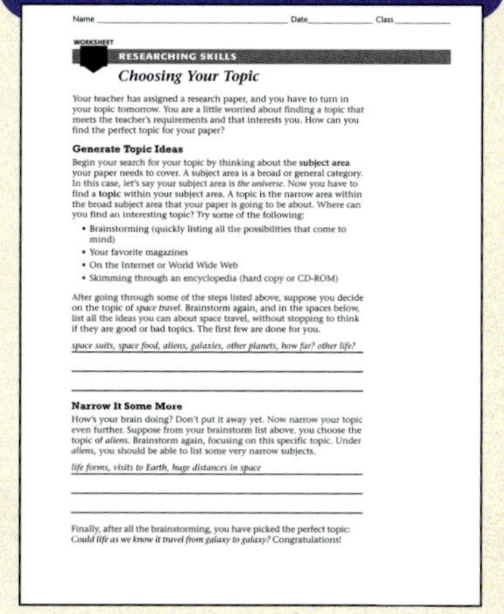

Choosing Your Topic

Your teacher has assigned a research paper, and you have to turn in your topic tomorrow. You are a little worried about finding a topic that meets the teacher's requirements and that interests you. How can you find the perfect topic for your paper?

Generate Topic Ideas

Begin your search for your topic by thinking about the **subject area** that your paper needs to cover. A subject area is a broad or general category. In this case, let's say your subject area is *the universe*. Now you have to find a topic within your subject area. A topic is the narrow area within the broad subject area that your paper is going to be about. Where can you find an interesting topic? Try some of the following:
- Brainstorming (quickly listing all the possibilities that come to mind)
- Your favorite magazines
- On the Internet or World Wide Web
- Skimming through an encyclopedia (hard copy or CD-ROM)

After going through some of the steps listed above, suppose you decide on the topic of *space travel*. Brainstorm again, and in the spaces below, list all the ideas you can about space travel, without stopping to think if they are good or bad topics. The first few are done for you.

space suits, space food, aliens, galaxies, other planets, how far? other life?

Narrow It Some More

How's your brain doing? Don't put it away yet. Now narrow your topic even further. Suppose from your brainstorm list above, you choose the topic of *aliens*. Brainstorm again, focusing on this specific topic. Under *aliens*, you should be able to list some very narrow subjects.

life forms, visits to Earth, huge distances in space

Finally, after all the brainstorming, you have picked the perfect topic:
Could life as we know it travel from galaxy to galaxy? Congratulations!

ORGANIZING YOUR RESEARCH

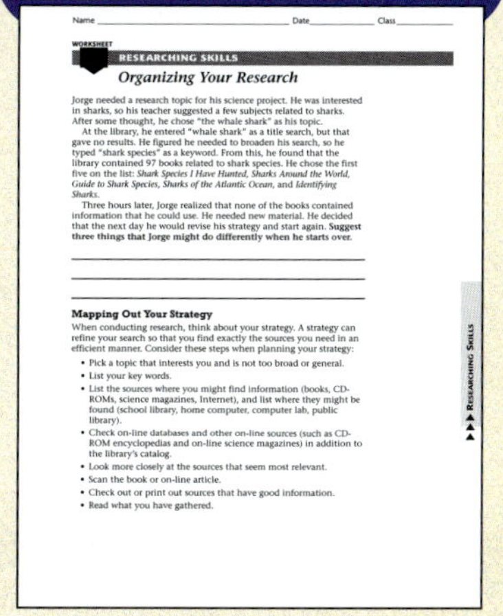

Organizing Your Research

Jorge needed a research topic for his science project. He was interested in sharks, so his teacher suggested a few subjects related to sharks. After some thought, he chose "the whale shark" as his topic.

At the library, he entered "whale shark" as a title search, but that gave no results. He figured he needed to broaden his search, so he typed "shark species" as a keyword. From this, he found that the library contained 97 books related to shark species. He chose the first five on the list: *Shark Species I Have Hunted, Sharks Around the World, Guide to Shark Species, Sharks of the Atlantic Ocean,* and *Identifying Sharks*.

Three hours later, Jorge realized that none of the books contained information that he could use. He needed new material. He decided that the next day he would revise his strategy and start again. **Suggest** three things that Jorge might do differently when he starts over.

Mapping Out Your Strategy

When conducting research, think about your strategy. A strategy can refine your search so that you find exactly the sources you need in an efficient manner. Consider these steps when planning your strategy:
- Pick a topic that interests you and is not too broad or general.
- List your key words.
- List the sources where you might find information (books, CD-ROMs, science magazines, Internet), and list where they might be found (school library, home computer, computer lab, public library).
- Check on-line databases and other on-line sources (such as CD-ROM encyclopedias and on-line science magazines in addition to the library's catalog.
- Look closely at those sources that seem most relevant.
- Scan the book or on-line article.
- Check out or print out sources that have good information.
- Read what you have gathered.

FINDING USEFUL SOURCES

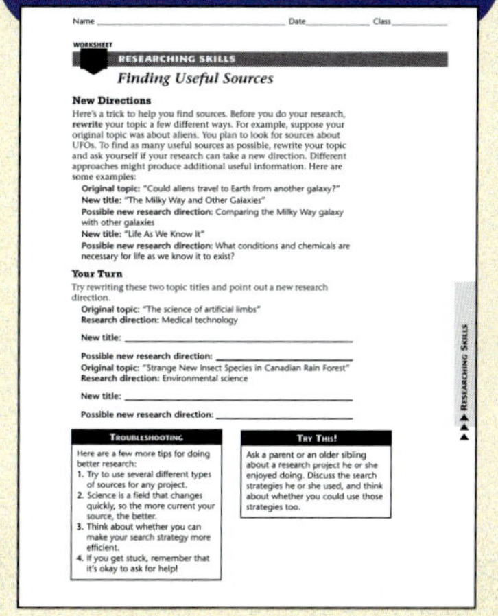

Finding Useful Sources

New Directions

Here's a trick to help you find sources. Before you do your research, **rewrite** your topic a few different ways. For example, suppose your original topic was about aliens. You plan to look for sources about UFOs. To find as many useful sources as possible, rewrite your topic and ask yourself if your research can take a new direction. Different approaches might produce additional useful information. Here are some examples:

- **Original topic:** "Could aliens travel to Earth from another galaxy?"
- **Possible new research direction:** Comparing the Milky Way galaxy with other galaxies
- **New title:** "The Milky Way and Other Galaxies"
- **New title:** "Life As We Know It"
- **Possible new research direction:** What conditions and chemicals are necessary for life as we know it to exist?

Your Turn

Try rewriting these two topic titles and point out a new research direction.

- **Original topic:** "The science of artificial limbs"
- **Research direction:** Medical technology
- **Possible new research direction:** _____
- **New title:** _____

- **Original topic:** "Strange New Insect Species in Canadian Rain Forest"
- **Research direction:** Environmental science
- **Possible new research direction:** _____
- **New title:** _____

TROUBLESHOOTING

If you need a few more tips for doing better research:
1. Try to use several different types of sources for any project.
2. Science is a field that keeps changing, so the more current your source, the better.
3. Think about whether you can make your search strategy more efficient.
4. If you get stuck, remember that it's okay to ask for help!

TRY THIS!

Ask a parent or an older sibling about a research project he or she enjoyed doing. Discuss the search strategies he or she used, and think about whether you could use those strategies too.

RESEARCHING ON THE WEB

Researching on the Web

Name one of your favorite Web sites on the World Wide Web.

How did you find that site?

The World Wide Web can be a great source of information. It is like an electronic library with information on almost every imaginable topic! Professional researchers rely heavily on the Web. In fact, some researchers find most of their sources on-line. Here are a few tips to help you conduct your searches effectively and efficiently.

Caught in the Web

As a Web researcher, you will face two major problems:
- too much information
- unreliable information

So, your challenge is to narrow your search to get the most useful information and to make sure the information is reliable and accurate.

Too Much Information

Here are a couple of hints to help you avoid too much information:
- Most search engines display the best matches at the top of the list. The ones you see first are the ones that are most likely to be useful.
- Do single-word searches like "Mars" and you type in a single keyword. For example, if your topic is "the search for life on Mars" and you type in "Mars" as your word, you may get more than 2,300 responses. If you narrow your search by typing in "'life on Mars' + evidence), you will get fewer responses, but they will be more focused on your topic.

You Try It

Go to the Internet. Type in the key word "armadillo." How many Web sites did you get in response? Are all of them about the animal called the nine-banded armadillo?

Now type in the key word "nine-banded armadillo." How many results did you get? Were they more closely related to the interesting mammal?

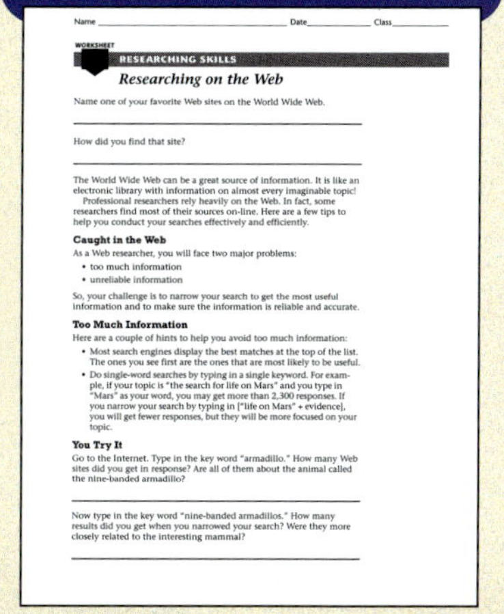

Science & Math Skills Worksheets (continued)

Science Skills Worksheets: Researching Skills (continued)

IDENTIFYING BIAS

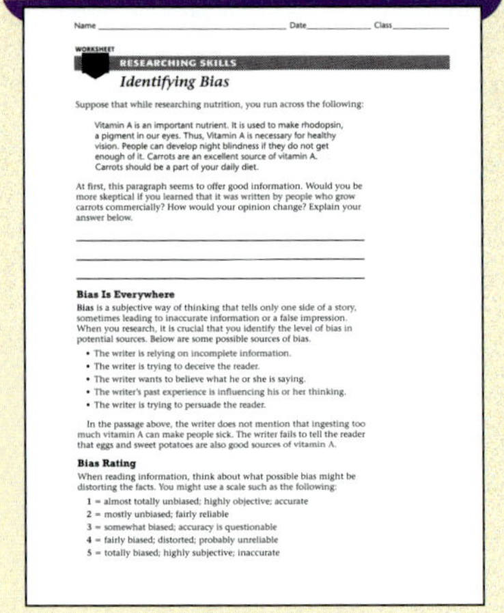

WORKSHEET
RESEARCHING SKILLS
Identifying Bias

Suppose that while researching nutrition, you run across the following:

Vitamin A is an important nutrient. It is used to make rhodopsin, a pigment in our eyes. Thus, Vitamin A is necessary for healthy vision. People can develop night blindness if they do not get enough of it. Carrots are an excellent source of vitamin A. Carrots should be a part of your daily diet.

At first, this paragraph seems to offer good information. Would you be more skeptical if you learned that it was written by people who grow carrots commercially? How would your opinion change? Explain your answer below.

Bias Is Everywhere
Bias is a subjective way of thinking that tells only one side of a story, sometimes leading to inaccurate information or a false impression. When you research, it is crucial that you identify the level of bias in potential sources. Below are some possible sources of bias.
• The writer is relying on incomplete information.
• The writer is trying to deceive you.
• The writer wants to believe what he or she is saying.
• The writer's past experience is influencing his or her thinking.
• The writer is trying to persuade the reader.

In the passage above, the writer does not mention that ingesting too much vitamin A can make people sick. The writer fails to tell the reader that eggs and sweet potatoes are also good sources of vitamin A.

Bias Rating
When reading information, think about what possible bias might be distorting the facts. You might use a scale such as the following:
1 = almost totally unbiased; highly objective; accurate
2 = mostly unbiased; fairly reliable
3 = somewhat biased; accuracy is questionable
4 = fairly biased; distorted; probably unreliable
5 = totally biased; highly subjective; inaccurate

TAKING NOTES

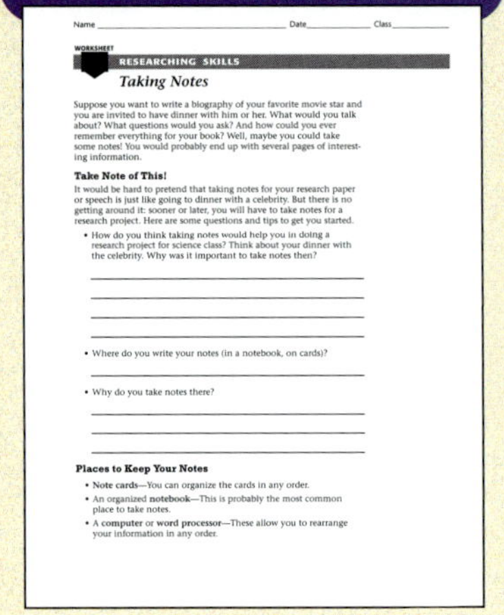

WORKSHEET
RESEARCHING SKILLS
Taking Notes

Suppose you want to write a biography of your favorite movie star and you are invited to have dinner with him or her. What would you talk about? What questions would you ask? And how could you ever remember everything for your book? Well, maybe you could take some notes! You would probably end up with several pages of interesting information.

Take Note of This!
It would be hard to pretend that taking notes for your research paper or speech is just like going to dinner with a celebrity. But there is no getting around it sooner or later, you will have to take notes for a research project. Here are some questions and tips to get you started.
• How do you think taking notes would help you in doing a research project for science class? Think about your dinner with the celebrity. Why was it important to take notes then?

• Where do you write your notes (in a notebook, on cards)?

• Why do you take notes there?

Places to Keep Your Notes
• Note cards—You can organize the cards in any order.
• An organized notebook—This is probably the most common place to take notes.
• A computer or word processor—These allow you to rearrange your information in any order.

SCIENCE WRITING

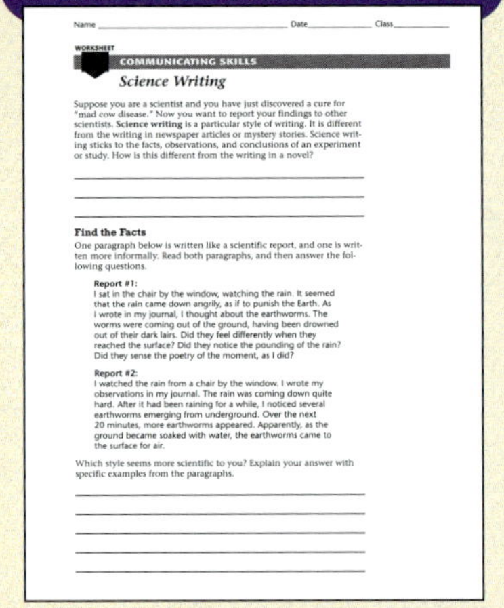

WORKSHEET
COMMUNICATING SKILLS
Science Writing

Suppose you are a scientist and you have just discovered a cure for "mad cow disease." Now you want to report your findings to other scientists. **Science writing** is a particular style of writing. It is different from the writing in newspaper articles or mystery stories. Science writing sticks to the facts, observations, and conclusions of an experiment or study. How is this different from the writing in a novel?

Find the Facts
One paragraph below is written like a scientific report, and one is written more informally. Read both paragraphs, and then answer the following questions.

Report #1:
I sat in the chair by the window, watching the rain. It seemed that the rain came down angrily, as if to punish the Earth. As I wrote in my journal, I thought about the earthworms. The worms were coming out of the ground, having been drowned out of their dark lairs. Did they feel differently when they reached the surface? Did they notice the pounding of the rain? Did they sense the poetry of the moment, as I did?

Report #2:
I watched the rain from a chair by the window. I wrote my observations in my journal. The rain was coming down quite hard. After it had been raining for a while, I noticed several earthworms emerging from underground. Over the next 20 minutes, more earthworms appeared. Apparently, as the ground became soaked with water, the earthworms came to the surface for air.

Which style seems more scientific to you? Explain your answer with specific examples from the paragraphs.

Science Skills Worksheets: Communicating Skills

SCIENCE DRAWING

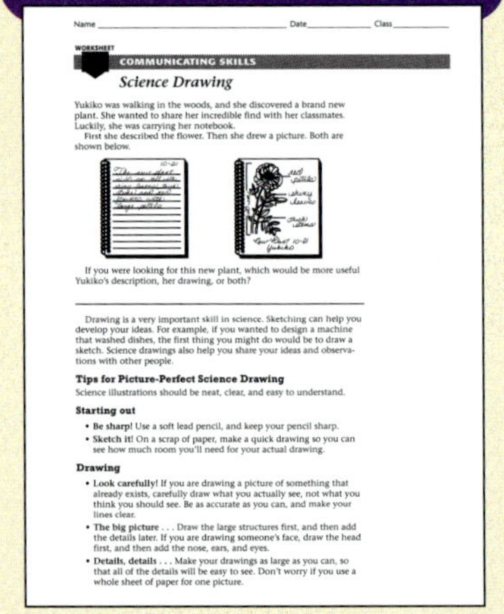

WORKSHEET
COMMUNICATING SKILLS
Science Drawing

Yukiko was walking in the woods, and she discovered a brand new plant. She wanted to share her incredible find with her classmates. Luckily, she was carrying her notebook.
First she described the flower. Then she drew a picture. Both are shown below.

If you were looking for this new plant, which would be more useful Yukiko's description, her drawing, or both?

Drawing is a very important skill in science. Sketching can help you develop your ideas. For example, if you wanted to design a machine that washed dishes, the first thing you might do would be to draw a sketch. Science drawings also help you share your ideas and observations with other people.

Tips for Picture-Perfect Science Drawing
Science illustrations should be neat, clear, and easy to understand.
Starting out
• **Be sharp!** Use a soft lead pencil, and keep your pencil sharp.
• **Sketch it!** On a scrap of paper, make a quick drawing so you can see how much room you'll need for your actual drawing.
Drawing
• **Look carefully!** If you are drawing a picture of something that already exists, carefully draw what you actually see, not what you think you should see. Be as accurate as you can, and make your lines clear.
• **The big picture . . .** Draw the large structures first, and then add the details later. If you are drawing someone's face, draw the head first, and then add the nose, ears, and eyes.
• **Details, details . . .** Make your drawings as large as you can, so that all of the details will be easy to see. Don't worry if you use a whole sheet of paper for one picture.

USING MODELS TO COMMUNICATE

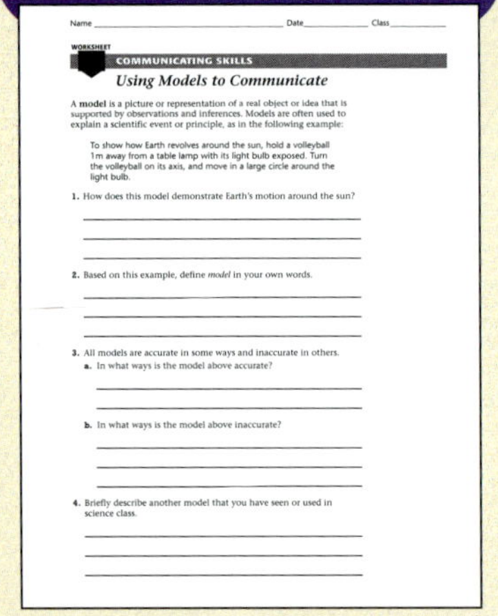

WORKSHEET
COMMUNICATING SKILLS
Using Models to Communicate

A **model** is a part or representation of a real object or idea that is supported by observations and inferences. Models are often used to explain a scientific event or principle, as in the following example:

To show how Earth revolves around the sun, hold a volleyball 1 m away from a table lamp with its light bulb exposed. Turn the volleyball on its axis, and move in a large circle around the light bulb.

1. How does this model demonstrate Earth's motion around the sun?

2. Based on this example, define *model* in your own words.

3. All models are accurate in some ways and inaccurate in others.
a. In what ways is the model above accurate?

b. In what ways is the model above inaccurate?

4. Briefly describe another model that you have seen or used in science class.

INTRODUCTION TO GRAPHS

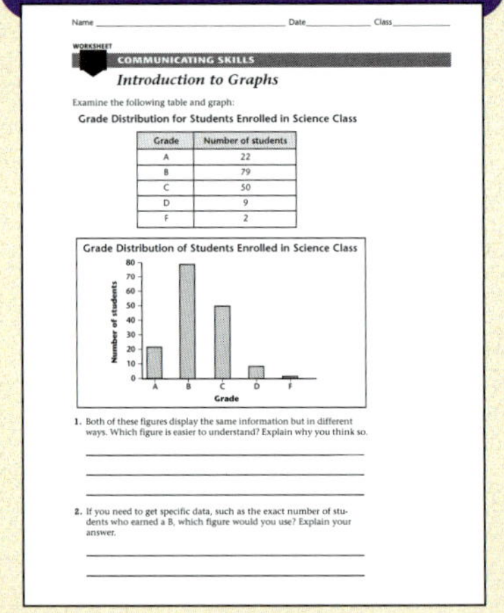

WORKSHEET
COMMUNICATING SKILLS
Introduction to Graphs

Examine the following table and graph.

Grade Distribution for Students Enrolled in Science Class

Grade	Number of students
A	22
B	79
C	50
D	9
F	2

Grade Distribution of Students Enrolled in Science Class

1. Both of these figures display the same information but in different ways. Which figure is easier to understand? Explain why you think so.

2. If you need to get specific data, such as the exact number of students who earned a B, which figure would you use? Explain your answer.

GRASPING GRAPHING

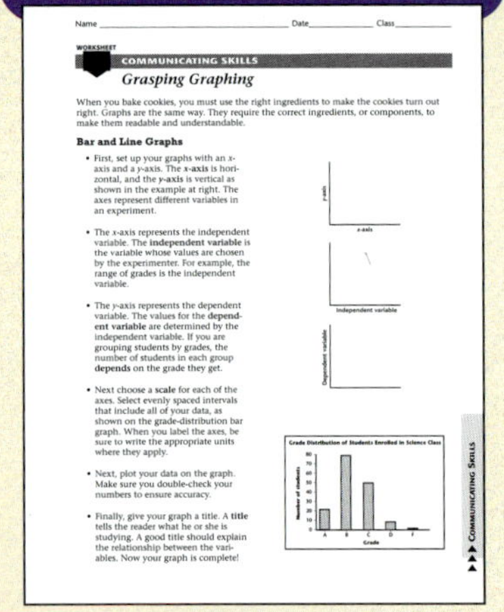

WORKSHEET
COMMUNICATING SKILLS
Grasping Graphing

When you bake cookies, you must use the right ingredients to make the cookies turn out right. Graphs are the same way. They require the correct ingredients, or components, to make them readable and understandable.

Bar and Line Graphs
• First, set up your graph with an x-axis and a y-axis. The x-axis is horizontal, and the y-axis is vertical as shown in the example at right. The axes represent different variables in an experiment.
• The x-axis represents the independent variable. The **independent variable** is the variable whose values are chosen by the experimenter. For example, the range of grades is the independent variable.
• The y-axis represents the dependent variable. The values for the **dependent variable** are determined by the independent variable. If you are grouping students by grades, the number of students in each group depends on the grade they get.
• Next choose a scale for each of the axes. Select evenly spaced intervals that include all of your data, as shown on the grade-distribution bar graph. When you label the axes, be sure to write the appropriate units where they apply.
• Next, plot your data on the graph. Make sure you double-check your numbers to ensure accuracy.
• Finally, give your graph a title. A **title** tells the reader what he or she is studying. A good title should explain the relationship between the variables. Now your graph is complete!

INTERPRETING YOUR DATA

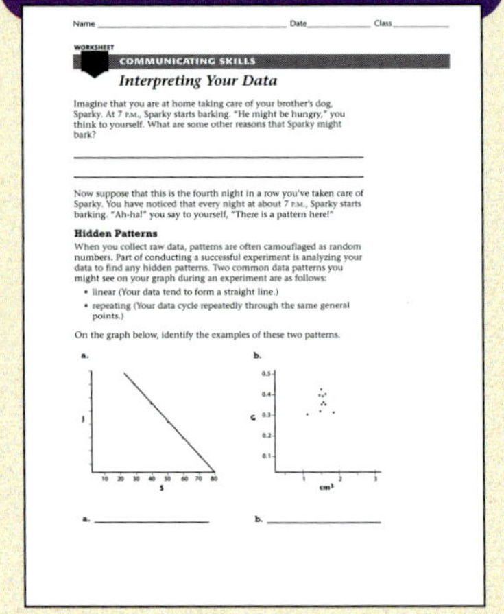

WORKSHEET
COMMUNICATING SKILLS
Interpreting Your Data

Imagine that you are at home taking care of your brother's dog, Sparky. At 7 P.M., Sparky starts barking. "He might be hungry," you think to yourself. What are some other reasons that Sparky might bark?

Now suppose that this is the fourth night in a row you've taken care of Sparky. You have noticed that each night at about 7 P.M., Sparky starts barking. "Ah-ha!" you say to yourself, "There is a pattern here!"

Hidden Patterns
When you collect raw data, patterns are often camouflaged as random numbers. Part of conducting a successful experiment is analyzing your data to find any hidden patterns. Two common data patterns you might see on your graph during an experiment are as follows:
• linear (Your data tend to form a straight line.)
• repeating (Your data cycle repeatedly through the same general points.)

On the graph below, identify the examples of these two patterns.

RECOGNIZING BIAS IN GRAPHS

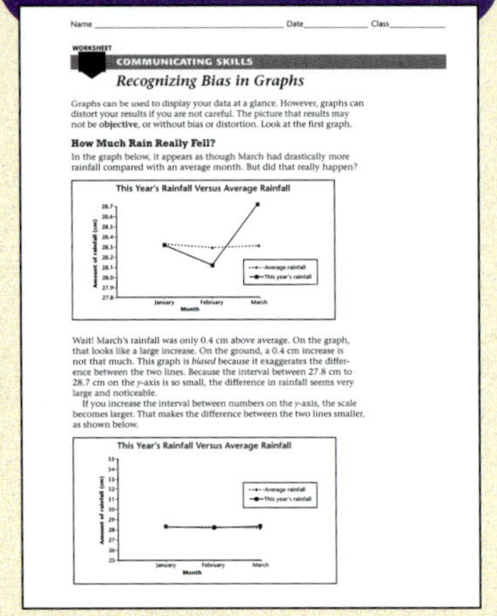

WORKSHEET
COMMUNICATING SKILLS
Recognizing Bias in Graphs

Graphs can be used to display your data at a glance. However, graphs can distort your results if you are not careful. The picture that results may not be objective, or without bias or distortion. Look at the first graph.

How Much Rain Really Fell?
In the graph below, it appears as though March had drastically more rainfall compared with an average month. But did that really happen?

This Year's Rainfall Versus Average Rainfall

Wait! March's rainfall was only 0.4 cm above average. On the graph, that looks like a large increase. On the ground, a 0.4 cm increase is not that much. This graph is biased because it exaggerates the difference between the two bars. Because the interval between 27.8 cm to 28.7 cm on the y-axis is so small, the difference in rainfall seems very large and noticeable.

If you increase the interval between numbers on the y-axis, the scale becomes larger. That makes the difference between the two bars smaller, as shown below.

This Year's Rainfall Versus Average Rainfall

MAKING DATA MEANINGFUL

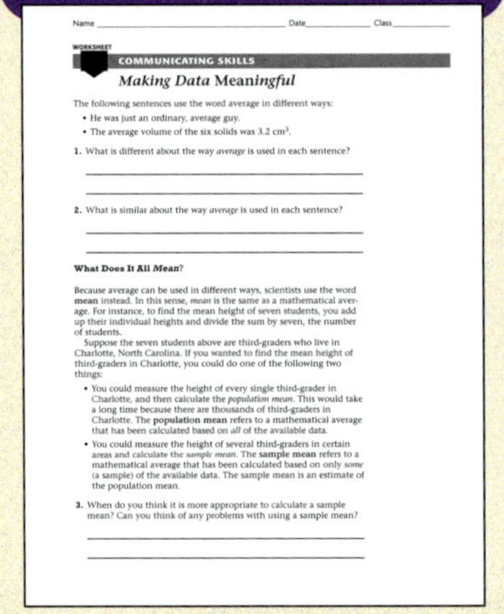

WORKSHEET
COMMUNICATING SKILLS
Making Data Meaningful

The following sentences use the word *average* in different ways.
• He was just an ordinary, average guy.
• The average volume of the six solids was 3.2 cm^3.

1. What is different about the way *average* is used in each sentence?

2. What is similar about the way *average* is used in each sentence?

What Does It All Mean?
Because *average* can be used in different ways, scientists use the word **mean** instead. In this sense, *mean* is the same as a mathematical average. For instance, to find the mean height of seven students, you add up their individual heights and divide the sum by seven, the number of students.

Suppose the seven students above are third-graders who live in Charlotte, North Carolina. If you wanted to find the mean height of third-graders in Charlotte, you could do one of the following two things:

• You could measure the height of every single third-grader in Charlotte, and then calculate the *population mean*. This would take a long time because there are thousands of third-graders in Charlotte. The **population mean** refers to a mathematical average that has been calculated based on all of the available data.
• You could measure the height of several third-graders in certain areas and calculate the *sample mean*. The **sample mean** refers to a mathematical average that has been calculated based on part (a sample) of the available data. The sample mean is an estimate of the population mean.

3. When do you think it is more appropriate to calculate a *sample mean*? Can you think of any problems with using a sample mean?

HINTS FOR ORAL PRESENTATIONS

WORKSHEET
COMMUNICATING SKILLS
Hints for Oral Presentations

Tomorrow, Gabe has to give a speech about pearls in his science class. Before going through this worksheet, he was very nervous, but now he is confident that he has a well-organized speech.

Giving a speech as an oral presentation is a real challenge to many people. This worksheet offers some hints for organizing your speech, controlling stage fright, and watching your language.

Organizing Your Speech
Just like a written report, a speech has three main parts: an introduction, a body, and a conclusion. Here are some hints about each of these parts to help you get organized.

■ **The Introduction**
The beginning of your speech is the **introduction**. The introduction can be as short as a few sentences. It is very important for the following three reasons:
• It gets the attention of your audience.
• It is a way for you to gain the audience's respect and "good feelings."
• It gives you the chance to build the audience's interest in your topic.

Gabe's introduction read, "Did you know that some jewelry is made by animals? It's true, only oysters make pearls."

What makes a good introduction?

Tell Them What You Are Going to Tell Them
Here are some hints for writing an interesting introduction. Choose one.
• Surprise your audience; begin your speech with a starting statement. Before going through this worksheet, he was very nervous, but now he is confident that he has a well-organized speech.
• Begin your speech with a question. Let the audience think about it for a few moments, and then answer the question. The audience will be listening for your answer.
• Begin your speech with a quotation that fits your topic.
• Begin your speech with a personal reference. If your speech is about how bicycles stay upright when being ridden, tell the story of how you learned to ride your bike.
• Begin your speech with an audio-visual presentation that supports your topic.

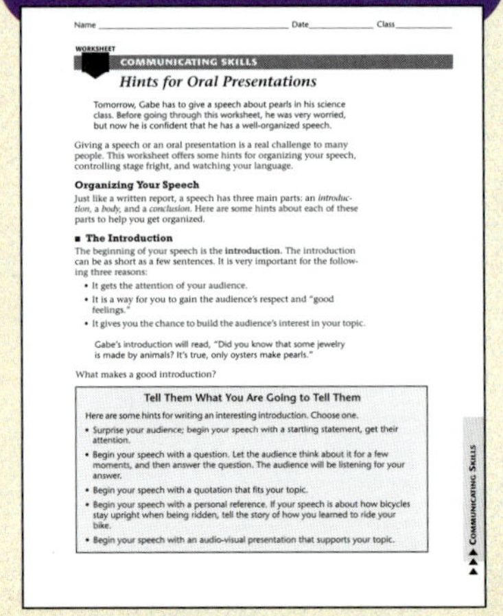

Math Skills for Science

ADDITION AND SUBTRACTION

Addition Review

Addition is used to find the total of two or more quantities. The answer to an addition problem is known as the *sum*.

PROCEDURE: To find the sum of a set of numbers, align the numbers vertically so that the ones digits are in the same column. Add each column, working from right to left.

SAMPLE PROBLEM: Find the sum of 317, 435, and 92.

Step 1: Add the ones. Don't forget to carry your numbers.	Step 2: Add the tens.	Step 3: Add the hundreds.

The sum is **844.**

Add It Up!

1. Find the sums of the following problems.

   a. 348 + 21   b. 98,125 + 233   c. 593 + 386   d. 36,186 + 27,309

2. Your doctor advises you to take 60 mg of vitamin C, 20 mg of niacin, and 15 mg of zinc every day. How many milligrams of nutrients will you take?

3. A chemistry experiment calls for 356 mL of water, 197 mL of saline solution, and 53 mL of vinegar. How much liquid is needed in all?

Subtraction Review

Subtraction is used to take one number from another number. The answer to a subtraction problem is known as the *difference*. The difference is how much larger or smaller one number is than the other.

PROCEDURE: To find the difference between two numbers, first align the numbers vertically so that the ones digits are in the same column, with the larger number above the smaller number. Subtract, working from right to left, one column at a time. Remember to borrow when necessary.

SAMPLE PROBLEM: Find the difference between 622 and 348.

The difference of the numbers is **274.**

Take It Away!

1. Find the difference in the following problems.

4. Mars has a diameter of 6790 km. The diameter of Jupiter is 142,984 km. How much larger is the diameter of Jupiter than the diameter of Mars?

5. A horse is born with a mass of 36 kg. It is expected to have a mass of 495 kg when fully grown. How much mass will it gain?

6. Traveling with the wind, a plane reaches a speed of 212 m/s. On the return trip, the same plane flies into the wind and achieves a speed of only 179 m/s. How much faster does the plane fly with the wind?

MULTIPLICATION

Multiplying Whole Numbers

Suppose every student in your class planted 5 seeds in your school's garden. How many seeds were planted? You could repeatedly add 5 seeds plus 5 seeds until every student's seeds had been added, but this would be pretty time consuming. **Multiplication**, which simplifies addition, is the process of calculating the total of a number that is added together a specific number of times. For example, 3 × 4 means adding 3 together 4 times, or 3 + 3 + 3 + 3 = 12. So 3 × 4 = 12. The answer to a multiplication problem is called the *product*.

PROCEDURE: To find the product of two whole numbers, align your numbers so that the ones digits are in the same column. Multiply each digit of the top number by the ones digit in the bottom number, carrying when necessary. Then multiply each digit in the top number by the tens in the bottom number, regrouping when necessary. Finally, add the partial products to find the final product.

SAMPLE PROBLEM: Find the product of 34 and 16.

The product is **544.**

Practice Your Skills!

1. Multiply. Don't forget to show all your work.

A Shortcut for Multiplying Large Numbers

Imagine that you are a doctor doing research on white blood cells. You know that there are approximately 80,000 white blood cells in 1 mL of blood. You have a sample of 50 mL of blood. How many white blood cells are in the sample? You could multiply to find the answer, of course, but it's a large number and you need an answer quickly. How can you make this easier? Read on to learn an easy way to find the product of large numbers.

PROCEDURE: To find the product of large numbers, remove the zeros at the end of each of your numbers. Next, multiply the non-zero numbers. Finally, at the end of the product, replace the same number of zeros that you removed from your multipliers.

SAMPLE PROBLEM: Multiply 80,000 by 50.

Step 1: Remove the zeros from the end of your numbers, and multiply the non-zero numbers.

Step 2: At the end of your product, replace the total number of zeros you removed from your multipliers, place five zeros after the product.

It's Your Turn!

Using the method above, find the products of the following problems, and write the corresponding letter from the correct answer on the line.

1. 300 × 90,000 ___   A. 11,720,000
2. 45 × 8500 ___   B. 3,524,000
3. 4400 × 7500 ___   C. 27,000,000
4. 52,000 × 610 ___   D. 33,000,000
5. 88,100 × 40 ___   E. 382,500

Challenge Yourself!

A super-fast chess computer can perform 200,000,000 calculations per second. How many calculations can it perform in the 3 minutes it is allowed for each move?

DIVISION

Dividing Whole Numbers with Long Division

Long division, which is used to divide numbers of more than one digit, is really just a series of simple division, multiplication, and subtraction problems. The number that you divide is called the *dividend*. The number you divide the dividend by is the *divisor*. The answer to a division problem is called a *quotient*.

Step 1: Because you cannot divide 12 into 5, you must start by dividing 12 into 56. To do this, ask yourself, "What number multiplied by 12 comes closest to 56 without going over?" 4 × 12 = 48, so place a 4 in the quotient.

Step 2: Multiply the 4 by the divisor and place the product under the 56. Then subtract this number from 56.

Step 3: Bring the next digit down from the dividend (4), and divide this new number (84) by the divisor. Because 12 divides into 84 seven times, write 7 in the quotient.

The quotient is **47.**

Divide It Up!

1. Fill in the blanks in the following long-division problems.

Checking Division with Multiplication

Multiplication and division "undo" one another. This means that when you ask yourself, "What is 12 divided by 3?" it is the same as asking, "What number *multiplied* by 3 gives 12?" You can use this method to catch mistakes in your division.

PROCEDURE: To check your division with multiplication, multiply the quotient of your division problem by the divisor and compare the result with the dividend. If the two are equal, your division was correct.

SAMPLE PROBLEM: Divide 564 by 47, and check your result with multiplication.

Step 1: Divide to find your quotient.	Step 2: Multiply the quotient by the divisor.	Step 3: Compare the product with the dividend.
		564 = 564 Correct!

Check It Out!

Complete the following divisions, and check your math by multiplying the quotient by your divisor. Are the product and dividend equal?

AVERAGES

What Is an Average?

Suppose that your class is doing an experiment to determine the boiling point of a particular liquid. Working in groups, your classmates come up with several answers that are all slightly different. Your teacher asks you to determine which temperature best represents all of the varying results from the class. A mathematical tool called an **average**, or *mean*, will help you solve the problem. An average allows you to simplify a list of numbers into a single number that *approximates* the value of all of them. Check it out!

PROCEDURE: To calculate the average of a set of numbers, first add all of the numbers to find the sum. Then divide the sum by the amount of numbers in your set. The result is the average of your numbers.

SAMPLE PROBLEM: Find the average of the following set of numbers: 5, 4, 7, 8

Step 1: Find the sum.

$$5 + 4 + 7 + 8 = 24$$

Step 2: Divide the sum by the amount of numbers in your set. Because there are four numbers in your set, divide the sum by 4.

$$24 \div 4 = 6$$

The average of the numbers is **6.**

Practice Your Skills!

Be sure to show your work for the following problems.

1. Find the average of each of the following sets of numbers.

   a. 19 m, 11 m, 29 m, 62 m, 14 m

   b. 12 cm, 16 cm, 23 cm, 13 cm

Average, Mode, and Median

Although an average, or mean, is the most common way to simplify a list of numbers, there are other mathematical tools that can help you work with lists of numbers. **Mode** is the number or value that appears most often in a particular set of numbers. **Median** is the number that falls in the *numerical center* of a list of numbers. Read on to find out how to find mode and median.

PROCEDURE: To find the mode, list your numbers in numerical order. Then determine which number appears most often in the set. That number is the mode. **Note:** A list of numbers may have more than one mode. If no number appears more often than the others, that series of numbers does not have a mode.

SAMPLE PROBLEM: Find the mode of 4, 3, 6, 10, and 3.

PROCEDURE: To find the median, list the numbers in numerical order. Next determine the number that appears in the middle of the set. **Note:** If more than one number falls in the middle, the median is the average of those numbers.

Get in the Mode!

1. Find the mode and median for the following sets of numbers.

POSITIVE AND NEGATIVE NUMBERS

Comparing Integers on a Number Line

An **integer** is any whole number (0, 1, 2, 3, . . .) or its opposite. A good way to compare integers is with a *number line*, which is used to represent positive and negative numbers in order. A number line looks like this:

The farther a number is to the right on a number line, the greater the number. The farther a number is to the left on a number line, the smaller the number.

PROCEDURE: To compare integers on a number line, simply place your values on the line, with positive numbers to the right of zero and negative numbers to the left of zero. The number that is the farthest to the right is the greatest number. The number that is the farthest to the left is the smallest number.

SAMPLE PROBLEM: Which is greater, −8 or −3?

Step 1: Draw your number line and select a point for 0. Then fill in the integer values on the line.

Step 2: Place the integers you are comparing on the number line. Because both numbers are negative, they will both be to the left of zero.

Because −3 is farther to the right than −8, −3 is greater than −8.

Practice Your Skills!

1. Locate the following integers on the number line. Then list them in order from smallest to greatest on the line below.

   4, 12, −2, 7, −5, 2, −7, 9, −13

2. Use a number line to correctly place the sign > (greater than) or < (less than) between

Arithmetic with Positive and Negative Numbers

The **absolute value** of a number is its distance from zero on the number line. For example, −7 (a negative number) and 7 (a positive number) are the same distance from zero on the number line, and have an absolute value of 7. Using absolute values simplifies the process of doing arithmetic with positive and negative numbers.

1. Find the absolute value of the following numbers.

   a. −7   b. 14
   c. 325,000   d. −475
   e. 230   f. −52

Part 1: Adding Positive and Negative Numbers

PROCEDURE: Determine if you are adding numbers that have the same or different signs. Then follow the appropriate set of directions below.

	Adding same signs	Example −3 + (−5)	Adding opposite signs	Example −5 + 3
Step 1	Add their absolute values.	3 + 5 = 8	Subtract the smaller absolute value from the larger.	5 − 3 = 2
Step 2	Make the sign of the answer the same as the sign of the original numbers.	Because −3 and −5 are both negative, the answer will be negative. Answer: −3 + (−5) = −8	Choose the sign of the number with the greater absolute value.	Because 5 has a greater absolute value than 3, and 5 is positive, your answer will be positive. Answer: −5 + 3 = 2

Add It Up!

2. Complete the following equations. When finished, go back and check your signs.

   a. 14 + (−17) = ___   b. −9 + (−23) = ___
   c. −16 + 21 = ___   d. −12 + 12 = ___
   e. 15 + (−4) = ___   f. −7 + (−7) = ___

FRACTIONS

What Is a Fraction?

Suppose that you are doing an experiment in your class on the benefits of sunlight to plants. Your teacher has asked you to put $\frac{1}{4}$ of the plants in the sun. What does that mean? While whole numbers, such as 1 and 879, are used to indicate *how many*, **fractions** are used to tell *how much of a whole*.

The number below the fraction bar in a fraction is called the *denominator*. This number indicates how many parts there are in the whole. The number above the fraction bar, called the *numerator*, tells you how many parts of that whole are represented.

SAMPLE PROBLEM: Your class has 24 plants. Your teacher instructs you to put 5 in a shady spot. What fraction does this represent?

Step 1: Write the total number of parts in the whole as the denominator.

$$\frac{}{24}$$

Step 2: Write the number of parts of the whole being represented as the numerator.

$$\frac{5}{24}$$ of the plants will be in the shade.

Constructing Fractions

1. What fraction of the whole does the shaded or patterned part represent?

Reducing Fractions to Lowest Terms

Suppose you have the fraction $\frac{20}{30}$. These are pretty big numbers to deal with. Is there a simpler way to write the same fraction? Well, one common method is to write the fraction in lowest terms. A fraction in lowest terms uses the smallest numbers possible that have the same relationship as the numbers in the original fraction. A fraction in lowest terms is the simplest form of that fraction. Read on to learn how to reduce a fraction to lowest terms.

PROCEDURE: To reduce a fraction to lowest terms, first find all the numbers that divide evenly into the numerator and the denominator. These numbers are known as *factors*. Find the largest factor that is common to both the numerator and the denominator. This is known as the Greatest Common Factor (GCF). Then divide both the numerator and the denominator by the GCF.

SAMPLE PROBLEM: Reduce the fraction $\frac{20}{30}$ to lowest terms.

Step 1: List the factors of the numerator and denominator, and determine which is the largest factor in both lists, or the GCF.

factors of the numerator: 1, 2, 4, 5, 10, 20
factors of the denominator: 45, 1, 3, 5, 9, 15, 45

Step 2: Divide the numerator and the denominator by the GCF, which is 15.

$$\frac{30}{45} \text{ reduced to lowest terms is } \frac{2}{3}.$$

How Low Can You Go?

1. Reduce each fraction to lowest terms.

2. Circle the fractions below that are already written in lowest terms.

Improper Fractions and Mixed Numbers

An **improper fraction** is a fraction whose numerator is greater than its denominator, such as $\frac{9}{5}$. An improper fraction can be changed to a **mixed number**, which is a whole number and a fraction, such as $2\frac{1}{2}$. Likewise, a mixed number can be changed to an improper fraction when it is necessary for doing mathematical operations with these numbers.

PROCEDURE: To change an improper fraction to a mixed number, divide the numerator by the denominator and write the quotient as the whole number. If there is a remainder, place it over the denominator to make the fraction of the mixed number.

SAMPLE PROBLEM A: Change $\frac{17}{5}$ to a mixed number.

Step 1: Divide the numerator by the denominator.

17 ÷ 5 = 3, remainder 2

Step 2: Write the quotient as the whole number, and put the remainder over the original denominator as the fraction.

$$\frac{17}{5} = 3\frac{2}{5}$$

PROCEDURE: To change a mixed number to an improper fraction, multiply the denominator of the fraction by the whole number. Then add that product to the numerator. Finally, write the sum over the denominator.

SAMPLE PROBLEM B: Change $4\frac{2}{3}$ to an improper fraction.

Step 1: Multiply denominator by the whole number.

3 × 4 = 12

Step 2: Add the product to the numerator.

12 + 2 = 14

$$4\frac{2}{3} = \frac{14}{3}$$

1. Write True or False next to each equation.

Adding and Subtracting Fractions

Part 1: Adding and Subtracting Fractions with the Same Denominator

To add fractions with the same denominator, add the numerators and put the sum over the original denominator. To subtract fractions with the same denominator, subtract the numerators and put the difference over the original denominator.

SAMPLE PROBLEM A: $\frac{3}{7} + \frac{1}{7} = ?$	**SAMPLE PROBLEM B:** $\frac{8}{11} - \frac{3}{11} = ?$
Add the numerators, and put the sum over the original denominator.	Subtract numerators and put the difference over the original denominator.

Practice What You've Learned!

1. Add and subtract to complete the following equations. Reduce your answers to lowest terms.

Part 2: Adding and Subtracting Fractions with Different Denominators

Sometimes you have to add or subtract fractions that have different denominators. To do this, you must first need to rewrite your fractions so that they DO have the same denominator. Figuring out the **least common denominator** (LCD) of your fractions is the first step.

PROCEDURE: To find the least common denominator of two fractions, find the least common multiple of the denominators. In other words, look at the multiples of the numbers, and find out which they have in common. The common multiple with the lowest value is your LCD.

SAMPLE PROBLEM: What is the LCD of $\frac{1}{4}$ and $\frac{2}{3}$?

Step 1: List the multiples of 4.

(4 × 1) = 4, (4 × 2) = 8, (4 × 3) = 16, etc.

Step 2: List the multiples of 3.

(3 × 1) = 3, (3 × 2) = 6, (3 × 3) = 9, (3 × 4) = 12, etc.

The least common denominator of $\frac{1}{4}$ and $\frac{2}{3}$ is 12.

Multiplying and Dividing Fractions

Compared with adding and subtracting fractions, multiplying and dividing fractions is quite simple. Just follow the steps below to see how it is done.

PROCEDURE 1: To multiply fractions, multiply the numerators and the denominators together and reduce the fraction (if necessary).

SAMPLE PROBLEM A: $\frac{5}{9} \times \frac{7}{10} = ?$

Step 1: Multiply your numerators and denominators.	Step 2: Reduce.

PROCEDURE 2: To divide fractions, switch the numerator and denominator of the divisor (the number you divide by) to make that fraction's *reciprocal*. Then multiply the fraction and the reciprocal, and reduce if necessary.

SAMPLE PROBLEM B: $\frac{5}{6} \div \frac{2}{3} = ?$

Step 1: Rewrite the divisor as its reciprocal.	Step 2: Multiply the dividend by the reciprocal.	Step 3: Reduce.

Practice Your Skills!

1. Multiply and divide to complete the equations. Give your answers in lowest terms.

2. You have $23\frac{1}{4}$ L of saline solution. Every student in the class needs $1\frac{1}{4}$ L for an experiment. How many students can do the experiment?

3. Because of differences in gravity, your weight on the moon would be $\frac{1}{6}$ what it is on Earth. If you weigh 72 N, what would be your weight on the moon?

Math Skills for Science (continued)

RATIOS AND PROPORTIONS

What Is a Ratio?

Using Proportions and Cross-Multiplication

DECIMALS

Decimals and Fractions

Arithmetic with Decimals

PERCENTAGES

Parts of 100: Calculating Percentages

Percentages, Fractions, and Decimals

Working with Percentages and Proportions

POWERS OF 10

Counting the Zeros

Creating Exponents

SCIENTIFIC NOTATION

What Is Scientific Notation?

Multiplying and Dividing in Scientific Notation

SI MEASUREMENT AND CONVERSION

What Is SI?

A Formula for SI Catch-up

Math Skills for Science (continued)

GEOMETRY

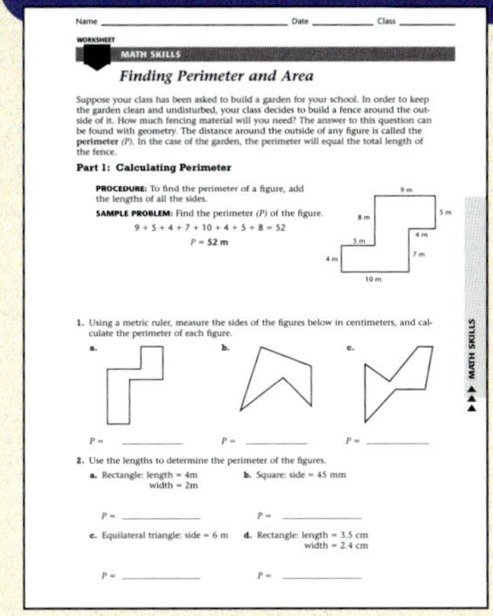

Finding Perimeter and Area

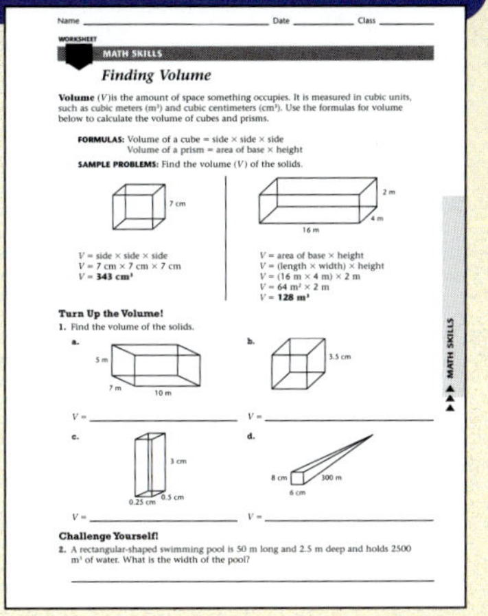

WORKSHEET
MATH SKILLS

Suppose your class has been asked to build a garden for your school. In order to keep the garden clean and undisturbed, your class decides to build a fence around the outside of it. How much fencing material will you need? The answer to this question can be found with geometry. The distance around the outside of any figure is called the **perimeter** (P). In the case of the garden, the perimeter will equal the total length of the fence.

Part 1: Calculating Perimeter

PROCEDURE: To find the perimeter of a figure, add the lengths of all the sides.

SAMPLE PROBLEM: Find the perimeter (P) of the figure.

$9 + 5 + 4 + 7 + 10 + 4 + 5 + 8 = 52$
$P = 52$ m

1. Using a metric ruler, measure the sides of the figures below in centimeters, and calculate the perimeter of each figure.

a. b. c.

$P =$ $P =$ $P =$

2. Use the lengths to determine the perimeter of the figures.

a. Rectangle: length = 4m
   width = 2m

b. Square: side = 45 mm

c. Equilateral triangle: side = 6 m

d. Rectangle: length = 3.5 cm
   width = 2.4 cm

$P =$ $P =$

$P =$ $P =$

Finding Volume

WORKSHEET
MATH SKILLS

Volume (V) is the amount of space something occupies. It is measured in cubic units, such as cubic meters (m^3) and cubic centimeters (cm^3). Use the formulas for volume below to calculate the volume of cubes and prisms.

FORMULAS: Volume of a cube = side × side × side
Volume of a prism = area of base × height

SAMPLE PROBLEMS: Find the volume (V) of the solids.

$V =$ side × side × side
$V = 7 cm × 7 cm × 7 cm$
$V = 343 cm^3$

$V =$ area of base × height
$V = $ (length × width) × height
$V = $ (16 m × 4 m) × 2 m
$V = 64 m^2 × 2 m$
$V = 128 m^3$

Turn Up the Volume!
1. Find the volume of the solids.

a. b.

$V =$ $V =$

c. d.

$V =$ $V =$

Challenge Yourself!
2. A rectangular-shaped swimming pool is 50 m long and 2.5 m deep and holds 2500 m^3 of water. What is the width of the pool?

THE UNIT FACTOR AND DIMENSIONAL ANALYSIS

The Unit Factor and Dimensional Analysis

WORKSHEET
MATH SKILLS

The measurements you take in science class, whether for time, mass, weight, or distance, are more than just numbers—they are also units. To make comparisons between measurements, it is convenient to have your measurements in the same units. A mathematical tool called a **unit factor** is used to convert back and forth between different kinds of units. A unit factor is a ratio that is equal to 1. Because it is equal to 1, multiplying a measurement by a unit factor changes the measurement's units but does not change its value. The skill of converting with a unit factor is known as **dimensional analysis**. Read on to see how it works.

Part 1: Converting with a Unit Factor

PROCEDURE: To convert units with a unit factor, determine the conversion factor between the units you have and the units you want to convert to. Then create the unit factor by making a ratio, in the form of a fraction, between the units you want to convert to in the numerator and the units you already have in the denominator. Finally, multiply your measurement by this unit factor to convert to the new units.

SAMPLE PROBLEM A: Convert 3.5 km to millimeters.

Step 1: Determine the conversion factor between kilometers and millimeters.
1 km = 1,000,000 mm

Step 2: Create the unit factor. Put the units you want to convert to in the numerator and the units you already have in the denominator.
$$\frac{1,000,000 \text{ mm}}{1 \text{ km}}$$

Step 3: Multiply the unit factor by the measurement. Notice that the original unit of the measurement cancels out with the unit in the denominator of the unit factor, leaving the units you are converting to.
$$3.5 \text{ km} \times \frac{1,000,000 \text{ mm}}{1 \text{ km}} = 3,500,000 \text{ mm}$$

On Your Own!
1. Convert the following measurements using a unit factor.

Conversion	Unit factor	Answer
a. 2.34 m = ? mm		
b. 54.6 mL = ? L		
c. 12 kg = ? g		

MATH IN SCIENCE: INTEGRATED SCIENCE

Density

WORKSHEET
MATH IN SCIENCE: INTEGRATED SCIENCE

Math Skills Used: Multiplication, Division, Decimals

Calculate density, and identify substances using a density chart.

Density is a measure of the amount of mass in a certain volume. This physical property is often used to identify and classify substances. It is usually measured in grams per cubic centimeters, or g/cm^3. The chart on the right lists the density of some common materials.

Densities of Substances

Substance	Density (g/cm³)
Gold	19.3
Mercury	13.5
Lead	11.4
Aluminum	2.7
Iron	7.87
Bone	1.7–2.0
Gasoline	0.66–0.69
Air (dry)	0.00119

FORMULA: $\text{density} = \frac{mass}{volume}$
$$D = \frac{m}{v}$$

SAMPLE PROBLEM: What is the density of a billiard ball that has a volume of 100 cm^3 and a mass of 250 g?
$$D = \frac{250 \text{ g}}{100 \text{ cm}^3}$$
$$D = 2.5 \text{ g/cm}^3$$

Your Turn!
1. A loaf of bread has a volume of 2270 cm^3 and a mass of 454 g. What is the density of the bread?

2. A liter of water has a mass of 1000 g. What is the density of water? (Hint: 1 mL = 1 cm^3)

3. A block of wood has a density of 0.6 g/cm^3 and a volume of 1.2 cm^3. What is the mass of the block of wood? Be careful!

4. Use the data below to calculate the density of each unknown substance. Then use the density chart above to determine the identity of each substance.

Mass (g)	Volume (cm³)	Density (g/cm³)	Substance
Example: 4725	350	4725 ÷ 350 = 13.5	mercury
a. 171	15		
b. 148	40		
c. 475	250		
d. 680	1000		

The Pressure Is On!

WORKSHEET
MATH IN SCIENCE: INTEGRATED SCIENCE

Math Skills Used: Multiplication, Division, Decimals, Percentages, Geometry

Use math to learn about force and pressure.

You are under pressure! Even though you may not be aware of it, the air above you presses down on every square centimeter of your body with the weight of a 1.03 kg mass! Because water is so much denser than air, pressure in water is many times greater than this. Pressure is defined as the force exerted on a particular area. The unit for pressure is the pascal (Pa), which is the force one newton (N) exerts on one square meter (m^2).

FORMULA: $\text{Pressure (Pa)} = \frac{\text{Force (N)}}{\text{Area (m}^2)}$ $Pa = \frac{N}{m^2}$

Apply Some Pressure!
Use the formula for pressure to answer the following questions:

1. An elephant that weighs 40,000 N stands on one leg during a circus performance. The area on the bottom of the elephant's foot is 0.4 m^2. How much pressure is exerted on the elephant's foot?

2. A carpenter hammers a nail with a force of 45 N with every stroke. The head of the nail has a surface area of 0.002 m^2. How much pressure is exerted on the nailhead with each stroke?

3. A brick falls from the third floor of a construction site. The brick hits the ground on its end, which measures 0.15 m by 0.25 m, with a force of 30 N. How much pressure is exerted by the brick on the ground? (Hint: Area of a rectangle = width × length)

Pressure in the Atmosphere
The air pressure we live under is about 101,000 Pa at sea level. Use this value to complete the following problems. Show all your work.

4. A mountain climber climbs to the top of Mt. Everest, which at 8848 m is the highest point on Earth. Because most of the air in the atmosphere is below this altitude, air pressure is about 30% less at the peak than at sea level. What is the air pressure exerted on the mountain climber?

5. A meteorologist reports that air pressure is reduced to 6,585 Pa by an approaching hurricane. What percentage change from normal air pressure does this represent?

Sound Reasoning

WORKSHEET
MATH IN SCIENCE: INTEGRATED SCIENCE

Math Skills Used: Multiplication, Division, Decimals

Use your math skills to understand dolphin echolocation.

Dolphins use echolocation to find their way through murky waters. They do this by emitting a clicking sound and listening for an echo. The direction and delay of the echo give the dolphins information about what objects are nearby and where the objects are located.

1. Sound travels about 1530 m/s in sea water. How many times faster does sound travel in sea water than in air? (The speed of sound in air at 25°C is about 345 m/s.)

2. A dolphin emits a click that is reflected off an object. If it takes 0.2 seconds for the sound to be sent and to come back, how far away is the object?

3. How long would it take the sound to be sent and returned from the same object in air?

4. Assume that the speed of sound decreases by 6 m/s for every 10°C decrease in water temperature. If a dolphin swam to the Arctic Ocean, where the water is about 5°C, how would the dolphin's ability to estimate the distance to an object be affected?

Using Temperature Scales

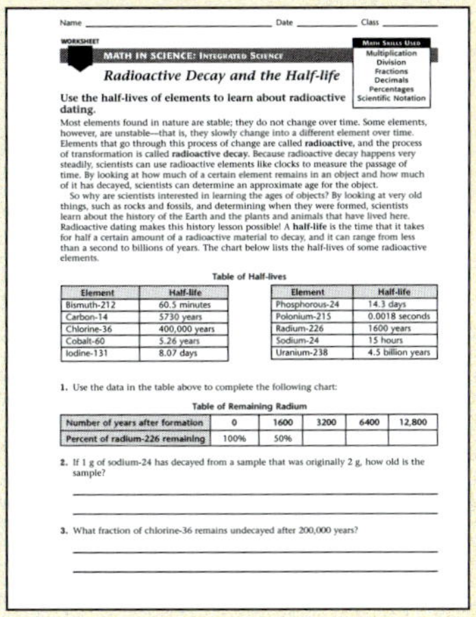

WORKSHEET
MATH IN SCIENCE: INTEGRATED SCIENCE

Math Skills Used: Addition, Multiplication, Fractions, Decimals, Percentages, SI Measurement and Conversion

Convert between degrees Fahrenheit and degrees Celsius.

Do you remember the last time you had your temperature taken? Your body temperature is usually about 98.6°F. This temperature is in degrees Fahrenheit (°F). The Fahrenheit temperature scale is a common temperature scale. In science class, however, a metric scale known as the Celsius (°C) scale is used. Temperatures in one scale can be mathematically converted to the other system using one of the formulas below.

FORMULAS: Conversion from Fahrenheit to Celsius: $\frac{5}{9} \times (°F − 32) = °C$
Conversion from Celsius to Fahrenheit: $\frac{9}{5} \times °C + 32 = °F$

SAMPLE PROBLEMS:
A. Convert 59°F to degrees Celsius.
$°C = \frac{5}{9} \times (°F − 32)$
$°C = \frac{5}{9} \times (59 − 32)$
$°C = \frac{5}{9} \times 27$
$°C = 15°C$

B. Convert 112°C to degrees Fahrenheit.
$°F = \frac{9}{5} \times °C + 32$
$°F = \frac{9}{5} \times 112 + 32$
$°F = 201\frac{3}{5} + 32$
$°F = 233\frac{3}{5}°F$

Turn Up the Temperature!
1. Convert the following temperatures from degrees Fahrenheit to degrees Celsius:

a. 98.6°F

b. 482°F

c. −4°F

2. Convert the following temperatures from degrees Celsius to degrees Fahrenheit:

a. 24°C

b. 17°C

c. 0°C

Challenge Yourself!
3. Convert 2.7×10^{14}°C to degrees Fahrenheit.

Radioactive Decay and the Half-life

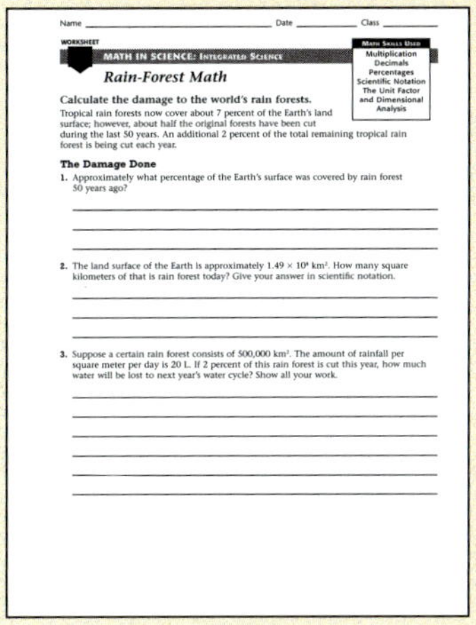

WORKSHEET
MATH IN SCIENCE: INTEGRATED SCIENCE

Math Skills Used: Multiplication, Division, Fractions, Decimals, Scientific Notation

Use the half-lives of elements to learn about radioactive dating.

Most elements found in nature are stable; they do not change over time. Some elements, however, are unstable—that is, they slowly change into a different element over time. Elements that go through this process of change are called **radioactive**, and the process of transformation is called **radioactive decay**. Because radioactive decay happens very steadily, scientists can use radioactive elements like clocks to measure the passage of time. By looking at how much of a certain element remains in an object and how much of it has decayed, scientists can determine an approximate age for the object.

So why are scientists interested in learning the ages of objects? By looking at very old things, such as rocks and fossils, and determining when they were formed, scientists learn about the history of the Earth and the plants and animals that have lived here. Radioactive dating makes this history lesson possible! A **half-life** is the time that it takes for half a certain amount of a radioactive material to decay, and it can range from less than a second to billions of years. The chart below lists the half-lives of some radioactive elements.

Table of Half-lives

Element	Half-life	Element	Half-life
Bismuth-212	60.5 minutes	Phosphorous-24	14.3 days
Carbon-14	5730 years	Polonium-213	0.0018 seconds
Chlorine-36	400,000 years	Radium-226	1600 years
Cobalt-60	5.26 years	Sodium-24	15 hours
Iodine-131	8.07 days	Uranium-238	4.5 billion years

1. Use the data in the table above to complete the following chart.

Table of Remaining Radium

Number of years after formation	0	1600	3200	6400	12,800
Percent of radium-226 remaining	100%	50%			

2. If 1 g of sodium-24 has decayed from a sample that was originally 2 g, how old is the sample?

3. What fraction of chlorine-36 remains undecayed after 200,000 years?

Rain-Forest Math

WORKSHEET
MATH IN SCIENCE: INTEGRATED SCIENCE

Math Skills Used: Multiplication, Decimals, Percentages, Scientific Notation, The Unit Factor and Dimensional Analysis

Calculate the damage to the world's rain forests.

Tropical rain forests now cover about 7 percent of the Earth's land surface; however, about half the original forests have been cut during the last 50 years. An additional 2 percent of the total remaining tropical rain forest is being cut each year.

The Damage Done
1. Approximately what percentage of the Earth's surface was covered by rain forest 50 years ago?

2. The land surface of the Earth is approximately 1.49 × 10⁸ km². How many square kilometers of that is rain forest today? Give your answer in scientific notation.

3. Suppose a certain rain forest consists of 500,000 km². The amount of rainfall per square meter per day is 20 L. If 2 percent of this rain forest is cut this year, how much water will be lost to next year's water cycle? Show all your work.

Math Skills for Science (continued)

MATH IN SCIENCE: PHYSICAL SCIENCE

Average Speed in a Pinewood Derby

Newton: Force and Motion

Momentum

Balancing Chemical Equations

Work and Power

A Bicycle Trip

Mechanical Advantage

Color at Light Speed

Assessment Checklists & Rubrics

The following is just a sample of over 50 checklists and rubrics contained in this booklet.

RUBRICS FOR WRITTEN WORK

RUBRIC FOR EXPERIMENTS

TEACHER EVALUATION OF COOPERATIVE LEARNING

TEACHER EVALUATION OF STUDENT PROGRESS

HOLT SCIENCE & TECHNOLOGY

PHYSICAL SCIENCE
NATIONAL SCIENCE EDUCATION STANDARDS CORRELATIONS

The following lists show the chapter correlation of **Holt Science and Technology: Introduction to Matter** with the *National Science Education Standards* (grades 5-8)

UNIFYING CONCEPTS AND PROCESSES

Standard	Chapter Correlation	
Systems, order, and organization Code: UCP 1	Chapter 2 Chapter 3 Chapter 4 Chapter 5	2.1, 2.2 3.1, 3.2, 3.3 4.1, 4.2 5.1
Evidence, models, and explanation Code: UCP 2	Chapter 2 Chapter 3 Chapter 4 Chapter 5	2.1, 2.2 3.1, 3.2, 3.3 4.1, 4.2 5.2
Change, constancy, and measurement Code: UCP 3	Chapter 2 Chapter 4 Chapter 5	2.1, 2.2 4.3 5.2

SCIENCE IN PERSONAL AND SOCIAL PERSPECTIVES

Standard	Chapter Correlation	
Personal health Code: SPSP 1	Chapter 1 Chapter 3 Chapter 5	1.1 3.3 5.2
Natural hazards Code: SPSP 3	Chapter 2 Chapter 3	2.2 3.3
Science and technology in society Code: SPSP 5	Chapter 1 Chapter 2 Chapter 3 Chapter 4 Chapter 5	1.1 2.2 3.2, 3.3 4.1 5.1, 5.2

SCIENCE AS INQUIRY

Standard	Chapter Correlation	
Abilities necessary to do scientific inquiry Code: SAI 1	Chapter 2 Chapter 3 Chapter 5	2.1, 2.2 3.1, 3.3 5.1
Understandings about scientific inquiry Code: SAI 2	Chapter 2 Chapter 3 Chapter 4	2.1, 2.2 3.2, 3.3 4.1

SCIENCE AND TECHNOLOGY

Standard	Chapter Correlation	
Abilities of technological design Code: ST 1	Chapter 1 Chapter 2	1.1 2.2
Understandings about science and technology Code: ST 2	Chapter 1 Chapter 3 Chapter 4 Chapter 5	1.1 3.3 4.1 5.1, 5.2

HISTORY AND NATURE OF SCIENCE

Standard	Chapter Correlation	
Science as a human endeavor Code: HNS 1	Chapter 4 Chapter 5	4.1 5.1
Nature of science Code: HNS 2	Chapter 4 Chapter 5	4.1, 4.2 5.1
History of science Code: HNS 3	Chapter 2 Chapter 4 Chapter 5	2.2 4.1 5.1, 5.2

PHYSICAL SCIENCE National Science Education Content Standards

PROPERTIES AND CHANGES OF PROPERTIES IN MATTER	
Standard	**Chapter Correlation**
A substance has characteristic properties, such as density, a boiling point, and solubility, all of which are independent of the amount of the sample. A mixture of substances often can be separated into the original substances using one or more of the characteristic properties. Code: PS 1a	**Chapter 1** 1.1, 1.2 **Chapter 2** 2.1, 2.2 **Chapter 3** 3.1, 3.3
Substances react chemically in characteristic ways with other substances to form new substances (compounds) with different characteristic properties. In chemical reactions, the total mass is conserved. Substances often are placed in categories or groups if they react in similar ways; metals is an example of such a group. Code: PS 1b	**Chapter 1** 1.2 **Chapter 3** 3.1, 3.2 **Chapter 4** 4.1, 4.2
Chemical elements do not break down during normal laboratory reactions involving such treatments as heating, exposure to electric current, or reaction with acids. There are more than 100 known elements that combine in a multitude of ways to produce compounds, which account for the living and nonliving substances that we encounter. Code: PS 1c	**Chapter 1** 1.1 **Chapter 3** 3.1, 3.2

TRANSFER OF ENERGY	
Standard	**Chapter Correlation**
Energy is a property of many substances and is associated with heat, light, electricity, mechanical motion, sound, nuclei, and the nature of a chemical. Energy is transferred in many ways. Code: PS 3a	**Chapter 2** 2.2
In most chemical and nuclear reactions, energy is transferred into or out of a system. Heat, light, mechanical motion, or electricity might all be involved in such transfers. Code: PS 3e	**Chapter 5** 5.2

Master Materials List

For added convenience, Science Kit® provides materials-ordering software on CD-ROM designed specifically for *Holt Science and Technology*. Using this software, you can order complete kits or individual items, quickly and efficiently.

CONSUMABLE MATERIALS	AMOUNT	PAGE
Bag, paper lunch	1	3, 120
Baking powder	1 tsp	20
Baking soda	1 tsp	20, 59
Balloon, round	1	138
Calcium chloride solution	approx. 10 mL	70
Can, aluminum	2	139
Cardboard, approx. 60 x 30 cm	1	79
Carton, egg, plastic-foam	1	20
Charcoal, activated	4 oz	142
Corn syrup, dark	20 mL	137
Cornstarch	1 tsp	20
Craft stick	1	70
Cream, heavy	4–6 fl oz	141
Cup, plastic-foam	1	109
Cup, plastic-foam	2	142
Dirt	1 cup	142

CONSUMABLE MATERIALS	AMOUNT	PAGE
Filter, coffee	1	53
Filter paper	2	142
Food coloring, dark red	1 bottle	137, 142
Food coloring, red, yellow, blue	1 bottle each	132
Gloves, protective	1 pair	70, 137, 142
Graphite for mechanical pencil	1	109
Hydrochloric acid 0.1 M (0.1N)	15 mL	70
Ice, crushed	3 cups	44
Ice cube	2–3	138
Iodine solution	5 drops	20
Marker, permanent, black	1	132
Marker, water-soluble, black	1	53
Oil, vegetable	10 mL	137
Paper, graphing	1 sheet	136
Paper, graphing	2 sheets	44, 120
Paper, white, approx. 60 x 30 cm	1 sheet	79

CONSUMABLE MATERIALS	AMOUNT	PAGE
Potassium chloride solution	approx. 10 mL	70
Rock salt	1 lb	44
Rubber band	1–2	142
Rubbing alcohol	25 mL	29
Sand, fine	7 oz	142
Sodium chloride solution	approx. 10 mL	70
Sugar cube	2	140
Sugar, granulated	1 tsp	20
Sugar, powdered	1 tsp	59
Swab, cotton	1	29
Tape, masking, 20–30 cm	1	70, 132
Tape, transparent, 20–30 cm	1	53
Toothpicks	20	94
Vinegar	5 mL	20
Vinegar	10 mL	59
Vinegar	50 mL	17

NONCONSUMABLE EQUIPMENT	AMOUNT	PAGE
Balance, metric	1	120, 133, 136
Ball, styrene, 2–3 cm diam.	10	94
Beaker, large (or bucket)	1	4
Beaker, 100 mL	1	137
Beaker, 100 mL	3	132
Beaker, 250 mL	1	44, 138
Beaker, 250 mL	2	70, 140
Beaker, 250 mL	4	142
Beaker, 1 L	1	139
Bottle, 2 L	1	134
Bunsen burner, adjustable	1	70
Can, coffee	1	44
Cup, clear plastic	1–2	4, 29, 53, 59
Dropper, plastic	1–3	20
Funnel	1–3	132, 134, 137
Gloves, heat-resistant	1 pair	44, 138, 139
Graduated cylinder, 10 mL	1	132

Nonconsumable Equipment	Amount	Page
Graduated cylinder, 10 mL	3	137
Graduated cylinder, 100 mL	1	44, 133, 134, 136, 140, 142
Hot plate	1	44, 138, 139
Igniter	1	70
Jar, clear plastic, with lid	1	141
Marble	1	79, 141
Marble	8–10	136
Nail, small	1	142
Pan, aluminum pie	1	134
Pan, aluminum, 2.5 in. deep	2	138
Penny, shiny	1–3	17
Penny, post-1982	5	133
Penny, pre-1982	5	133

Nonconsumable Equipment	Amount	Page
Pie plate, small	1	17
Ruler, metric	1	120, 138, 142
Scissors	1	142
Spatula	4	20
Spoon, plastic	2	142
Stirring rod	1	20
Stopwatch	1	44, 140, 141
Syringe, small	1	41
Test tube	2–6	70, 132
Test-tube rack	1	70, 132
Thermometer	1	44
Tongs, beaker with plain jaws	1	139
Wire, bare copper, 25 cm	1	44, 109
Wire, bare Nichrome®, 30 cm	1	70

Answers to Concept Mapping Questions

The following pages contain sample answers to all of the concept mapping questions that appear in the Chapter Reviews. Because there is more than one way to do a concept map, your students' answers may vary.

CHAPTER 1 The Properties of Matter

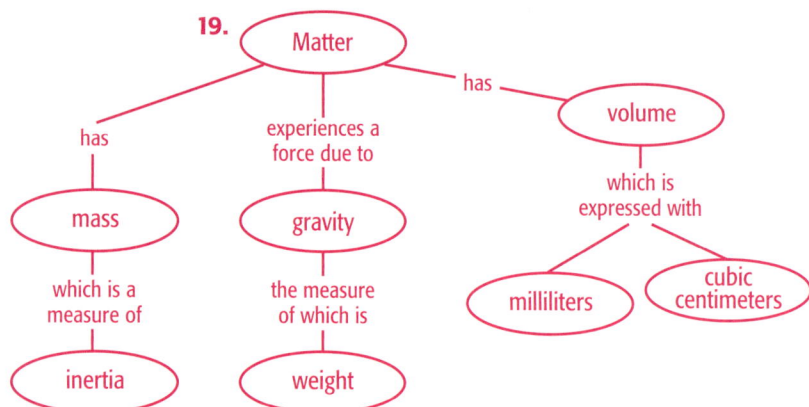

19.

Matter
- has → mass → which is a measure of → inertia
- experiences a force due to → gravity → the measure of which is → weight
- has → volume → which is expressed with → milliliters / cubic centimeters

CHAPTER 2 States of Matter

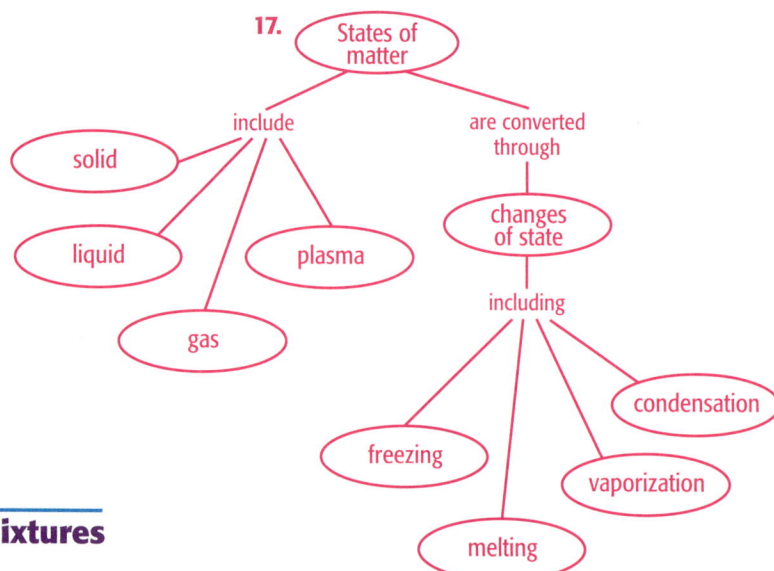

17.

States of matter
- include → solid, liquid, plasma, gas
- are converted through → changes of state → including → freezing, melting, vaporization, condensation

CHAPTER 3 Elements, Compounds, and Mixtures

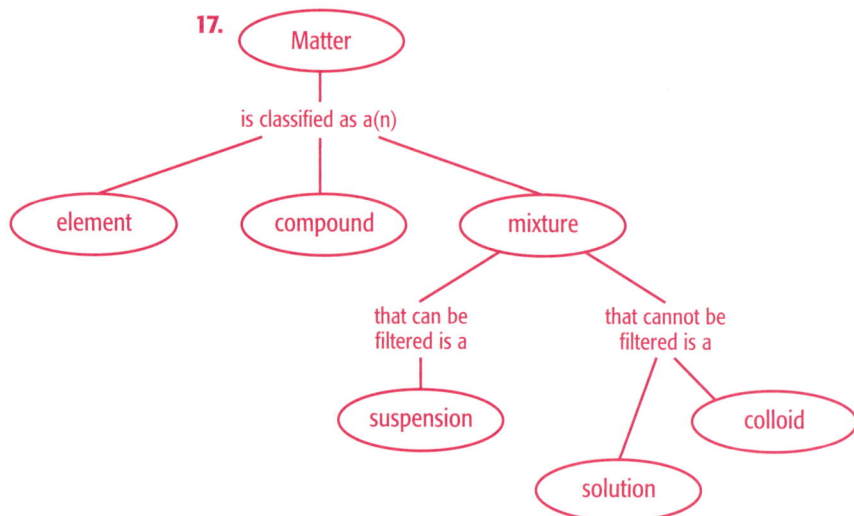

17.

Matter
- is classified as a(n) → element, compound, mixture
  - mixture that can be filtered is a → suspension
  - mixture that cannot be filtered is a → solution, colloid

17.

An atom

contains a

nucleus

surrounded by

composed of

electrons

protons

neutrons

which determine the

which are
different in

which are added
together to find the

atomic
number

which is the
same in

isotopes

mass number

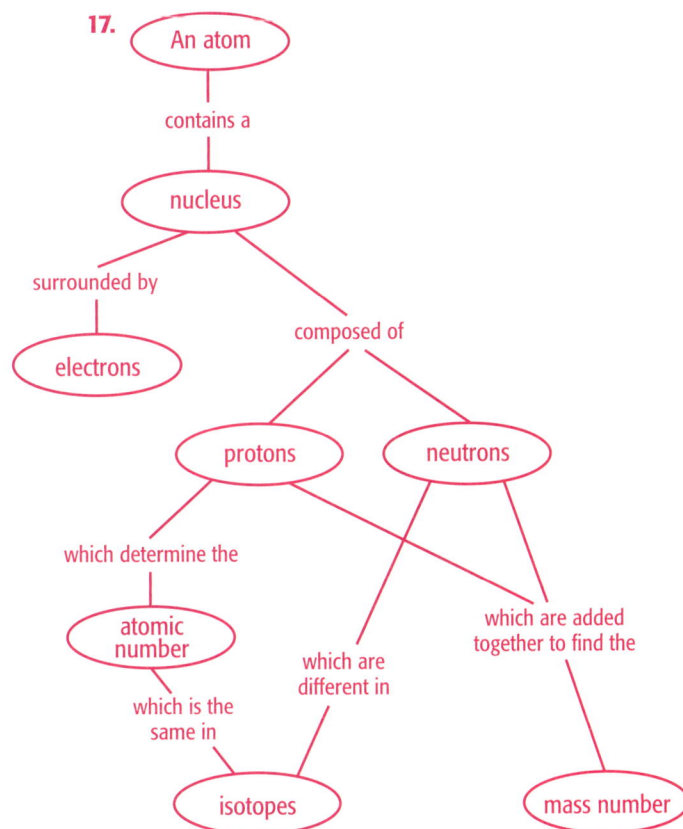

CHAPTER 5 The Periodic Table

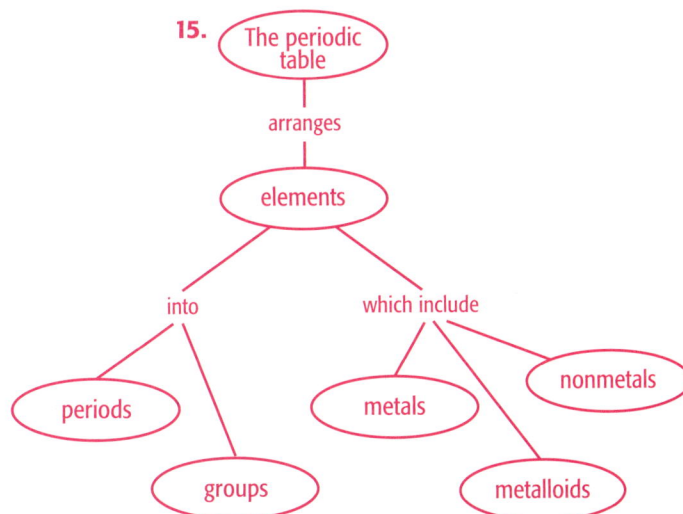

15.

The periodic
table

arranges

elements

into

which include

periods

metals

nonmetals

groups

metalloids

To the Student

This book was created to make your science experience interesting, exciting, and fun!

Go for It!

Science is a process of discovery, a trek into the unknown. The skills you develop using *Holt Science & Technology*— such as observing, experimenting, and explaining observations and ideas— are the skills you will need for the future. There is a universe of exploration and discovery awaiting those who accept the challenges of science.

Science & Technology

You see the interaction between science and technology every day. Science makes technology possible. On the other hand, some of the products of technology, such as computers, are used to make further scientific discoveries. In fact, much of the scientific work that is done today has become so technically complicated and expensive that no one person can do it entirely alone. But make no mistake, the creative ideas for even the most highly technical and expensive scientific work still come from individuals.

Activities and Labs

The activities and labs in this book will allow you to make some basic but important scientific discoveries on your own. You can even do some exploring on your own at home! Here's your chance to use your imagination and curiosity as you investigate your world.

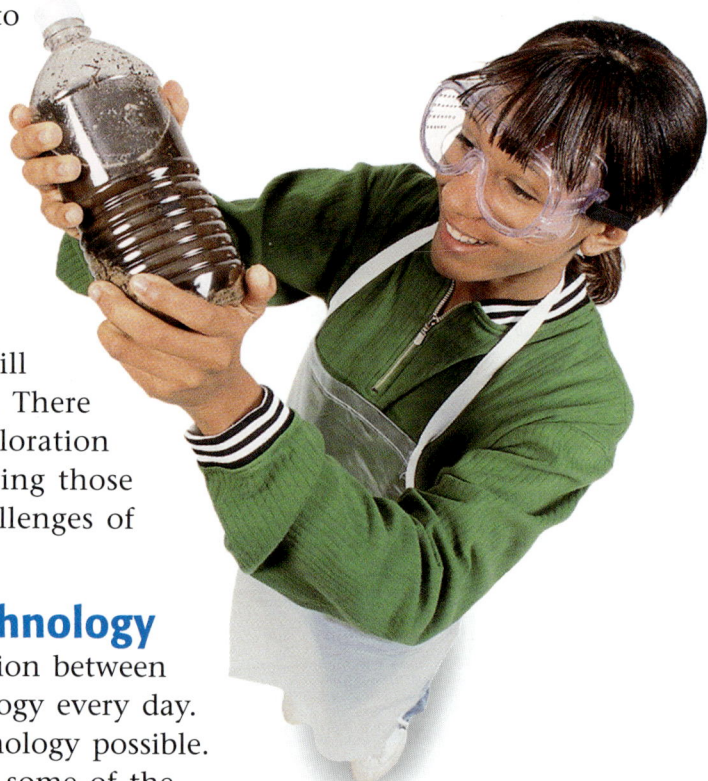

Keep a ScienceLog

In this book, you will be asked to keep a type of journal called a ScienceLog to record your thoughts, observations, experiments, and conclusions. As you develop your ScienceLog, you will see your own ideas taking shape over time. You'll have a written record of how your ideas have changed as you learn about and explore interesting topics in science.

Know "What You'll Do"

The "What You'll Do" list at the beginning of each section is your built-in guide to what you need to learn in each chapter. When you can answer the questions in the Section Review and Chapter Review, you know you are ready for a test.

Check Out the Internet

You will see this *sciLINKS* logo throughout the book. You'll be using *sci*LINKS as your gateway to the Internet. Once you log on to *sci*LINKS using your computer's Internet link, type in the *sci*LINKS address. When asked for the keyword code, type in the keyword for that topic. A wealth of resources is now at your disposal to help you learn more about that topic.

In addition to *sci*LINKS you can log on to some other great resources to go with your text. The addresses shown below will take you to the home page of each site.

internet connect

This textbook contains the following on-line resources to help you make the most of your science experience.

go. hrw .com

Visit **go.hrw.com** for extra help and study aids matched to your textbook. Just type in the keyword HB2 HOME.

SCi LINKS NSTA

Visit **www.scilinks.org** to find resources specific to topics in your textbook. Keywords appear throughout your book to take you further.

Smithsonian Institution® Internet Connections

Visit **www.si.edu/hrw** for specifically chosen on-line materials from one of our nation's premier science museums.

CNNfyi.com

Visit **www.cnnfyi.com** for late-breaking news and current events stories selected just for you.

Chapter Organizer

CHAPTER ORGANIZATION	TIME MINUTES	OBJECTIVES	LABS, INVESTIGATIONS, AND DEMONSTRATIONS	
Chapter Opener pp. 2–3	45	National Standards: SAI 2, PS 1a	**Start-Up Activity,** Sack Secrets, p. 3	
Section 1 **What Is Matter?**	90	▶ Name the two properties of all matter. ▶ Describe how volume and mass are measured. ▶ Compare mass and weight. ▶ Explain the relationship between mass and inertia. ST 1, 2, SPSP 1, 5, PS 1a, 1c; Labs PS 1a	**QuickLab,** Space Case, p. 4 **Demonstration,** p. 4 in ATE **Skill Builder,** Measuring Liquid Volume, p. 132 **Datasheets for LabBook,** Measuring Liquid Volume **Demonstration,** p. 6 in ATE **Skill Builder,** Volumania! p. 134 **Datasheets for LabBook,** Volumania!	
Section 2 **Describing Matter**	90	▶ Give examples of matter's different properties. ▶ Describe how density is used to identify different substances. ▶ Compare physical and chemical properties. ▶ Explain what happens to matter during physical and chemical changes. PS 1a, 1b; Labs PS 1a, 1b	**Demonstration,** p. 11 in ATE **Skill Builder,** Coin Operated, p. 133 **Datasheets for LabBook,** Coin Operated **QuickLab,** Changing Change, p. 17 **Skill Builder,** Determining Density, p. 136 **Datasheets for LabBook,** Determining Density **Discovery Lab,** Layering Liquids, p. 137 **Datasheets for LabBook,** Layering Liquids **Skill Builder,** White Before Your Eyes, p. 20 **Datasheets for LabBook,** White Before Your Eyes **Inquiry Labs,** Whatever Floats Your Boat **Whiz-Bang Demonstrations,** Curious Cubes **Whiz-Bang Demonstrations,** The Dancing Toothpicks **Whiz-Bang Demonstrations,** Does 2 + 2 = 4? **Long-Term Projects & Research Ideas,** And We Have Thales to Thank	

See page **T23** *for a complete correlation of this book with the*

NATIONAL SCIENCE EDUCATION STANDARDS.

TECHNOLOGY RESOURCES

Guided Reading Audio CD
English or Spanish, Chapter 1

Science Discovery Videodiscs
Image and Activity Bank with Lesson Plans: In Search of Chemical Change, Material Science

CNN **Scientists in Action,** Neutrino Breakthrough, Segment 4

One-Stop **Planner CD-ROM with Test Generator**

Chapter 1 • The Properties of Matter

CLASSROOM WORKSHEETS, TRANSPARENCIES, AND RESOURCES	SCIENCE INTEGRATION AND CONNECTIONS	REVIEW AND ASSESSMENT
Directed Reading Worksheet **Science Puzzlers, Twisters & Teasers**		
Directed Reading Worksheet, Section 1 **Math Skills for Science Worksheet,** The Unit Factor and Dimensional Analysis **Transparency 95,** Growth Chart **Transparency 203,** How Mass and Distance Affect Gravity Between Objects **Transparency 204,** Differences Between Mass and Weight **Science Skills Worksheet,** Measuring	**Cross-Disciplinary Focus,** p. 5 in ATE **MathBreak,** Calculating Volume, p. 6 **Math and More,** p. 6 in ATE **Biology Connection,** p. 8 **Connect to Life Science,** p. 8 in ATE **Multicultural Connection,** p. 9 in ATE **Apply,** p. 10 **Across the Sciences:** In the Dark About Dark Matter, p. 26	**Self-Check,** p. 9 **Section Review,** p. 10 **Quiz,** p. 10 in ATE **Alternative Assessment,** p. 10 in ATE
Directed Reading Worksheet, Section 2 **Math Skills for Science Worksheet,** Density **Transparency 205,** Examples of Chemical Changes **Reinforcement Worksheet,** A Matter of Density **Critical Thinking Worksheet,** As a Matter of Fact!	**Real-World Connection,** p. 12 in ATE **MathBreak,** Density, p. 13 **Math and More,** p. 13 in ATE **Apply,** p. 14 **Real-World Connection,** p. 15 in ATE **Environment Connection,** p. 19 **Health Watch:** Building a Better Body, p. 27	**Homework,** p. 12 in ATE **Section Review,** p. 14 **Section Review,** p. 19 **Quiz,** p. 19 in ATE **Alternative Assessment,** p. 19 in ATE

internet connect

go.hrw.com
Holt, Rinehart and Winston On-line Resources
go.hrw.com
For worksheets and other teaching aids related to this chapter, visit the HRW Web site and type in the keyword: **HSTMAT**

sciLINKS NSTA
National Science Teachers Association
www.scilinks.org
Encourage students to use the sciLINKS numbers listed in the internet connect boxes to access information and resources on the **NSTA** Web site.

END-OF-CHAPTER REVIEW AND ASSESSMENT

Chapter Review in Study Guide
Vocabulary and Notes in Study Guide
Chapter Tests with Performance-Based Assessment, Chapter 1 Test
Chapter Tests with Performance-Based Assessment, Performance-Based Assessment 1
Concept Mapping Transparency 2

Chapter Resources & Worksheets

Visual Resources

TEACHING TRANSPARENCIES

#203

How Mass and Distance Affect Gravity Between Objects

Holt Science and Technology

Teaching Transparency 203

#204

Holt Science and Technology — Teaching Transparency 204

Differences Between Mass and Weight

Mass is . . .
- a measure of the amount of matter in an object.
- always constant for an object no matter where the object is in the universe.
- measured with a balance (shown below).
- expressed in kilograms (kg), grams (g), and milligrams (mg).

Weight is . . .
- a measure of the gravitational force on an object.
- varied depending on where the object is in relation to the Earth (or any other large body in the universe).
- measured with a spring scale (shown above).
- expressed in newtons (N).

#205

Examples of Chemical Changes

Holt Science and Technology

Teaching Transparency 205

TEACHING TRANSPARENCIES

#95

Growth Chart

Infant   4 years   7 years   11 years   Adult

Holt Science and Technology

95

LINK TO LIFE SCIENCE

CONCEPT MAPPING TRANSPARENCY

#2

Holt Science and Technology — Concept Mapping Transparency 2

The Properties of Matter

Use the following terms to complete the concept map below: weight, milliliters, mass, cubic centimeters, matter, motion, volume, gravity

has

which is used to measure

which is measured in

SI units

inertia

which tends to resist any change in

which is the force on an object exerted by

such as

Meeting Individual Needs

DIRECTED READING

#1

DIRECTED READING WORKSHEET

The Properties of Matter

Chapter Introduction

As you begin this chapter, answer the following.
1. Read the title of the chapter. List three things that you already know about this subject.

2. Write two questions about this subject that you would like answered by the time you finish this chapter.

Section 1: What Is Matter?
3. What do a human, hot soup, and a neon sign have in common?

Everything Is Made of Matter (p. 4)
4. Anything that has _____ and _____ is called matter.

REINFORCEMENT & VOCABULARY REVIEW

#1

REINFORCEMENT WORKSHEET

A Matter of Density

Complete this worksheet after you finish reading Chapter 2, Section 2.

Imagine that you work at a chemical plant. This morning, four different liquid chemicals accidentally spilled into the same tank. Luckily, none of the liquids reacted with each other! The sides of the tank are made of steel, so you can only see the surface of what's inside. But you need to remove the red chemical to use in a reaction later this afternoon. How will you find and remove the red chemical? By finding the chemicals' different densities, of course!

The following liquids were spilled into the tank.
- a green liquid that has a volume of 48 m³ and a mass of 36 kg.
- a red liquid that has a volume of 96 m³ and a mass of 115.2 kg.
- a blue liquid that has a volume of 144 m³ and a mass of 129.6 kg.
- a black liquid that has a volume of 120 m³ and a mass of 96 kg.
1. Calculate the density of each liquid.
   Green liquid: _____   Red liquid: _____
   Blue liquid: _____   Black liquid: _____
2. Determine the order in which the liquids have settled in the tank.
   First (bottom): _____   Third: _____
   Second: _____   Fourth (top): _____
3. Use colored pencils to sketch the liquid layers in the container.
4. Now that you know where the red chemical is inside the tank, how could you remove it?

REINFORCEMENT & VOCABULARY REVIEW

#1

VOCABULARY REVIEW WORKSHEET

Search for Matter

Complete the puzzle after you've finished Chapter 2.
Fill in each blank with the correct word. Then find the words in the puzzle. Words in the puzzle can be spelled forward or backward, and can be vertical, horizontal, or diagonal.
1. The tendency of an object to resist any change in its motion is called _____.
2. When water is in a container, the edge of the surface of the water is curved. This curve is called the _____.
3. The amount of space taken up by an object is its _____.
4. Iron _____ is also known as fool's gold.
5. The _____ of an object is the amount of matter in the object. The SI unit for measuring this quantity is the _____.
6. The force that pulls any object toward Earth is called _____. The measure of this force is the object's _____. The SI unit for measuring this force is the _____.
7. _____ is anything that has volume and mass.
8. The _____ of an object or liquid is its mass divided by volume.
9. A _____ change occurs when one or more substances are changed into entirely new substances.
10. Examples of _____ properties are color and odor.
11. The properties that are most useful in identifying a substance are its _____ properties.

W	P	F	X	D	E	N	S	I	T	Y	E	P	C
R	E	V	Q	C	J	N	D	Q	W	M	I	I	J
B	P	I	N	E	W	T	O	N	U	A	T	G	K
A	E	F	G	E	X	J	O	L	N	S	I	K	I
G	X	C	J	H	H	P	O	D	I	I	I	K	L
R	Y	M	H	R	T	V	V	R	C	N	Q	P	O
A	S	A	K	E	T	S	E	M	A	E	X	H	G
V	T	S	L	D	M	T	N	F	M	R	U	Y	R
I	W	S	N	N	C	I	M	V	X	T	Z	S	A
T	Y	U	K	A	C	G	C	A	X	I	N	I	M
Y	O	D	R	J	I	N	T	A	T	A	Q	C	M
P	T	A	P	Y	R	I	T	E	L	T	R	A	W
C	H	Z	M	M	P	V	Q	P	B	Z	E	L	B
C	T	Z	C	M	E	N	I	S	C	U	S	R	P

SCIENCE PUZZLERS, TWISTERS & TEASERS

#1

SCIENCE PUZZLERS, TWISTERS & TEASERS

The Properties of Matter

Let's Go Bowling
1. The 7-10 Splits, a bowling team, won their Tuesday night league championship. To celebrate, they threw a party. Here are some events that occurred during the party.

   a. Randy, the captain of the bowling team, set up the ten pins in his basement bowling lane and got out his two favorite bowling balls. After a quick spit-polish, his teammate, Nigel, rolled the purple ball down the lane toward the pins. Nigel knocked down all ten pins. He then tossed the green ball to his friend Basil, who tossed it back, declaring, "I never use a green ball on a full moon. It's bad luck." So Nigel tossed him the purple ball. Nigel noted that the green ball was more difficult to throw than the purple ball. Basil noticed that the green ball was more difficult to stop than the purple ball. Both balls are exactly the same size. Why might the purple ball be easier to set into motion and stop moving than the green ball?

   b. For the party, Basil baked a victory cake shaped like a huge bowling ball. On the cake he placed 10 candles, one for each team victory in the championship tournament. The candles burned for 15 minutes before the team blew them out, leaving puddles of wax on top of the cake. What kind of change did the wax undergo?

   c. What kind of change occurred with the wick?

   d. Basil's enormous cat Binkie ate three large pieces of cake. Basil's tiny dog Booboo ate two. Binkie became sleepy from overeating, so he lay down on the floor to take a nap. Booboo ran around the basement at a constant speed, yapping loudly. Booboo had run four laps before he tripped over Binkie and landed on the floor.

   What stopped Booboo? (circle one)
   a. Binkie's inertia
   b. Binkie's momentum
   c. Booboo's inertia
   d. Booboo's momentum

Chapter 1 • The Properties of Matter

Review & Assessment

STUDY GUIDE

#1 VOCABULARY & NOTES WORKSHEET
The Properties of Matter

By studying the Vocabulary and Notes listed for each section below, you can gain a better understanding of this chapter.

SECTION 1
Vocabulary
In your own words, write a definition for each of the following terms in the space provided.

1. matter

2. volume

3. meniscus

4. mass

5. gravity

6. weight

#1 CHAPTER REVIEW WORKSHEET
The Properties of Matter

USING VOCABULARY
For each pair of terms, explain the difference in their meanings.

1. mass/volume

2. mass/weight

3. inertia/mass

4. volume/density

5. physical property/chemical property

6. physical change/chemical change

CHAPTER TESTS WITH PERFORMANCE-BASED ASSESSMENT

#1 THE PROPERTIES OF MATTER
Chapter 1 Test

USING VOCABULARY
To complete the following sentences, choose the correct term from each pair of terms listed, and write the term in the blank.

1. The amount of matter in an object is its _____. (volume or mass)

2. Due to _____, all objects resist a change in motion. (inertia or gravity)

3. Cheese melting on a hamburger is an example of a _____ change. (chemical or physical)

4. An object's _____ is affected by gravitational force. (mass or weight)

5. A copper penny can turn green as it reacts with CO_2 and water. This is a _____ change. (physical or chemical)

UNDERSTANDING CONCEPTS

Multiple Choice
Circle the correct answer.

6. An amount of vinegar would be expressed in _____ for scientific experiments.
   a. centimeters or meters
   b. grams or milligrams
   c. liters or milliliters
   d. ounces or gallons

7. Which of the following is not a physical property of matter?
   a. ductility
   b. color
   c. thermal conductivity
   d. reactivity to water

8. A _____ object has more inertia than a 20 kg object.
   a. 2 mg
   b. 5 kg
   c. 2,000 g
   d. 30,000 g

9. During physical changes, matter always retains its
   a. size.
   b. identity.
   c. state.
   d. texture.

10. One way to compare the densities of oil and water is to pour the liquids into a glass container and observe how they
   a. change color.
   b. evaporate quickly.
   c. separate into layers.
   d. create an odor.

#1 THE PROPERTIES OF MATTER
Chapter 1 Performance-Based Assessment

SKILL BUILDER

Objective
Every metal has a unique density. You will measure the densities to find out what metals coins are made of.

Know the Score!
As you work through the activity, keep in mind that you will be earning a grade for the following:
• how well you work with the materials and equipment (30%)
• how well you make observations and test the hypothesis (40%)
• how accurately you identify test objects (30%)

MATERIALS
• 10 pennies
• 10 nickels
• 10 dimes
• 5 quarters
• copper sample
• zinc sample
• balance
• graduated cylinder (1.0 mL increments)
• 1 L of water
• calculator

METHOD • Ask a Question
How can I determine which metal I have?

Make Observations
1. Compare the luster and appearance of the coins with the luster and appearance of the copper and zinc samples.

Form a Hypothesis
2. Which metals do you think the coins are mostly made of? Support your hypothesis with observations.

Test the Hypothesis
3. Place the copper sample on the balance. Measure and record the mass in the table on the next page.
4. Pour water into the graduated cylinder until it is half full. Record the starting volume of the water in the table.
5. Place the copper sample in the cylinder. Record the ending volume of the water in the table.

Lab Worksheets

INQUIRY LABS

#1 STUDENT WORKSHEET
DESIGN YOUR OWN
Whatever Floats Your Boat

You are the dreaded Captain Sly of the pirate ship *Revenge*. On your recent excursion to Morocco, you bumped into a few vacationing royal families and relieved them of their "excess" gold. You thought you did them a favor, but the ungrateful families have sent an armada after you. Your only hope is to get back to London, where pirates and scoundrels are a dime a dozen. They'll never find you there!

The safest route is to take the Atlantic Ocean northward around Portugal, Spain, and France, go through the English Channel toward the North Sea, and then sail straight up the Thames, where you'll be home, sweet home.

There is one problem: your vessel is too heavy. Being a savvy sailor, you know that as you sail from Morocco to England, the waters become less salty and therefore less buoyant. With the market, make a line indicating the level of the liquid on the body of the hydrometer. The density of the blue liquid is 1.0 g/mL. Write a "1" beside the line you made.

4. Repeat step 3 for the three remaining solutions. Your teacher will give you the densities of each liquid. Mark very accurate, fine lines on your hydrometer so that you will get the correct measurements. Be sure to rinse off your hydrometer after each measurement.

MATERIALS
• plastic drinking straw
• modeling clay
• scissors
• pencils
• thumbtacks
• 4 large test tubes each filled with one of the following: blue liquid, brown liquid, soy sauce, "Atlantic sea water"
• fine-tipped permanent marker
• 60 mL plastic cup
• triple-beam balance
• 35 pennies
• paper towels

Objective
Build and calibrate a hydrometer to help determine the minimum amount of cargo to remove from a ship in order to keep it afloat.

Oh Buoy, a Hydrometer!
1. Choose a straw or pencil to form the body of your hydrometer. A *hydrometer* is a device used to determine the density of a liquid.

2. Now experiment with the available materials until you find a way to construct a hydrometer that floats vertically. If your hydrometer doesn't float vertically, it will not give accurate readings. The hydrometer should be stable but light enough so that it doesn't sink. Adjust the mass so that the water level is half-way up the body of the hydrometer.

3. Once the hydrometer floats properly, it must be calibrated. To calibrate the hydrometer, place it in the test tube filled with blue liquid. With the market, make a line indicating

WHIZ-BANG DEMONSTRATIONS

#1 TEACHER-LED DEMONSTRATION
DISCOVERY LAB
Curious Cubes

#1 Purpose

#1 Time Required
10–15 minutes

#1
Concept Level:
Clean Up □

MATERIALS
• 250 mL beakers (2)
• 200 mL of tap water
• food coloring
• 200 mL of rubbing alcohol or a nontoxic liquid with a density less than 0.92 g/cm³
• rubber gloves
• 2 ice cubes

Explanation
The density of water is 1.00 g/cm³. The density of ice is 0.92 g/cm³. The density of rubbing alcohol is 0.79 g/cm³. Therefore, ice is more dense than alcohol and less dense than water. Ice floats on water and sinks in alcohol because a substance floats only if it is less dense than the fluid that surrounds it. Water is one of the few substances that is less dense in its solid form than as a liquid. As water freezes, its crystalline structure becomes more open and the ice expands, making it less dense.

What to Do
1. Fill one beaker with 200 mL of water. Fill the other with 200 mL of rubbing alcohol.

2. Add two drops of food coloring to each beaker to improve visibility.

3. Ask a student wearing safety goggles and rubber gloves to place an ice cube in each of the beakers.

4. Ask students to observe what happens. (The cube in the alcohol will sink, while the cube in the water will float.)

Discussion
Review the concept of density with students. Explain that a cork floats because it is less dense than water, not because it is lighter. Encourage class discussion using the following questions as a guide:
• Why did one ice cube sink while the other floated? (The ice cube was more dense than the rubbing alcohol, so it sank. It floated in the water because it was less dense than the water.)
• How does the fact that water expands as it freezes explain the floating ice cube? (Because the same mass of water occupies a greater volume when it freezes, the ice cube must be less dense. Why is this is floats in liquid water.)

SAFETY ALERT
Be sure the appliance is unplugged before you take it apart.

LONG-TERM PROJECTS & RESEARCH IDEAS

#1 STUDENT WORKSHEET
DESIGN YOUR OWN
And We Have Thales To Thank

Thales of Miletus (585 B.C.) was a Greek philosopher and a very inquisitive fellow. Thales asked the question, "What is the world really made of?" After careful thought, he concluded that all the matter in the world originated from water! Thales was wrong, but by asking questions about the origin and properties of matter, he provided an early foundation for our modern scientific method.

A Philosophical Matter
1. Other Greek philosophers, such as Anaximedes (525 B.C.), Heraclitus (500 B.C.), and Aristotle (380 B.C.), had their own theories about matter. Find out how these philosophers classified matter and which properties of matter they described. For example, according to Aristotle's system, how would you describe an apple? How did each philosopher develop his theory? How is each theory different from modern theories of matter? Write a play centered around a discussion these philosophers might have had if they all participated in one debate about the nature of matter.

Another Research Idea
2. The space shuttle is protected by an outer covering of ceramic tiles with certain physical properties. What are these properties? What factors did the engineers consider when designing the tiles for the shuttle? How could the ceramic tiles be improved? What other uses do you think ceramic tiles would have on Earth? Write an article describing your findings.

HELPFUL HINT
Some useful properties of matter are color, reactivity, strength, heat conductivity, malleability, ductility, and density.

Long-Term Project Idea
3. Have you ever wondered why some toasters have a metal covering and some have a plastic one? Go to a junk shop or second-hand store, and purchase an old countertop appliance, such as a toaster, coffee maker, or blender. Take it apart and try to determine what material each of its parts is made of. List the physical and chemical properties of these materials. You may need to measure some of the properties yourself or look them up in a reference book. Why do you think the appliance is made out of materials with those properties? You might want to consider things such as: durability, practicality, consumer interest, and manufacturing cost. Make a chart that summarizes the properties of each material and the benefits of using it in this appliance.

PHYSICAL SCIENCE

DATASHEETS FOR LABBOOK

#1 Measuring Liquid Volume

#1 Coin Operated

#1 Volumania!

#1 Determining Density

#1 Layering Liquids

#1 White Before Your Eyes

Applications & Extensions

CRITICAL THINKING & PROBLEM SOLVING

#1 CRITICAL THINKING WORKSHEET
As a Matter of Fact!

From the journal of Captain Jane P. Fleet

LOG 2551
I have sent two of my best science officers to explore the planet Xerxes. Their mission is to collect samples of matter from the planet's surface.

LOG 2552
The science officers brought back a small cube of space matter from Xerxes. The cube is white, odorless, and grapefruit-sized, and it glows in the dark. We will observe the cube for a few days.

LOG 2553
Last night, the space cube expanded to three times its normal size. It is now about the size of a packing crate. Its mass did not change.

LOG 2554
Today the cube divided into four smaller cubes. Each new cube has more mass than the original cube.

LOG 2555
Lab officers applied electricity to one of the cubes. The cube burst into flames and exploded, covering the room and our science officers with a green paste. The paste reacted with the surfaces, and now the walls and the science officers are permanently green.

Comprehending Ideas
1. What changes in the properties of the space cube were recorded in Log 2553?

2. Explain how the changes recorded in Log 2553 affected the space cube's density.

SCIENTISTS IN ACTION

#4 in the News: Critical Thinking Worksheets

Segment 4
Neutrino Breakthrough

1. Why do you think scientists previously believed that neutrinos had no mass?

2. Explain why the discovery of the neutrino's mass could change scientists' theories about the expansion of the universe.

3. Why is a neutrino with mass important to solving the dark matter mystery?

4. Why are scientists other than those mentioned in the video doing mass experiments?

Chapter Background

SECTION 1

What Is Matter?

▶ Measuring Volume

Body measurements probably provided the basis for many early measurements. The Babylonian liquid measure, the *ka*, was the volume of a cube with sides of one hand-breadth (between 99 and 102 mm). Three hundred *ka* equaled 3,000 *gin* or 1 *gur*. The *gur* was equal to a volume of approximately 50 L. The basic Roman unit of volume was the *sextarius*. It had several subdivisions and multiples. The largest multiple, the *amphora*, was equal to 48 *sextarii*. The *amphora* was equal to 25.5 L.

▶ Weight on Other Planets

The weight of a person on any given planet depends on the attraction between the person and the planet. The more massive the planet, the greater the gravitational force on the person, and the greater the person's weight. A person who weighs 445 N on Earth would have different weights on other planets. On Mercury, the person would weigh about 164.6 N, on Venus 400.3 N, on Mars 169 N, on Jupiter 1,169.8 N, on Saturn 502.6 N, on Uranus 351.4 N, on Neptune 498.2 N, and on Pluto about 22.2 N.

IS THAT A FACT!

◆ A balance is a freely suspended beam that is balanced by known and unknown masses. Balances have been used for almost 3,000 years.

▶ Knife-Edge Balances

The modern knife-edge balance was developed during the sixteenth and seventeenth centuries. At the end of the seventeenth century, balances were developed in which the mass and goods plates were positioned above the balance beam, allowing the goods or masses to be placed anywhere on the plates without affecting accuracy.

▶ Spring Scales

Another type of device used to measure weight is the spring scale. It uses the relationship between a spring's deflection and the weight of the object on the scale. Spring scales are not as accurate as balances, which compare a pair of masses.

SECTION 2

Describing Matter

▶ **Physical and Chemical Properties**

The color and the density of a substance are physical properties. So are the temperatures at which a substance changes state. For example, chlorine is a greenish yellow gas with a density of 0.00321 g/cm^3. It can be changed to a liquid by cooling it to −34.6°C.

- The ability of chlorine to react explosively with sodium to form sodium chloride (table salt) is a chemical property. When chlorine reacts with sodium, an entirely different substance is formed.

IS THAT A FACT!

■ By comparing the density of King Hieron II's crown with that of a bar of pure gold, the Greek inventor and mathematician Archimedes was able to prove that the crown was not made of pure gold and that a goldsmith had cheated the king.

▶ **Dealing with Density**

Density is often a difficult concept for some middle school students to grasp. Discuss with students the principles that the world we know is made up of a variety of matter and that all matter has mass and volume.

- Not all matter is the same. Ask students which they would rather carry around all day, a backpack full of feathers or a backpack full of sand. Ask them to explain why. Lead them to the idea that even though the backpack is fixed in size (volume), there is more mass in it when it is full of sand because sand is more dense than feathers.

For background information about teaching strategies and issues, refer to the *Professional Reference for Teachers.*

CHAPTER
1

The Properties
of Matter

The Properties of
Matter

Pre-Reading Questions

Students may not know the answers to these questions before reading the chapter, so accept any reasonable response.

Suggested Answers

1. Matter is anything that has volume and has mass.

2. A physical property of matter can be observed without changing the identity of the matter. A chemical property of matter can be observed only when a chemical change might occur that would change the identity of the matter.

3. A physical change changes the shape or form of the matter without changing its identity. It is still the same matter as before the change and has most of the same properties. In a chemical change, the matter changes its identity and becomes a new form of matter with a different identity and different properties.

Sections

Pre-Reading Questions

1. What is matter?
2. What is the difference between a physical property and a chemical property?
3. What is the difference between a physical change and a chemical change?

Nice Ice

You've seen water in many forms: steam rising from a kettle, dew collecting on grass, and tiny crystals of frost forming on the windows in winter. But no matter what its form, water is still water. In this chapter, you'll learn more about the many different properties of matter, such as water. You'll also learn about changes in matter that take place all around you.

2

internet connect

HRW On-line Resources

go.hrw.com
For worksheets and other teaching aids, visit the HRW Web site and type in the keyword: **HSTMAT**

SCiLINKS NSTA

www.scilinks.com
Use the sciLINKS numbers at the end of each chapter for additional resources on the **NSTA** Web site.

Smithsonian Institution®

www.si.edu/hrw
Visit the Smithsonian Institution Web site for related on-line resources.

CNNfyi.com

www.cnnfyi.com
Visit the CNN Web site for current events coverage and classroom resources.

SACK SECRETS

In this activity, you will test your skills in determining the identity of an object based on its properties.

Procedure

1. You and two or three of your classmates will receive a **sealed paper sack** containing a **mystery object.** Do not open the sack!

2. For 5 minutes, make as many observations as you can about the object. You may touch, smell, or listen to the object through the sack; shake the sack; and so on. Be sure to record your observations.

Analysis

3. At the end of 5 minutes, discuss your findings with your partners.

4. In your ScienceLog, list the object's properties. Make a conclusion about the object's identity.

5. Share your observations, your list of properties, and your conclusion with the class. Now you are ready to open the sack.

6. Did you properly identify the object? If so, how? If not, why not? Write your answers in your ScienceLog. Share them with the class.

SACK SECRETS

MATERIALS

FOR EACH GROUP:
Anything that fits in the sack can be used for the object. Objects with interesting shapes, odors, and textures are preferable. Some objects to consider are a rubber ball, a jack, a pink school eraser, a piece of chalk, an orange, and a potato. Almost anything that is not sharp, corrosive, or prone to spoilage will work. Giving each group a different object will add to the mystery.

Answers to START-UP Activity

6. Students may or may not be able to identify the object, but their observations should demonstrate an attempt to identify various properties of the object, such as mass, shape, sound when shaken, and so on.

3

Focus

What Is Matter?

This section explains that matter is anything that has volume and mass. Students explore how the volume of solids, liquids, and gases are measured. Students learn the difference between mass and weight and learn how both are measured. Finally, students learn about inertia.

🔔 Bellringer

Ask students to list in their ScienceLog what they think some of the components might be for the following items:

loaf of bread, textbook, bicycle

Discuss the variety of answers students come up with.

1 Motivate

DEMONSTRATION

Display the following objects for students to view:

a rock, a paper clip, a book, a pencil, and a large cardboard box

Point out that the objects are alike because they all take up space. Discuss with students which objects are largest and smallest, and what those terms mean. Then discuss the connection between "taking up space" and volume. This is what makes the game of musical chairs so much fun. Point out that the amount of space something takes up is its volume.

`Sheltered English`

📄 **Directed Reading Worksheet** Section 1

Terms to Learn

matter	gravity
volume	weight
meniscus	newton
mass	inertia

What You'll Do

- Name the two properties of all matter.
- Describe how volume and mass are measured.
- Compare mass and weight.
- Explain the relationship between mass and inertia.

⏱ QuickLab

Space Case

1. Crumple a **piece of paper,** and fit it tightly in the bottom of a **cup** so that it won't fall out.
2. Turn the cup upside down. Lower the cup straight down into a **large beaker or bucket** half-filled with **water** until the cup is all the way underwater.
3. Lift the cup straight out of the water. Turn the cup upright and observe the paper. Record your observations in your ScienceLog.
4. Now punch a small hole in the bottom of the cup with the point of a **pencil.** Repeat steps 2 and 3.
5. How do these results show that air has volume? Record your explanation in your ScienceLog.

4

⏱ QuickLab

MATERIALS

FOR EACH STUDENT:
- piece of paper
- cup

FOR EACH GROUP:
- large beaker or bucket
- water
- pencil

What Is Matter?

Here's a strange question: What do you have in common with a toaster?

Give up? Okay, here's another question: What do you have in common with a steaming bowl of soup or a bright neon sign?

You are probably thinking these are trick questions. After all, it is hard to imagine that a human—you—has anything in common with a kitchen appliance, some hot soup, or a glowing neon sign.

Everything Is Made of Matter

From a scientific point of view you have at least one characteristic in common with these things. You, the toaster, the bowl, the soup, the steam, the glass tubing, and the glowing gas are all made of matter. But what is matter exactly? If so many different kinds of things are made of matter, you might expect the definition of the word *matter* to be complicated. But it is really quite simple. **Matter** is anything that has volume and mass.

Matter Has Volume

All matter takes up space. The amount of space taken up, or occupied, by an object is known as the object's **volume.** The sun, shown in **Figure 1,** has volume because it takes up space at the center of our solar system. Your fingernails, the Statue of Liberty, the continent of Africa, and a cloud all have volume. And because these things have volume, they cannot share the same space at the same time. Even the tiniest speck of dust takes up space, and there's no way another speck of dust can fit into that space without somehow bumping the first speck out of the way. Try the QuickLab on this page to see for yourself that matter takes up space.

Figure 1 *The volume of the sun is about 1,000,000 (1 million) times larger than the volume of the Earth.*

---

4. **Teacher Notes:** Students can feel the air being "moved out of the way" by the water if they hold a finger above the hole while they submerge the cup.

Answer to QuickLab

5. The water could not enter the cup because air occupied the space. Once the hole was punched, the water could force the air out of the cup and occupy the space in the cup.

Liquid Volume   Lake Erie, the smallest of the Great Lakes, has a volume of approximately 483,000,000,000,000 (483 trillion) liters of water. Can you imagine that much liquid? Well, think of a 2 liter bottle of soda. The water in Lake Erie could fill more than 241 trillion of those bottles. That's a lot of water! On a smaller scale, a can of soda has a volume of only 355 milliliters, which is approximately one-third of a liter. You can read the volume printed on the soda can. Or you can check the volume by pouring the soda into a large measuring cup from your kitchen, as shown in **Figure 2.**

Figure 2   *If the measurement is accurate, the volume measured should be the same as the volume printed on the can.*

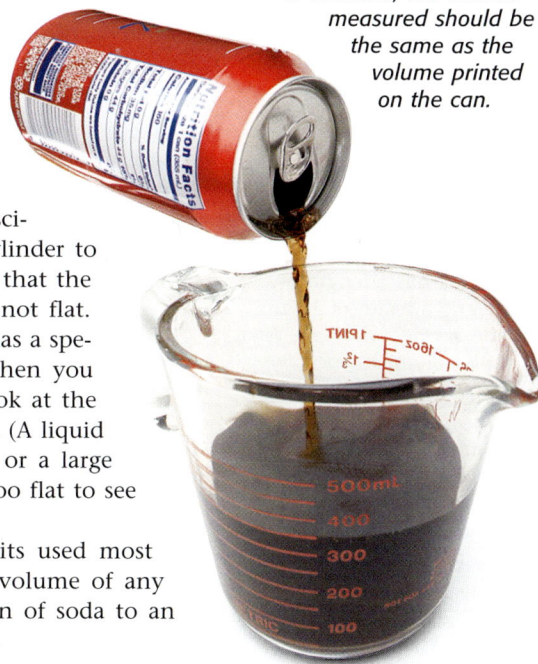

Measuring the Volume of Liquids   In your science class, you'll probably use a graduated cylinder to measure the volume of liquids. Keep in mind that the surface of a liquid in a graduated cylinder is not flat. The curve that you see at the liquid's surface has a special name—the **meniscus** (muh NIS kuhs). When you measure the volume of a liquid, you must look at the bottom of the meniscus, as shown in **Figure 3.** (A liquid in any container, including a measuring cup or a large beaker, has a meniscus. The meniscus is just too flat to see in a wider container.)

Liters (L) and milliliters (mL) are the units used most often to express the volume of liquids. The volume of any amount of liquid, from one raindrop to a can of soda to an entire ocean, can be expressed in these units.

Figure 3   *To measure volume correctly, read the scale at the lowest part of the meniscus (as indicated) at eye level.*

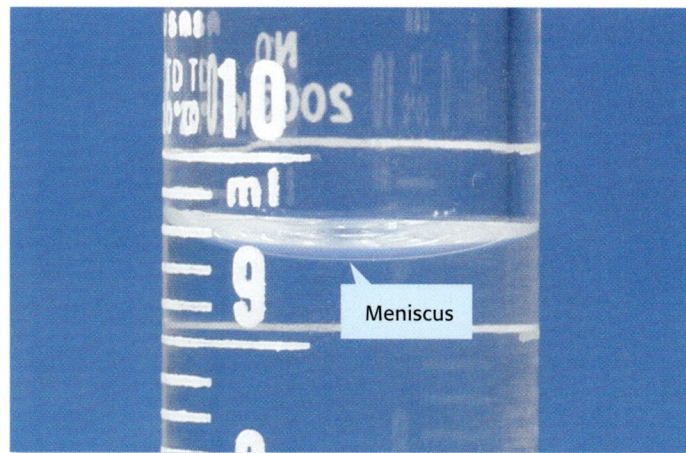

Meniscus

BRAIN FOOD

The volume of a typical raindrop is approximately 0.09 mL, which means that it would take almost 4,000 raindrops to fill a soda can.

READING STRATEGY

Prediction Guide   Before students read the next four pages, ask them whether these statements are true or false.

- The volume of a gas can be measured with a graduated cylinder. (false)
- If you know the volume of the container a gas is in, you know the volume of the gas. (true)
- Volumes of solids can be expressed in liters or milliliters. (false)
- Weight and mass are the same thing. (false)

GROUP ACTIVITY

Provide small groups with a variety of jars, bottles, cans, and cartons. Allow students to use a measuring cup or a graduated cylinder to determine the volume of each container. A volume reading is always made at the flattest part of the meniscus— the bottom of the curve for water and the top for mercury.

DISCUSSION

Discuss with students the fact that we are surrounded by matter. Have students give examples of matter around them. Ask them to list some characteristics of matter. Discuss with them the notion that they are actually pushing air aside as they walk "through" it.

CROSS-DISCIPLINARY FOCUS

Music   Provide each small group with several identical glass containers. Tell students to add a different amount of water to each container, then lightly strike each container with a pen to hear the pitch of the sound it makes. Ask students to arrange the containers in order from the lowest pitch to highest pitch. Finally, have students measure the volume of water in each container, and ask them to determine the relationship between pitch and the volume of water. Sheltered English

2) Teach, continued

MATH and MORE

In 1997 Americans consumed an average of 204 L of soft drinks per person. How many cans of soft drinks would that be? Assume that a can holds 355 mL and remember that 1 L = 1,000 mL. (more than 574 cans)

The total volume of soft drinks consumed by Americans in 1997 was approximately 53 billion liters. How many cans of soft drinks would that be? (more than 149 billion cans)

Math Skills Worksheet
"The Unit Factor and Dimensional Analysis"

Answers to MATHBREAK

1. 1,800 cm^3
2. 95,000 cm^3, or 0.095 m^3
3. Answers will vary with the objects chosen. Students should show the units in each measurement and in the final answer.

LabBook PG 134
Volumania!

DEMONSTRATION

Display a variety of classroom objects, such as pencils, books, and notebook paper. Ask students which objects contain the largest amount of matter and thus have the greatest mass. Then ask which contain the smallest amount of matter and thus have the smallest mass. Ask them if the objects with the greatest volume always have the most mass. Sheltered English

---

MATH BREAK

Calculating Volume

A typical compact disc (CD) case has a length of 14.2 cm, a width of 12.4 cm, and a height of 1.0 cm. The volume of the case is the length multiplied by the width multiplied by the height:

14.2 cm × 12.4 cm × 1.0 cm = 176.1 cm^3

Now It's Your Turn

1. A book has a length of 25 cm, a width of 18 cm, and a height of 4 cm. What is its volume?
2. What is the volume of a suitcase with a length of 95 cm, a width of 50 cm, and a height of 20 cm?
3. For additional practice, find the volume of other objects that have square or rectangular sides. Compare your results with those of your classmates.

LabBook

How would you measure the volume of this strangely shaped object? To find out, turn to page 134 in the LabBook.

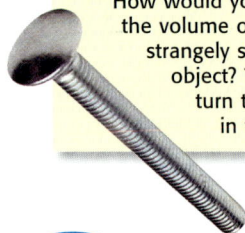

MISCONCEPTION ALERT

Students may not understand that a gas not only will expand to fill its container but also can be compressed, or squeezed, to fill a smaller container. Scuba divers' tanks contain compressed air. A 2.24 m^3 scuba tank holds enough air for an average adult to breathe underwater for 30 to 45 minutes.

Solid Volume The volume of any solid object is expressed in cubic units. *Cubic* means "having three dimensions." One cubic unit, a cubic meter, is shown in **Figure 4.** In science, cubic meters (m^3) and cubic centimeters (cm^3) are the units most often used to express the volume of solid items. The *3* in these unit abbreviations shows that three quantities were multiplied to get the final result. For a rectangular object, these three quantities are length, width, and height. Try this for yourself in the MathBreak at left.

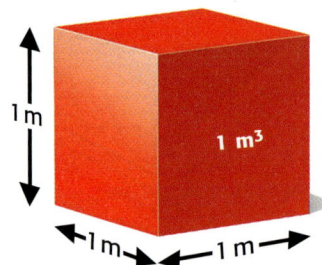

Figure 4 *A cubic meter has a height of 1 m, a length of 1 m, and a width of 1 m, so its volume is 1 m × 1 m × 1 m = 1 m^3.*

Comparing Solid and Liquid Volumes Suppose you want to determine whether the volume of an ice cube is equal to the volume of water that is left when the ice cube melts. Because 1 mL is equal to 1 cm^3, you can express the volume of the water in cubic centimeters and compare it with the volume of the ice cube. The volume of any liquid can be expressed in cubic units in this way. (However, in SI, volumes of solids are never expressed in liters or milliliters.)

Measuring the Volume of Gases How do you measure the volume of a gas? You can't hold a ruler up to a gas, and you can't pour a gas into a graduated cylinder. So it's impossible, right? Wrong! A gas expands to fill its container, so if you know the volume of the container the gas is in, then you know the volume of the gas.

Matter Has Mass

Another characteristic of all matter is mass. **Mass** is the amount of matter that something is made of. For example, the Earth is made of a very large amount of matter and therefore has a large mass. A peanut is made of a much smaller amount of matter and thus has a smaller mass. Remember, even something as small as a speck of dust is made of matter and therefore has mass.

WEIRD SCIENCE

Mauna Loa, in Hawaii, is the world's most active volcano. The volume of lava that has flowed from the volcano is enough to pave a four-lane highway that reaches around the world 30 times.

Is a Puppy Like a Bowling Ball? An object's mass can be changed only by changing the amount of matter in the object. Consider the bowling ball shown in **Figure 5.** Its mass is constant because the amount of matter in the bowling ball never changes (unless you use a sledgehammer to remove a chunk of it!). Now consider the puppy. Does its mass remain constant? No, because the puppy is growing. If you measured the puppy's mass next year or even next week, you'd find that it had increased. That's because more matter—more puppy—would be present.

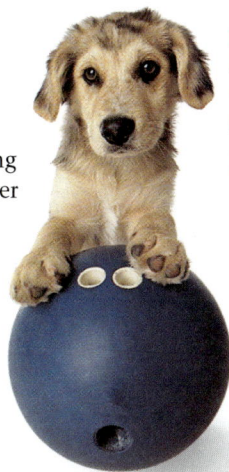

Figure 5 *The mass of the bowling ball does not change. The mass of the puppy increases as more matter is added—that is, as the puppy grows.*

The Difference Between Mass and Weight

Weight is different from mass. To understand this difference, you must first understand gravity. **Gravity** is a force of attraction between objects that is due to their masses. This attraction causes objects to exert a pull on other objects. Because all matter has mass, all matter experiences gravity. The amount of attraction between objects depends on two things—the masses of the objects and the distance between them, as shown in **Figure 6.**

Figure 6   How Mass and Distance Affect Gravity Between Objects

a Gravitational force (represented by the width of the arrows) is large between objects with large masses that are close together.

b Gravitational force is smaller between objects with smaller masses that are close together than between objects with large masses that are close together (as shown in **a**).

c An increase in distance reduces gravitational force between two objects. Therefore, gravitational force between objects with large masses (such as those in **a**) is less if they are far apart.

Activity

Imagine the following items resting side by side on a table: an elephant, a tennis ball, a peanut, a bowling ball, and a housefly. In your ScienceLog, list these items in order of their attraction to the Earth due to gravity, from least to greatest amount of attraction. Follow your list with an explanation of why you arranged the items in the order that you did.

TRY at HOME

7

Science Bloopers

In July 1983, an Air Canada Boeing 767 airliner with 69 people on board ran out of fuel in mid-flight. Before leaving their previous stop, the crew had calculated that they had enough fuel to make it to their final destination. However, the plane's crew miscalculated the mass of the fuel because they had used the wrong units; they used pounds instead of kilograms! Fortunately, the plane was able to make a safe landing at an abandoned air force base in Manitoba, Canada.

CONNECT TO
LIFE SCIENCE

Although calcium makes up about 70 percent of the mass of the human skeleton, it is not the most abundant substance in the human body. The majority of the human body—between 70 and 80 percent—is composed of water.

RETEACHING

The gravitational force exerted on an astronaut on the moon is one-sixth that exerted on the astronaut by the Earth. Ask students to compare the mass and weight of an astronaut on Earth with the mass and weight of the astronaut on the moon. (The astronaut's mass would not change at all. The astronaut's weight would be only one-sixth his or her weight on Earth.)

GUIDED PRACTICE

Writing | Ask students to write a paragraph or make a poster explaining how the weight of an object can change even though its mass does not change.

Teacher Notes: An object at sea level may weigh more than it would high above sea level, but the difference would be very small. For all practical purposes, mass and weight remain constant everywhere on Earth.

PORTFOLIO

Biology
CONNECTION

The mineral calcium is stored in bones, and it accounts for about 70 percent of the mass of the human skeleton. Calcium strengthens bones, helping the skeleton to remain upright against the strong force of gravity pulling it toward the Earth.

May the Force Be with You Gravitational force is experienced by all objects in the universe all the time. But the ordinary objects you see every day have masses so small (relative to, say, planets) that their attraction toward each other is hard to detect. Therefore, the gravitational force experienced by objects with small masses is very slight. However, the Earth's mass is so large that the gravitational force between objects, such as our atmosphere or the space shuttle, and the Earth is great. Gravitational force is what keeps you and everything else on Earth from floating into space.

So What About Weight? A measure of the gravitational force exerted on an object is called **weight.** Consider the brick in **Figure 7.** The brick has mass. The Earth also has mass. Therefore, the brick and the Earth are attracted to each other. A force is exerted on the brick because of its attraction to the Earth. The weight of the brick is a measure of this gravitational force.

Now look at the sponge in Figure 7. The sponge is the same size as the brick, but its mass is much less. Therefore, the sponge's attraction toward the Earth is not as great, and the gravitational force on the sponge is not as great. Thus, the *weight* of the sponge is less than the *weight* of the brick.

Figure 7 *This brick and sponge may be the same size, but their masses, and therefore their weights, are quite different.*

At a Distance The attraction between objects decreases as the distance between them increases. As a result, the gravitational force exerted on objects also decreases as the distance increases. For this reason, a brick floating in space would weigh less than it does resting on Earth's surface. However, the brick's mass would stay the same.

Massive Confusion Back on Earth, the gravitational force exerted on an object is about the same everywhere, so an object's weight is also about the same everywhere. Because mass and weight remain constant everywhere on Earth, the terms *mass* and *weight* are often used as though they mean the same thing. But using the terms interchangeably can lead to confusion. So remember, weight depends on mass, but weight is not the same thing as mass.

WEIRD SCIENCE

In 1993, while on its way to Jupiter, the *Galileo* spacecraft passed close enough to asteroid 243 to photograph it. This asteroid, named Ida, is approximately 52 × 24 × 21 km. When scientists analyzed the photo from *Galileo*, they noticed that Ida had a small moon circling it. The moon, which is approximately 1.5 km in diameter, is held in orbit by Ida's gravitational force.

Measuring Mass and Weight

The SI unit of mass is the kilogram (kg), but mass is often expressed in grams (g) and milligrams (mg) as well. These units can be used to express the mass of any object, from a single cell in your body to the entire solar system. Weight is a measure of gravitational force and must be expressed in units of force. The SI unit of force is the **newton (N).** So weight is expressed in newtons.

A newton is approximately equal to the weight of a 100 g mass on Earth. So if you know the mass of an object, you can calculate its weight on Earth. Conversely, if you know the weight of an object on Earth, you can determine its mass. **Figure 8** summarizes the differences between mass and weight.

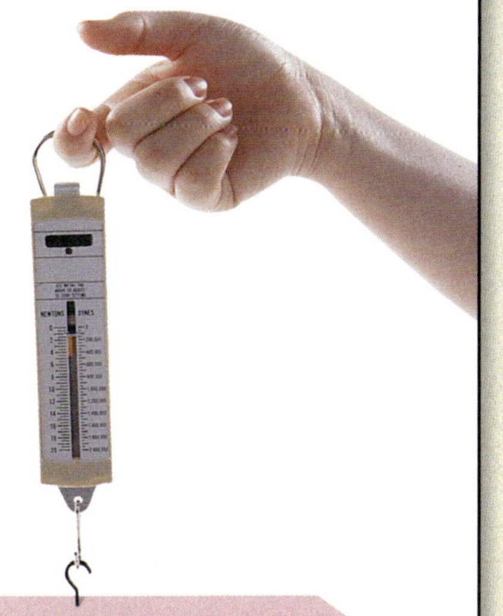

Figure 8 **Differences Between Mass and Weight**

Mass is . . .

- a measure of the amount of matter in an object.
- always constant for an object no matter where the object is in the universe.
- measured with a balance (shown below).
- expressed in kilograms (kg), grams (g), and milligrams (mg).

Weight is . . .

- a measure of the gravitational force on an object.
- varied depending on where the object is in relation to the Earth (or any other large body in the universe).
- measured with a spring scale (shown above).
- expressed in newtons (N).

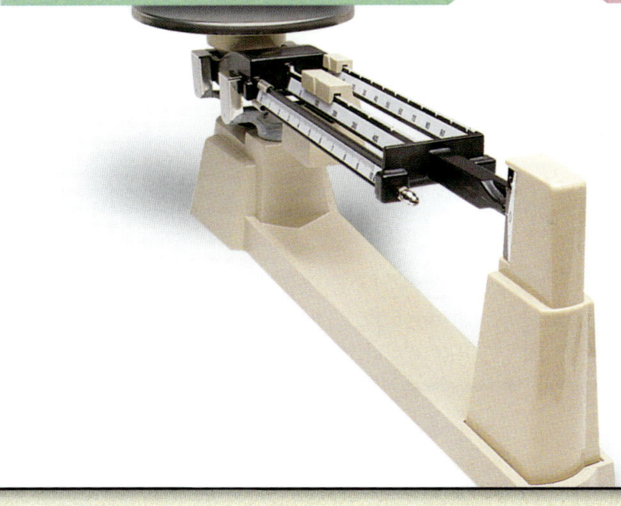

Self-Check

If all of your school books combined have a mass of 3 kg, what is their total weight in newtons? Remember that 1 kg = 1,000 g. *(See page 168 to check your answer.)*

3 Extend

Multicultural CONNECTION

Cultures throughout history have developed different units to express mass, volume, and weight. Have students make a poster that shows or describes the units of measurement used by a particular culture.

RESEARCH

PORTFOLIO

Have students research the different kinds of instruments used to measure mass and weight. Have students draw or diagram an instrument and explain how it works. Sheltered English

MEETING INDIVIDUAL NEEDS

Students Having Difficulty

Rather than having students explain how volume, weight, and mass are measured, have them demonstrate the concepts. Provide them with instruments such as graduated cylinders and triple-beam balances. Have them explain step-by-step how to measure the volume and the mass of a given substance. Provide students with spring scales, and help them observe the difference between mass and weight.

Answer to Self-Check

approximately 30 N

SCIENTISTS AT ODDS

The official standard kilogram is a cylinder made of platinum-iridium alloy. The mass of the cylinder is supposed to equal the mass of 1 dL^3 of pure water at 4°C. Some scientists believe that this cylinder is imprecise and needs to be changed.

In fact, the kilogram is the only SI unit based on a single physical standard that can be destroyed or altered. Some scientists now suggest redefining the kilogram as the mass of an exact number of atoms of a particular element.

Teaching Transparency 204
"Differences Between Mass and Weight"

Science Skills Worksheet
"Measuring"

Quiz

Allow students to use a balance and a spring scale to compare the masses and the weights of a variety of small objects, such as a stone, a marble, a washer, and a nail.

ALTERNATIVE ASSESSMENT

Writing Have students write a short science-fiction story in which matter does not behave in the ways described in this section. Ask them to consider what life would be like in this kind of universe. Some examples: a universe where gravity does not exist; a universe where objects repulse rather than attract one another; a universe where inertia becomes greater as mass decreases. Ask volunteers to share their stories with the class.

INDEPENDENT PRACTICE

Concept Mapping Have each student make a concept map comparing a golf ball with a table-tennis ball. Tell students they must use the terms *matter, mass, volume, weight,* and *inertia* in their map.

internet connect

SCI LINKS
NSTA

TOPIC: What Is Matter?
GO TO: www.scilinks.org
*sci*LINKS NUMBER: HSTP030

APPLY

Mass, Weight, and Bathroom Scales

Ordinary bathroom scales are spring scales. Many scales available today show a reading in both pounds (a common, though not SI, unit of weight) and kilograms. How does such a reading contribute to the confusion between mass and weight?

Mass Is a Measure of Inertia

Imagine trying to kick a soccer ball that has the mass of a bowling ball. It would be painful! The reason has to do with inertia (in UHR shuh). **Inertia** is the tendency of all objects to resist any change in motion. Because of inertia, an object at rest will remain at rest until something causes it to move. Likewise, a moving object continues to move at the same speed and in the same direction unless something acts on it to change its speed or direction.

Mass is a measure of inertia because an object with a large mass is harder to start in motion and harder to stop than an object with a smaller mass. This is because the object with the large mass has greater inertia. For example, imagine that you are going to push a grocery cart that has only one potato in it. No problem, right? But suppose the grocery cart is filled with potatoes, as in **Figure 9.** Now the total mass—and the inertia—of the cart full of potatoes is much greater. It will be harder to get the cart moving and harder to stop it once it is moving.

Figure 9 *Why is a cartload of potatoes harder to get moving than a single potato? Because of inertia, that's why!*

internet connect

SCI LINKS
NSTA

TOPIC: What Is Matter?
GO TO: www.scilinks.org
*sci*LINKS NUMBER: HSTP030

SECTION REVIEW

1. What are the two properties of all matter?
2. How is volume measured? How is mass measured?
3. **Analyzing Relationships** Do objects with large masses always have large weights? Explain your reasoning.

10

▼ **Answers to Section Review**

1. All matter has mass and volume.

2. Volumes of liquids are measured with graduated cylinders and are expressed in liters and milliliters. Volumes of rectangular solids are calculated by multiplying length, width, and height measurements and are expressed in cubic units, such as m^3 or cm^3. Mass is measured with a balance and is expressed in kilograms, grams, and milligrams.

3. Not all objects with large masses have large weights because the weight of an object can change depending on where it is located in the universe. Mass remains the same everywhere in the universe. So a massive object in space may not have a large weight.

Terms to Learn

physical property   physical change
density   chemical change
chemical property

What You'll Do

◆ Give examples of matter's different properties.
◆ Describe how density is used to identify different substances.
◆ Compare physical and chemical properties.
◆ Explain what happens to matter during physical and chemical changes.

Describing Matter

Have you ever heard of the game called "20 Questions"? In this game, your goal is to determine the identity of an object that another person is thinking of by asking questions about the object. The other person can respond with only a "yes" or a "no." If you can identify the object after asking 20 or fewer questions, you win! If you still can't figure out the object's identity after asking 20 questions, you may not be asking the right kinds of questions.

What kinds of questions should you ask? You might find it helpful to ask questions about the properties of the object. Knowing the properties of an object can help you determine the object's identity, as shown below.

Could I hold it in my hand? Yes.

Does it have an odor? Yes.

Is it safe to eat? Yes.

Is it an apple? Yes.

Is it orange? No. Yellow? No. Red? Yes.

Activity

With a partner, play a game of 20 Questions. One person will think of an object, and the other person will ask yes/no questions about it. Write the questions in your ScienceLog as you go along. Put a check mark next to the questions asked about physical properties. When the object is identified or when the 20 questions are up, switch roles. Good luck!

Physical Properties

Some of the questions shown above help the asker gather information about *color* (Is it orange?), *odor* (Does it have an odor?), and *mass* and *volume* (Could I hold it in my hand?). Each of these properties is a physical property of matter. A **physical property** of matter can be observed or measured without changing the identity of the matter. For example, you don't have to change what the apple is made of to see that it is red or to hold it in your hand.

11

IS THAT A FACT!

The element bromine gets its name from the Greek word *bromos*, which means "stench" or "bad smell."

Focus

Describing Matter

This section introduces students to the properties of matter and compares physical properties and chemical properties of substances. It also explains how density is used to identify different substances. The section concludes with an explanation of what happens to matter during physical and chemical changes.

Bellringer

Ask your students the following question:

If you were asked to describe an orange to someone who had never seen an orange, what would you tell the person?

Write your description in your ScienceLog.

1 Motivate

DEMONSTRATION

Display several objects that have differences in color, odor, texture, size, shape, and state. Allow students to examine the objects. Then ask them to describe each object in terms of its color, odor, texture, size, shape, and state. Ask students why it is important to describe objects using a variety of properties. Sheltered English

📄 **Directed Reading Worksheet** Section 2

DISCUSSION

Draw students' attention to the chart of physical properties. After reading through the definition and example of each property, ask volunteers to give another example of the same property.

REAL-WORLD CONNECTION

Electronic coin testers in vending machines can instantly identify the properties of real coins and reject fake coins. First, an electric current passes through the coin to measure its metal content and size. Only proper coins conduct the right amount of electricity. Next a magnet and light sensors are used to detect the coin's value. Incorrect coins are rejected.

Homework

Writing | Have students write a description of the properties of a favorite object from home, such as a bicycle, a pet, a type of food, or an article of clothing. Let students read their description aloud, and have other class members try to guess the object being described. Sheltered English

LabBook PG 136
Determining Density

Physical Properties Identify Matter You rely on physical properties all the time. For example, physical properties help you determine whether your socks are clean (odor), whether you can fit all your books into your backpack (volume), or whether your shirt matches your pants (color). The table below lists some more physical properties that are useful in describing or identifying matter.

More Physical Properties		
Physical property	**Definition**	**Example**
Thermal conductivity	the ability to transfer thermal energy from one area to another	Plastic foam is a poor conductor, so hot chocolate in a plastic-foam cup will not burn your hand.
State	the physical form in which a substance exists, such as a solid, liquid, or gas	Ice is water in its solid state.
Malleability (MAL ee uh BIL uh tee)	the ability to be pounded into thin sheets	Aluminum can be rolled or pounded into sheets to make foil.
Ductility (duhk TIL uh tee)	the ability to be drawn or pulled into a wire	Copper is often used to make wiring.
Solubility (SAHL yoo BIL uh tee)	the ability to dissolve in another substance	Sugar dissolves in water.
Density	mass per unit volume	Lead is used to make sinkers for fishing line because lead is more dense than water.

Spotlight on Density Density is a very helpful property when you need to distinguish different substances. Look at the definition of density in the table above—mass per unit volume. If you think back to what you learned in Section 1, you can define density in other terms: **density** is the amount of matter in a given space, or volume, as shown in **Figure 10.**

Figure 10 *A golf ball is more dense than a table-tennis ball because the golf ball contains more matter in a similar volume.*

12

SCIENCE HUMOR

Two fish swim by a fisherman's baited hook. One fish says to the other, "You know, I never could figure out why those worms always go swimming with lead weights tied around their necks."

The other fish replies, "Yeah, they must be pretty dense."

To find an object's density (D), first measure its mass (m) and volume (V). Then use the following equation:

$$D = \frac{m}{V}$$

Units for density are expressed using a mass unit divided by a volume unit, such as g/cm^3, g/mL, kg/m^3, and kg/L.

Using Density to Identify Substances Density is a useful property for identifying substances for two reasons. First, the density of a particular substance is always the same at a given pressure and temperature. For example, the helium in a huge airship has a density of 0.0001663 g/cm^3 at 20°C and normal atmospheric pressure. You can calculate the density of any other sample of helium at that same temperature and pressure—even the helium in a small balloon—and you will get 0.0001663 g/cm^3. Second, the density of one substance is usually different from that of another substance. Check out the table below to see how density varies among substances.

Densities of Common Substances*

Substance	Density (g/cm³)	Substance	Density (g/cm³)
Helium (gas)	0.0001663	Copper (solid)	8.96
Oxygen (gas)	0.001331	Silver (solid)	10.50
Water (liquid)	1.00	Lead (solid)	11.35
Iron pyrite (solid)	5.02	Mercury (liquid)	13.55
Zinc (solid)	7.13	Gold (solid)	19.32

* at 20°C and normal atmospheric pressure

Mass = 96.6 g
Volume = 5.0 cm³

The nugget in **Figure 11** looks like gold. But is it? It might be fool's gold instead. Fool's gold is another name for iron pyrite (PIE RIET), a mineral that looks like gold. How can you tell what the nugget really is? You could compare the density of the nugget with the known densities for gold and iron pyrite at the same temperature and pressure. By comparing densities, you'd know whether you've struck gold or you've been fooled!

Figure 11 Is this gold or fool's gold?

MATH BREAK

Density

You can rearrange the equation for density to find mass and volume as shown below:

$$D = \frac{m}{V}$$

$$m = D \times V \qquad V = \frac{m}{D}$$

1. Find the density of a substance with a mass of 5 kg and a volume of 43 m³.

2. Suppose you have a lead ball with a mass of 454 g. What is its volume? (Hint: Use the table at left.)

3. What is the mass of a 15 mL sample of mercury? (Hint: Use the table at left.)

BRAIN FOOD

Pennies minted before 1982 are made mostly of copper and have a density of 8.85 g/cm^3. In 1982, a penny's worth of copper began to cost more than one cent, so the U.S. Department of the Treasury began producing pennies using mostly zinc with a copper coating. Pennies minted after 1982 have a density of 7.14 g/cm^3. Check it out for yourself!

13

MATH and MORE

Ask students to use the equation for density to solve the following problems.

1. A block of pine wood has a mass of 120 g and a volume of 300 cm³. What is the density of the wood? (0.4 g/cm³)

   Extension Ask students to predict whether this block of pine would float in a pool of water. Why? (Yes; it is less dense than water.)

2. A sample of metal has a mass of 4,059 g and a volume of 453 cm³. What metal is it? (copper)

Math Skills Worksheet "Density"

Answers to MATHBREAK

1. $D = \frac{m}{V}$, so $D = \frac{5 \text{ kg}}{43 \text{ m}^3} =$ 0.12 kg/m³

2. $V = \frac{m}{D}$, so $V = \frac{454 \text{ g}}{11.35 \text{ g/cm}^3} =$ 40 cm³ (Students must use the density of lead from the table on this page.)

3. $m = D \times V$, so $m =$ 13.55 g/mL × 15 mL = 203 g (Students must use the density of mercury from the table on this page.)

MISCONCEPTION ALERT

The density of a substance changes as temperature and pressure change. This is especially true at the point at which a substance changes state. Remind students that densities such as those listed in the table on this page are valid only at the given temperature and pressure.

IS THAT A FACT!

The density of a fresh egg is about 1.2 g/mL, while the density of a spoiled egg is about 0.9 g/mL. So don't eat an egg that floats. It's spoiled!

WEIRD SCIENCE

At one time, a person who was suspected of being a witch was tossed into a lake. It was believed that a witch would float, while a person who was not a witch would sink.

MEETING INDIVIDUAL NEEDS

Learners Having Difficulty
Visual learners and students with limited English proficiency may benefit from handling sealed bottles of oil and vinegar to observe how the different densities cause the oil and vinegar to separate in layers. Sheltered English

MISCONCEPTION ///// ALERT \\\\\

Factors unrelated to density determine whether liquids are miscible (will dissolve in each other). Liquids of different densities may be miscible, so liquid layers *will not* form. For example, ethyl alcohol and water have different densities, but they dissolve in each other. Conversely, liquids of identical density may be immiscible, so liquid layers *will* form. However, because the density of one substance is usually different from that of another, layering liquids is still a useful way to compare the relative densities of immiscible liquids.

LabBook PG 137
Layering Liquids

Answer to APPLY

The grease and the meat juices have different densities. The top layer is the grease. The spout is at the bottom so the meat juices can be poured out while leaving the grease behind.

Liquid Layers What do you think causes the liquid in **Figure 12** to look the way it does? Is it magic? Is it trick photography? No, it's differences in density! There are actually four different liquids in the jar. Each liquid has a different density. Because of these differences in density, the liquids do not mix together but instead separate into layers, with the densest layer on the bottom and the least dense layer on top. The order in which the layers separate helps you determine how the densities of the liquids compare with one another.

The Density Challenge Imagine that you could put a lid on the jar in the picture and shake up the liquids. Would the different liquids mix together so that the four colors would blend into one interesting color? Maybe for a minute or two. But if the liquids are not soluble in one another, they would start to separate, and eventually you'd end up with the same four layers.

The same thing happens when you mix oil and vinegar to make salad dressing. When the layers separate, the oil is on top. But what do you think would happen if you added more oil? What if you added so much oil that there was several times as much oil as there was vinegar? Surely the oil would get so heavy that it would sink below the vinegar, right? Wrong! No matter how much oil you have, it will always be less dense than the vinegar, so it will always rise to the top. The same is true of the four liquids shown in Figure 12. Even if you add more yellow liquid than all of the other liquids combined, all of the yellow liquid will rise to the top. That's because density does not depend on how much of a substance you have.

Figure 12 *The yellow liquid is the least dense, and the green liquid is the densest.*

APPLY

Density and Grease Separators

The grease separator shown here is a kitchen device that cooks use to collect the best meat juices for making gravies. Based on what you know about density, describe how a grease separator works. Be sure to explain why the spout is at the bottom.

SECTION REVIEW

1. List three physical properties of water.

2. Why does a golf ball feel heavier than a table-tennis ball?

3. Describe how you can determine the relative densities of liquids.

4. **Applying Concepts** How could you determine that a coin is not pure silver?

▼ Answers to Section Review

1. Water is colorless, is liquid at room temperature, has a density of 1.00 g/mL, is odorless, has a melting point of 0°C, has a boiling point of 100°C, and can dissolve table salt and sugar.

2. A golf ball feels heavier than a table-tennis ball because it is more dense; that is, it has more mass in a similar volume.

3. Sample answer: Pour the liquids into a container, and allow the container to stand undisturbed. The liquids will form layers with the least dense liquid on top and the densest liquid on the bottom.

4. Sample answer: Measure the mass and volume of the coin. Calculate the coin's density. Compare this density with the known density of silver.

Chemical Properties

Physical properties are not the only properties that describe matter. **Chemical properties** describe a substance based on its ability to change into a new substance with different properties. For example, a piece of wood can be burned to create new substances (ash and smoke) with properties different from the original piece of wood. Wood has the chemical property of *flammability*—the ability to burn. A substance that does not burn, such as gold, has the chemical property of nonflammability. Other common chemical properties include reactivity with oxygen, reactivity with acid, and reactivity with water. (The word *reactivity* just means that when two substances get together, something can happen.)

Observing Chemical Properties Chemical properties can be observed with your senses. However, chemical properties aren't as easy to observe as physical properties. For example, you can observe the flammability of wood only while the wood is burning. Likewise, you can observe the nonflammability of gold only when you try to burn it and it won't burn. But a substance always has its chemical properties. A piece of wood is flammable even when it's not burning.

Some Chemical Properties of Car Maintenance Look at the old car shown in **Figure 13.** Its owner calls it Rust Bucket. Why has this car rusted so badly while some other cars the same age remain in great shape? Knowing about chemical properties can help answer this question.

Most car bodies are made from steel, which is mostly iron. Iron has many desirable physical properties, including strength, malleability, and a high melting point. Iron also has many desirable chemical properties, including nonreactivity with oil and gasoline. All in all, steel is a good material to use for car bodies. It's not perfect, however, as you can probably tell from the car shown here.

Paint doesn't react with oxygen, so it provides a barrier between oxygen and the iron in the steel.

This hole started as a small chip in the paint. The chip exposed the iron in the car's body to oxygen. The iron rusted and eventually crumbled away.

Figure 13 **Rust Bucket**
One unfavorable chemical property of iron is its reactivity with oxygen. When iron is exposed to oxygen, it rusts.

This bumper is rust free because it is coated with a barrier of chromium, which is nonreactive with oxygen.

15

READING STRATEGY

Prediction Guide Before students read this page, ask them:

What is one unfavorable chemical property of iron?

 a. its high melting point

 b. its nonreactivity with oil and gasoline

 c. its reactivity with oxygen

 d. its nonflammability
  (c)

GUIDED PRACTICE

Show students a sheet of paper, and ask them to write a list of the physical and chemical properties of paper. (physical properties: white color, smooth texture, no odor; chemical properties: flammability)

DISCUSSION

After students have studied the picture of the car, have them describe where they have seen other examples of rusting.
Sheltered English

REAL-WORLD CONNECTION

Rustproofing is one way to help protect cars from rust. The process involves treating the car's underside and panels—such as the doors, the trunk, and the hood—with sealants. The sealants penetrate all the seams, cracks, and holes to keep out air and moisture, which can increase the rate at which rust forms.

Science Bloopers

In 1870, John and Isaiah Hyatt patented a plastic they called celluloid. It was to be used as a substitute for the ivory used in billiard balls. Unfortunately, one of the chemicals in celluloid was explosive, and billiard balls made from the Hyatt brothers' celluloid often blew up when struck with a pool cue.

IS THAT A FACT!

Galvanized steel is steel that is coated with zinc to prevent rusting. It is used in buckets and nails. And steel plated with tin was used in food cans and containers. Today, aluminum cans have replaced most steel cans.

Advanced Learners Most automobile manufacturers use many plastic parts in their automobiles. A new car contains up to 180 kg of plastic. Back in 1970, cars contained only about 27 kg of plastic. Ask students to research the use of plastic parts in the manufacture of cars, to list as many areas in a car where plastic is used, and to describe the advantages of using plastic parts instead of metal parts.

MISCONCEPTION ALERT

Although **Figure 14** shows bleach being mixed with food coloring, remind students never to mix household chemicals. Mixing chemicals, even cleaning products, can cause a chemical reaction that produces poisonous or flammable fumes or even an explosion.

Learners Having Difficulty Divide the class into small groups. Give each group a banana, a lump of clay, and a sheet of paper. Ask each group to brainstorm about ways to change the physical properties of each of the objects. Then select one person from each group for each item. Have them demonstrate the changes. Guide students to observe that each object remained the same substance after the physical change. **Sheltered English**

Figure 14 *Substances have different physical and chemical properties.*

a Helium is used in airships because it is less dense than air and is nonflammable.

b If you add bleach to water that is mixed with red food coloring, the red color will disappear.

Physical Vs. Chemical Properties

You can describe matter by both physical and chemical properties. The properties that are most useful in identifying a substance, such as density, solubility, and reactivity with acids, are its characteristic properties. The *characteristic properties* of a substance are always the same whether the sample you're observing is large or small. Scientists rely on characteristic properties to identify and classify substances. **Figure 14** describes some physical and chemical properties.

It is important to remember the differences between physical and chemical properties. For example, you can observe physical properties without changing the identity of the substance. You can observe chemical properties only in situations in which the identity of the substance could change.

Comparing Physical and Chemical Properties		
Substance	**Physical property**	**Chemical property**
Helium	less dense than air	nonflammable
Wood	grainy texture	flammable
Baking soda	white powder	reacts with vinegar to produce bubbles
Powdered sugar	white powder	does not react with vinegar
Rubbing alcohol	clear liquid	flammable
Red food coloring	red color	reacts with bleach and loses color
Iron	malleable	reacts with oxygen

Physical Changes Don't Form New Substances

A **physical change** is a change that affects one or more physical properties of a substance. For example, if you break a piece of chalk in two, you change its physical properties of size and shape. But no matter how many times you break it, chalk is still chalk. The chemical properties of the chalk remain unchanged. Each piece of chalk would still produce bubbles if you placed it in vinegar.

Science Bloopers

The German zeppelin *Hindenburg,* which was filled with hydrogen, caught fire upon landing in 1937. The entire airship was engulfed in an orange fireball and burned in less than 32 seconds. For decades, most people believed the fire started when a spark ignited the flammable hydrogen. But hydrogen burns with a near-colorless flame, not an orange one. Scientists now think that the spark actually ignited the airship's highly flammable outer covering.

Examples of Physical Changes Melting is a good example of a physical change, as you can see in **Figure 15.** Still another physical change occurs when a substance dissolves into another substance. If you dissolve sugar in water, the sugar seems to disappear into the water. But the identity of the sugar does not change. If you taste the water, you will also still taste the sugar. The sugar has undergone a physical change. See the chart below for more examples of physical changes.

Figure 15 *A physical change turned a stick of butter into the liquid butter that makes popcorn so tasty, but the identity of the butter did not change.*

More Examples of Physical Changes	
■ Freezing water for ice cubes	■ Crushing an aluminum can
■ Sanding a piece of wood	■ Bending a paper clip
■ Cutting your hair	■ Mixing oil and vinegar

Can Physical Changes Be Undone? Because physical changes do not change the identity of substances, they are often easy to undo. If you leave butter out on a warm counter, it will undergo a physical change—it will melt. Putting it back in the refrigerator will reverse this change. Likewise, if you create a figure from a lump of clay, you change the clay's shape, causing a physical change. But because the identity of the clay does not change, you can crush your creation and form the clay back into its previous shape.

Chemical Changes Form New Substances

A **chemical change** occurs when one or more substances are changed into entirely new substances with different properties. Chemical changes will or will not occur as described by the chemical properties of substances. But chemical changes and chemical properties are not the same thing. A chemical property describes a substance's ability to go through a chemical change; a chemical change is the actual process in which that substance changes into another substance. You can observe chemical properties only when a chemical change might occur.

QuickLab

Changing Change

1. Place a folded **paper towel** in a **small pie plate.**

2. Pour **vinegar** into the pie plate until the entire paper towel is damp.

3. Place **two or three shiny pennies** on top of the paper towel.

4. Put the pie plate in a place where it won't be bothered, and wait 24 hours.

5. Describe the chemical change that took place.

6. Write your observations in your ScienceLog.

17

DISCUSSION

Encourage students to think of ways to cause physical changes in the following objects:

- a pencil (break, sharpen, grind, or sand it)
- hair (comb, cut, curl, or wash it)
- salt (crush it or dissolve it in water)

QuickLab

MATERIALS

FOR EACH STUDENT:
- paper towel
- pie plate
- vinegar
- 2 or 3 shiny pennies

Answers to QuickLab

5. The shiny copper surface became coated with a dull, green substance. The color change and change in the appearance of the coin indicated that a chemical change took place.

ACTIVITY

MATERIALS

FOR EACH STUDENT:
- spoon
- baking soda
- 2 plastic cups
- vinegar
- water

Safety Caution: Students should wear safety goggles, gloves, and an apron.

Have students place a teaspoon of baking soda in each of two plastic cups. Tell them to pour a small amount of water into one cup and a small amount of vinegar into the other. Ask students to describe what happens and to determine in which cup a chemical change occurred. (The chemical change occurred in the cup containing vinegar; gas bubbles were produced in this cup.)

3 Extend

GOING FURTHER

MATERIALS

FOR EACH STUDENT:
- self-sealing plastic bag
- small plastic pill bottle
- hydrogen peroxide (3% solution)
- clean steel wool

Safety Caution: Caution students to wear safety goggles, gloves, and an apron when doing this activity. Also caution them to handle the hydrogen peroxide with care.

Instruct students to fill the pill bottle halfway with hydrogen peroxide. Then have them place a small piece of steel wool and the pill bottle into the plastic bag, being careful not to spill the hydrogen peroxide. Tell them to force the air out of the bag and seal it tightly. Then instruct them to tip the bottle over so that the hydrogen peroxide comes in contact with the steel wool. Tell them to feel the bag and to observe what happens. Ask them how they know a chemical change has occurred. (A gas was formed that inflated the bag, and the bag heated up.)

MEETING INDIVIDUAL NEEDS

Advanced Learners Have students explore why desalination is so expensive. Then have them try to determine under what conditions it makes sense to use desalination on a large scale. (in places where it is less expensive than piping or shipping fresh water from distant sources)

A fun (and delicious) way to see what happens during chemical changes is to bake a cake. When you bake a cake, you combine eggs, flour, sugar, butter, and other ingredients as shown in **Figure 16.** Each ingredient has its own set of properties. But if you mix them together and bake the batter in the oven, you get something completely different. The heat of the oven and the interaction of the ingredients cause a chemical change. As shown in **Figure 17,** you get a cake that has properties completely different to any of the ingredients. Some more examples of chemical changes are shown below.

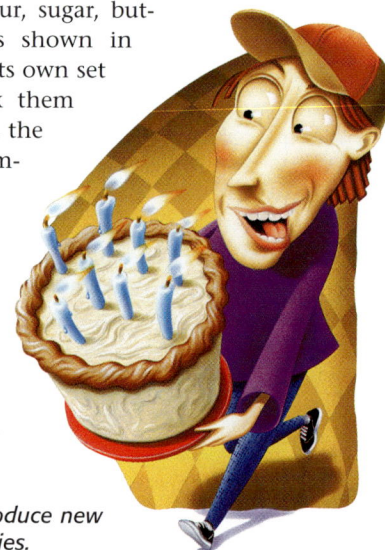

Figure 16 Each of these ingredients has different physical and chemical properties.

Figure 17 Chemical changes produce new substances with different properties.

Examples of Chemical Changes

Soured milk smells bad because bacteria have formed new substances in the milk.

Effervescent tablets bubble when the citric acid and baking soda in them react with water.

The hot gas formed when hydrogen and oxygen join to make water helps blast the space shuttle into orbit.

The Statue of Liberty is made of shiny, orange-brown copper. But the metal's interaction with carbon dioxide and water has formed a new substance, copper carbonate, and made this landmark lady green over time.

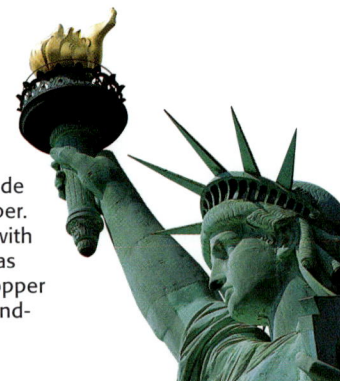

18

internet connect

SCI LINKS
NSTA

TOPIC: Describing Matter
GO TO: www.scilinks.org
sciLINKS NUMBER: HSTP035

Clues to Chemical Changes Look back at the bottom of the previous page. In each picture, there is at least one clue that signals a chemical change. Can you find the clues? Here's a hint: chemical changes often cause color changes, fizzing or foaming, heat, or the production of sound, light, or odor.

In the cake example, you would probably smell the sweet aroma of the cake as it baked. If you looked into the oven, you would see the batter rise and turn brown. When you cut the finished cake, you would see the spongy texture created by gas bubbles that formed in the batter (if you baked it right, that is!). All of these yummy clues are signals of chemical changes. But are the clues and the chemical changes the same thing? No, the clues just result from the chemical changes.

Can Chemical Changes Be Undone? Because new substances are formed, you cannot reverse chemical changes using physical means. In other words, you can't uncrumple or iron out a chemical change. Imagine trying to un-bake the cake shown in **Figure 18** by pulling out each ingredient. No way! Most of the chemical changes you see in your daily life, such as a cake baking or milk turning sour, would be difficult to reverse. However, some chemical changes can be reversed under the right conditions by other chemical changes. For example, the water formed in the space shuttle's rockets could be split back into hydrogen and oxygen using an electric current.

Environment CONNECTION

When fossil fuels are burned, a chemical change takes place involving sulfur (a substance in fossil fuels) and oxygen (from the air). This chemical change produces sulfur dioxide, a gas. When sulfur dioxide enters the atmosphere, it undergoes another chemical change by interacting with water and oxygen. This chemical change produces sulfuric acid, a contributor to acid precipitation. Acid precipitation can kill trees and make ponds and lakes unable to support life.

Figure 18 *Looking for the original ingredients? You won't find them—their identities have changed.*

SECTION REVIEW

1. Classify each of the following properties as either physical or chemical: reacts with water, dissolves in acetone, is blue, does not react with hydrogen.

2. List three clues that indicate a chemical change might be taking place.

3. **Comparing Concepts** Describe the difference between physical changes and chemical changes in terms of what happens to the matter involved in each kind of change.

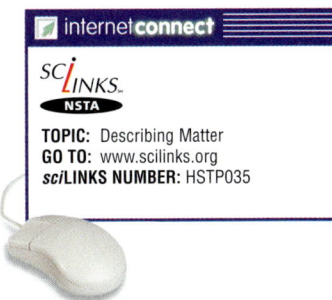

internet**connect**

SC**LINKS**
NSTA

TOPIC: Describing Matter
GO TO: www.scilinks.org
*sci***LINKS NUMBER:** HSTP035

---

4 Close

Quiz

1. You have two objects, both about the size of an orange. Object A has a mass of 1,487 g, and object B has a mass of 878 g. Which object do you think has the greater density? Explain your answer. (Both objects have the same volume, so the object with more mass in the same volume has the greater density. Object A has the greater density.)

2. Give an example of a chemical change that occurs during the preparation of a meal. (Possible answers: burning of gas in an oven or a stove burner; cooking an egg; baking a pie or cake)

ALTERNATIVE ASSESSMENT

Concept Mapping Ask students what they could do to a sugar cube to cause it to undergo a physical change. (crush it, grind it, dissolve it in water)

Then ask what they could do that would cause the sugar cube to undergo a chemical change. (burn it, eat it, cause it to react with another chemical)

Have students make a concept map that shows physical changes and chemical changes to sugar.

▼ *Answers to Section Review*

1. reacts with water: chemical; dissolves in acetone: physical; is blue: physical; does not react with hydrogen: chemical

2. Acceptable answers include color change; bubbling; fizzing or foaming; heat; and the production of light, sound, or odor.

3. In a physical change, the material does not change its identity. It is still the same matter it was before the change and has most of the same properties. In a chemical change, the initial matter changes its identity and becomes a new form of matter with a different identity and different properties.

Reinforcement Worksheet
"A Matter of Density"

Critical Thinking Worksheet
"As a Matter of Fact!"

White Before Your Eyes
Teacher's Notes

Time Required

One or two 45-minute class periods

Lab Ratings

EASY ——————→ HARD

Teacher Prep ♦♦
Student Set-Up ♦
Concept Level ♦♦
Clean Up ♦♦

MATERIALS

Use an iodine solution that contains no more than 1.0% iodine in water. Instead of using an egg carton, you may wish to use a 24-well spot plate or test tubes. The containers used for the solids can be baby food jars or even plastic film canisters. Canisters may be available from a local film processing store. If you do not have small medicine-dropper bottles, you can use small soft-drink bottles for liquids and solutions. Label all containers. A small test tube taped to the bottle makes a great holder for a dropper or pipette and decreases the chances of contamination. A drinking straw cut in half at an angle works well as a spatula; the pointed end is a great scoop, and its large size makes it easy to handle.

Safety Caution

When iodine is being used, be certain that a functioning eyewash is available in case of a splash. Caution students that iodine can stain skin and clothes. Students should wash their face and hands when finished. Clean up any spills immediately to avoid slips and falls.

Skill Builder Lab

White Before Your Eyes

You have learned how to describe matter based on its physical and chemical properties. You also have learned some clues that can help you determine whether a change in matter is a physical change or a chemical change. In this lab, you'll use what you have learned to describe four substances, based on their properties and the changes they undergo.

MATERIALS

- 4 spatulas
- baking powder
- plastic-foam egg carton
- 3 eyedroppers
- water
- stirring rod
- vinegar
- iodine solution
- baking soda
- cornstarch
- sugar

Procedure

1. Copy Tables 1 and 2, shown on the next page, into your ScienceLog. Be sure to leave plenty of room in each box to write down your observations. Before you start the lab, put on your safety goggles.

2. Use a spatula to place a small amount of baking powder (just enough to cover the bottom of the cup) into three cups of your egg carton. Look closely at the baking powder. Record your observations about its appearance, such as color and texture, in Table 1 in the column titled "Unmixed."

3. Use an eyedropper to add 60 drops of water to the baking powder in the first cup. Stir the mixture with the stirring rod. Record your observations in Table 1 in the column titled "Mixed with water." Clean your stirring rod.

4. Use a clean dropper to add 20 drops of vinegar to the second cup of baking powder. Stir. Record your observations in Table 1 in the column titled "Mixed with vinegar." Clean your stirring rod.

5. Use a clean dropper to add 5 drops of iodine solution to the third cup of baking powder. Stir. Record your observations in Table 1 in the column titled "Mixed with iodine solution." Clean your stirring rod.
Caution: Be careful when using iodine. Iodine will stain your skin and clothes.

6. Repeat steps 2–5 for each of the other substances (baking soda, cornstarch, and sugar). Use a clean spatula for each substance.

20

Datasheets for LabBook

CLASSROOM TESTED & APPROVED

Joseph Price
H. M. Browne Junior High
Washington, D.C.

Table 1 Observations

Substance	Unmixed	Mixed with water	Mixed with vinegar	Mixed with iodine solution
Baking powder				
Baking soda				
Cornstarch				
Sugar				

DO NOT WRITE IN BOOK

Table 2 Changes and Properties

Substance	Mixed with water		Mixed with vinegar		Mixed with iodine solution	
	Change	Property	Change	Property	Change	Property
Baking powder						
Baking soda						
Cornstarch						
Sugar						

DO NOT WRITE IN BOOK

Analysis

7 What physical properties do all four substances share?

8 In Table 2, write the type of change you observed, and state the property that the change demonstrates.

9 Classify the four substances you tested by their chemical properties. For example, which substances are reactive with vinegar (acid)?

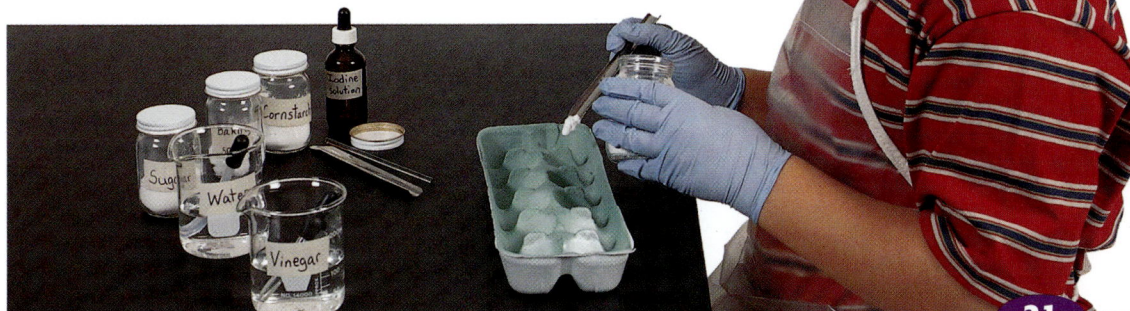

21

Skill Builder Lab

Answers

7. Baking powder, baking soda, cornstarch, and sugar are all white powders.

8. See the table at the bottom of the page.

9. Baking soda, cornstarch, and sugar do not react with water, and baking powder reacts with water. Baking powder and baking soda react with vinegar, and cornstarch and sugar do not react with vinegar. Cornstarch reacts with iodine, and baking powder, baking soda, and sugar do not react with iodine.

Lab Notes

Remind students that vinegar is an acid.

Disposal Information

Dispose of any unreacted iodine solution by combining all student solutions. Decolorize if necessary by adding 1.0 M $Na_2S_2O_3$ while stirring until the dark color disappears. Dilute the mixture with at least 10 times its volume of water. Then pour down the drain.

Science Skills Worksheet
"Using Your Senses"

Teacher's Note: Although Table 2 states that baking powder reacts with water, baking powder is actually not reactive with water. Baking powder can contain baking soda (a base), a weak acid, and a starch. The baking soda and the weak acid in baking powder will react with each other when the baking powder is dissolved in water. This reaction produces the bubbles that students observe in this experiment.

Answers

8.

Substance	Mixed with water		Mixed (vinegar)		Mixed (iodine solution)	
	Change	Property	Change	Property	Change	Property
Baking powder	chemical	reactivity with water	chemical	reactivity with acid	physical	solubility
Baking soda	physical	solubility	chemical	reactivity with acid	physical	solubility
Cornstarch	physical	solubility	physical	solubility	chemical	reactivity with iodine
Sugar	physical	solubility	physical	solubility	physical	solubility

SECTION 1

matter anything that has volume and mass

volume the amount of space that something occupies or the amount of space that something contains

meniscus the curve at a liquid's surface by which you measure the volume of the liquid

mass the amount of matter that something is made of; its value does not change with the object's location in the universe

gravity a force of attraction between objects that is due to their masses

weight a measure of the gravitational force exerted on an object, usually by Earth

newton (N) the SI unit of force

inertia the tendency of all objects to resist any change in motion

Chapter Highlights

SECTION 1

Vocabulary

matter *(p. 4)*
volume *(p. 4)*
meniscus *(p. 5)*
mass *(p. 6)*
gravity *(p. 7)*
weight *(p. 8)*
newton *(p. 9)*
inertia *(p. 10)*

Section Notes

- Matter is anything that has volume and mass.
- Volume is the amount of space taken up by an object.
- The volume of liquids is expressed in liters and milliliters.
- The volume of solid objects is expressed in cubic units, such as cubic meters.
- Mass is the amount of matter that something is made of.
- Mass and weight are not the same thing. Weight is a measure of the gravitational force exerted on an object, usually in relation to the Earth.
- Mass is usually expressed in milligrams, grams, and kilograms.
- The newton is the SI unit of force, so weight is expressed in newtons.
- Inertia is the tendency of all objects to resist any change in motion. Mass is a measure of inertia. The more massive an object is, the greater its inertia.

Labs

Volumania! *(p. 134)*
Measuring Liquid Volume *(p. 132)*
Coin Operated *(p. 133)*

☑ Skills Check

Math Concepts

DENSITY To calculate an object's density, divide the mass of the object by its volume. For example, the density of an object with a mass of 45 g and a volume of 5.5 cm³ is calculated as follows:

$$D = \frac{m}{V}$$
$$D = \frac{45 \text{ g}}{5.5 \text{ cm}^3}$$
$$D = 8.2 \text{ g/cm}^3$$

Visual Understanding

MASS AND WEIGHT Mass and weight are related, but they're not the same thing. Look back at Figure 8 on page 9 to learn about the differences between mass and weight.

PHYSICAL AND CHEMICAL PROPERTIES All substances have physical and chemical properties. You can compare some of those properties by reviewing the table on page 16.

22

Lab and Activity Highlights

White Before Your Eyes PG 20

Measuring Liquid Volume PG 132

Coin Operated PG 133

Volumania! PG 134

Determining Density PG 136

Layering Liquids PG 137

Datasheets for LabBook
(blackline masters for these labs)

SECTION 2

Vocabulary
physical property (p. 11)
density (p. 12)
chemical property (p. 15)
physical change (p. 16)
chemical change (p. 17)

Section Notes
- Physical properties of matter can be observed without changing the identity of the matter.
- Density is the amount of matter in a given space, or the mass per unit volume.
- The density of a substance is always the same at a given pressure and temperature regardless of the size of the sample of the substance.

- Chemical properties describe a substance based on its ability to change into a new substance with different properties.
- Chemical properties can be observed only when one substance might become a new substance.
- The characteristic properties of a substance are always the same whether the sample observed is large or small.
- When a substance undergoes a physical change, its identity remains the same.
- A chemical change occurs when one or more substances are changed into new substances with different properties.

Labs
Determining Density (p. 136)
Layering Liquids (p. 137)

23

VOCABULARY DEFINITIONS, continued

SECTION 2

physical property a property of matter that can be observed or measured without changing the identity of the matter

density the amount of matter in a given space; mass per unit volume

chemical property a property of matter that describes a substance based on its ability to change into a new substance with different properties

physical change a change that affects one or more physical properties of a substance; most physical changes are easy to undo

chemical change a change that occurs when one or more substances are changed into entirely new substances with different properties; cannot be reversed using physical means

Vocabulary Review Worksheet

Blackline masters of these Chapter Highlights can be found in the **Study Guide.**

Lab and Activity Highlights

LabBank

Inquiry Labs, Whatever Floats Your Boat

Whiz-Bang Demonstrations
- Curious Cubes
- The Dancing Toothpicks
- Does 2 + 2 = 4?

Long-Term Projects & Research Ideas, And We Have Thales to Thank

Chapter Review Answers

USING VOCABULARY

1. Mass is the amount of matter in an object; volume is the amount of space the object occupies.
2. Mass is the amount of matter in an object and is always constant. Weight is a measure of the gravitational force, and it will change, depending on the object's distance from the Earth or other bodies.
3. Mass is a measure of inertia. The more massive an object is, the more inertia it has.
4. Volume is the amount of space occupied by an object, and density is the amount of mass in a given volume.
5. physical property: can be observed without changing the identity of the matter; chemical property: can be observed only when a chemical change might occur that would change the matter into something new
6. When matter undergoes a physical change, its shape or form changes, but its identity remains the same. When matter undergoes a chemical change, its identity and properties change.

UNDERSTANDING CONCEPTS

Multiple Choice

7. c	11. a
8. d	12. a
9. b	13. c
10. d	14. c

Short Answer

15. Sample answer: Measure the volume of a liquid by pouring it into a graduated cylinder and reading the scale at the bottom of the meniscus. The volume of a rectangular solid is determined by multiplying the object's length, width, and height.
16. Sample answer: Mass is a measure of the inertia of an object. The more an object resists a change in motion, the greater its mass.

USING VOCABULARY

For each pair of terms, explain the difference in their meanings.

1. mass/volume
2. mass/weight
3. inertia/mass
4. volume/density
5. physical property/chemical property
6. physical change/chemical change

UNDERSTANDING CONCEPTS

Multiple Choice

7. Which of these is *not* matter?
   - a. a cloud
   - b. your hair
   - c. sunshine
   - d. the sun
8. The mass of an elephant on the moon would be
   - a. less than its mass on Mars.
   - b. more than its mass on Mars.
   - c. the same as its weight on the moon.
   - d. None of the above
9. Which of the following is *not* a chemical property?
   - a. reactivity with oxygen
   - b. malleability
   - c. flammability
   - d. reactivity with acid
10. Your weight could be expressed in which of the following units?
    - a. pounds
    - b. newtons
    - c. kilograms
    - d. both (a) and (b)

17. density = $\frac{mass}{volume}$
18. Characteristic properties include density, solubility, reactivity with acids, melting point, and boiling point.

Concept Mapping

19. An answer to this exercise can be found at the front of this book.

---

11. You accidentally break your pencil in half. This is an example of
    - a. a physical change.
    - b. a chemical change.
    - c. density.
    - d. volume.
12. Which of the following statements about density is true?
    - a. Density depends on mass and volume.
    - b. Density is weight per unit volume.
    - c. Density is measured in milliliters.
    - d. Density is a chemical property.
13. Which of the following pairs of objects would have the greatest attraction toward each other due to gravity?
    - a. a 10 kg object and a 10 kg object, 4 m apart
    - b. a 5 kg object and a 5 kg object, 4 m apart
    - c. a 10 kg object and a 10 kg object, 2 m apart
    - d. a 5 kg object and a 5 kg object, 2 m apart
14. Inertia increases as ___?___ increases.
    - a. time
    - b. length
    - c. mass
    - d. volume

Short Answer

15. In one or two sentences, explain the different processes in measuring the volume of a liquid and measuring the volume of a solid.
16. In one or two sentences, explain the relationship between mass and inertia.
17. What is the formula for calculating density?
18. List three characteristic properties of matter.

CRITICAL THINKING AND PROBLEM SOLVING

20. Sample answer: Cooking eggs involves a chemical change. I cannot change the cooked eggs back to raw eggs in order to poach them.
21. Sample answer: If my neighbor has trouble lifting a small box, I would conclude that the box's inertia is large. The box resists my neighbor's attempt to move it. A large inertia means that the item(s) in the box has a large mass.

Concept Mapping

19. Use the following terms to create a concept map: matter, mass, inertia, volume, milliliters, cubic centimeters, weight, gravity.

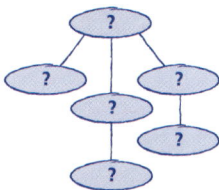

CRITICAL THINKING AND PROBLEM SOLVING

20. You are making breakfast for your picky friend, Filbert. You make him scrambled eggs. He asks, "Would you please take these eggs back to the kitchen and poach them?" What scientific reason do you give Filbert for not changing his eggs?

Poach these, please!

21. You look out your bedroom window and see your new neighbors moving in. Your neighbor bends over to pick up a small cardboard box, but he cannot lift it. What can you conclude about the item(s) in the box? Use the terms *mass* and *inertia* to explain how you came to this conclusion.

22. You may sometimes hear on the radio or on television that astronauts are "weightless" in space. Explain why this is not true.

23. People commonly use the term *volume* to describe the capacity of a container. How does this definition of volume differ from the scientific definition?

MATH IN SCIENCE

24. What is the volume of a book with the following dimensions: a width of 10 cm, a length that is two times the width, and a height that is half the width? Remember to express your answer in cubic units.

25. A jar contains 30 mL of glycerin (mass = 37.8 g) and 60 mL of corn syrup (mass = 82.8 g). Which liquid is on top? Show your work, and explain your answer.

INTERPRETING GRAPHICS

Examine the photograph below, and answer the following questions:

26. List three physical properties of this can.

27. Did a chemical change or a physical change cause the change in this can's appearance?

28. How does the density of the metal in the can compare before and after the change?

29. Can you tell what the chemical properties of the can are just by looking at the picture? Explain.

Reading Check-up

Take a minute to review your answers to the Pre-Reading Questions found at the bottom of page 2. Have your answers changed? If necessary, revise your answers based on what you have learned since you began this chapter.

22. Sample answer: An astronaut weighs less in orbit than on Earth because of the astronaut's increased distance from the Earth. However, an astronaut is not weightless because there are still gravitational forces between the astronaut and all other objects in the universe.

23. Sample answer: The scientific definition of *volume* is the amount of space that matter, such as a container, occupies. The *capacity* of a container describes the maximum amount of space that matter inside the container can occupy.

MATH IN SCIENCE

24. volume = length × width × height = 20 cm × 10 cm × 5 cm = 1,000 cm³

25. Density of glycerin = $\frac{37.8 \text{ g}}{30 \text{ mL}}$ = 1.26 g/mL. Density of corn syrup = $\frac{82.8 \text{ g}}{60 \text{ mL}}$ = 1.38 g/mL. The glycerin will be on top because it is less dense than corn syrup.

INTERPRETING GRAPHICS

26. sample answer: crushed shape, somewhat shiny, metallic

27. a physical change

28. The density before and after the change is the same because density is a characteristic property of matter.

29. No; chemical properties cannot be determined simply by looking at a substance. Chemical properties can only be observed when a chemical change might occur.

25

ACROSS THE SCIENCES

In the Dark About Dark Matter

Background

Keep an eye out for new information about MACHOs and WIMPs. Bring relevant articles from newspapers or magazines to class to discuss with your students. Astronomy magazines, general-science magazines such as *Discover* and *Scientific American,* and science Web sites are a good source of up-to-date information. Web searches using the keywords "MACHO" and "WIMP" will produce results.

You may wish to review with students the types of astronomical bodies that are now known. You may also want to discuss the life cycle of a star.

Astronomers have proposed several candidates for MACHOs. These include brown dwarfs (stars that do not have enough mass to undergo combustion), neutron stars, and black holes.

In 1997, 1998, and 1999, scientists found evidence of MACHOs. They are still looking for convincing evidence of WIMPs.

ACROSS THE SCIENCE

PHYSICAL SCIENCE • ASTRONOMY

In the Dark About Dark Matter

What is the universe made of? Believe it or not, when astronomers try to answer this question, they still find themselves in the dark. Surprisingly, there is more to the universe than meets the eye.

A Matter of Gravity

Astronomers noticed something odd when studying the motions of galaxies in space. They expected to find a lot of mass in the galaxies. Instead, they discovered that the mass of the galaxies was not great enough to explain the large gravitational force causing the galaxies' rapid rotation. So what was causing the additional gravitational force? Some scientists think the universe contains matter that we cannot see with our eyes or our telescopes. Astronomers call this invisible matter *dark matter.*

Dark matter doesn't reveal itself by giving off any kind of electromagnetic radiation, such as visible light, radio waves, or gamma radiation. According to scientific calculations, dark matter could account for between 90 and 99 percent of the total mass of the universe! What is dark matter? Would you believe MACHOs and WIMPs?

MACHOs

Scientists recently proved the existence of *MAssive Compact Halo Objects* (MACHOs) in our Milky Way galaxy by measuring their gravitational effects. Even though scientists know MACHOs exist, they aren't sure what MACHOs are made of. Scientists suggest that MACHOs may be brown dwarfs, old white dwarfs, neutron stars, or black holes. Others suggest they

▲ *The Large Magellanic Cloud, located 180,000 light-years from Earth*

are some type of strange, new object whose properties still remain unknown. Even though the number of MACHOs is apparently very great, they still do not represent enough missing mass. So scientists offer another candidate for dark matter—WIMPs.

WIMPs

Theories predict that *Weakly Interacting Massive Particles* (WIMPs) exist, but scientists have never detected them. WIMPs are thought to be massive elementary particles that do not interact strongly with normal matter (which is why scientists have not found them).

More Answers Needed

So far, evidence supports the existence of MACHOs, but there is little or no solid evidence of WIMPs or any other form of dark matter. Scientists who support the idea of WIMPs are conducting studies of the particles that make up matter to see if they can detect WIMPs. Other theories are that gravity acts differently around galaxies or that the universe is filled with things called "cosmic strings." Scientists admit they have a lot of work to do before they will be able to describe the universe—and all the matter in it.

On Your Own

▶ What is microlensing, and what does it have to do with MACHOs? How might the neutrino provide valuable information to scientists who are interested in proving the existence of WIMPs? Find out on your own!

26

Answer to On Your Own

Students will find that microlensing is the occasional amplification of light from distant stars outside the Milky Way by the gravitational force of certain massive objects in the Milky Way. Neutrinos are important because some scientists think certain neutrinos are good candidates for WIMPs.

Health WATCH

Building a Better Body

Have you ever broken an arm or a leg? If so, you probably wore a cast while the bone healed. But what happens when a bone is too badly damaged to heal? In some cases, a false bone made from a metal called titanium can take the original bone's place. Could using titanium bone implants be the first step in creating bionic body parts? Think about it as you read about some of titanium's amazing properties.

▲ *Titanium bones—even better than the real thing?*

Imitating the Original

Why would a metal like titanium be used to imitate natural bone? Well, it turns out that a titanium implant passes some key tests for bone replacement. First of all, real bones are incredibly lightweight and sturdy, and healthy bones last for many years. Therefore, a bone-replacement material has to be lightweight but also very durable. Titanium passes this test because it is well known for its strength, and it is also lightweight.

Second, the human body's immune system is always on the lookout for foreign substances. If a doctor puts a false bone in place and the patient's immune system attacks it, an infection can result. Somehow, the false bone must be able to chemically trick the body into thinking that the bone is real. Does titanium pass this test? Keep reading!

Accepting Imitation

By studying the human body's immune system, scientists found that the body accepts certain metals. The body almost always accepts one metal in particular. Yep, you guessed it—titanium! This turned out to be quite a discovery.

Doctors could implant pieces of titanium into a person's body without triggering an immune reaction. A bond can even form between titanium and existing bone tissue, fusing the bone to the metal!

Titanium is shaping up to be a great bone-replacement material. It is lightweight and strong, is accepted by the body, can attach to existing bone, and resists chemical changes, such as corrosion. But scientists have encountered a slight problem. Friction can wear away titanium bones, especially those used near the hips and elbows.

Real Success

An unexpected surprise, not from the field of medicine but from the field of nuclear physics, may have solved the problem. Researchers have learned that by implanting a special form of nitrogen on the surface of a piece of metal, they can create a surface layer on the metal that is especially durable and wear-resistant. When this form of nitrogen is implanted in titanium bones, the bones retain all the properties of pure titanium bones but also become very wear-resistant. The new bones should last through decades of heavy use without needing to be replaced.

Think About It

▶ What will the future hold? As time goes by, doctors become more successful at implanting titanium bones. What do you think would happen if the titanium bones were to eventually become better than real bones?

HEALTH WATCH
Building a Better Body

Background

You may want to tell your students that the nitrogen implanted in the surface of the titanium bones is actually nitrogen ions that have been blasted into the surface of the titanium by a particle accelerator. The ions bond to the titanium atoms, resulting in a titanium surface that is harder and smoother than nonimplanted titanium. The nitrogen-titanium bond also attracts and holds a thin film of joint fluid in the body. This fluid lubricates moving parts of the joint and increases the durability of the bone replacement.

Students may want to investigate the most recent advances in materials used for replacement bones. Surgeons and scientists are also making replacement bones that use a titanium-cobalt alloy or a cobalt-chrome alloy and a plastic socket. Other scientists are pursuing the use of ceramic materials for replacement bones.

internetconnect

SCiLINKS
NSTA
TOPIC: Building a Better Body
GO TO: www.scilinks.org
*sci*LINKS NUMBER: HSTP045

Answer to Think About It
During the discussion about the future of better bones, accept all reasonable answers.

Chapter Organizer

CHAPTER ORGANIZATION	TIME MINUTES	OBJECTIVES	LABS, INVESTIGATIONS, AND DEMONSTRATIONS
Chapter Opener pp. 28–29	45	National Standards: UCP 2, 3, SAI 1, PS 1a, 3a	**Start-Up Activity,** Vanishing Act, p. 29
Section 1 Four States of Matter	60	▶ Describe the properties shared by particles of all matter. ▶ Describe the four states of matter discussed here. ▶ Describe the differences between the states of matter. ▶ Predict how a change in pressure or temperature will affect the volume of a gas. UCP 1, 2, PS 1a; Labs UCP 2, 3, SAI 1, 2, PS 1a	**Demonstration,** p. 30 in ATE **Demonstration,** p. 34 in ATE **Discovery Lab,** Full of Hot Air! p. 138 **Datasheets for LabBook,** Full of Hot Air! **Whiz-Bang Demonstrations,** Demonstration with a CRUNCH!
Section 2 Changes of State	90	▶ Describe how substances change from state to state. ▶ Explain the difference between an exothermic change and an endothermic change. ▶ Compare the changes of state. UCP 1, 3, SAI 1, 2, ST 1, SPSP 3, 5, HNS 3, PS 1a, 3a; Labs UCP 2, 3, SAI 1, 2, PS 1a, 3a	**Demonstration,** p. 38 in ATE **QuickLab,** Boiling Water Is Cool, p. 41 **Skill Builder,** Can Crusher, p. 139 **Datasheets for LabBook,** Can Crusher **Discovery Lab,** A Hot and Cool Lab, p. 44 **Datasheets for LabBook,** A Hot and Cool Lab **Calculator-Based Labs,** A Hot and Cool Lab **Labs You Can Eat,** How Cold Is Ice-Cream Cold? **Long-Term Projects & Research Ideas,** Episode IV: Sam and His Elephants Get That Sinking Feeling

*See page **T23** for a complete correlation of this book with the*

NATIONAL SCIENCE EDUCATION STANDARDS.

TECHNOLOGY RESOURCES

Guided Reading Audio CD
English or Spanish, Chapter 2

Science Discovery Videodiscs
Image and Activity Bank with Lesson Plans: States of Change

CNN **Scientists in Action,** In Search of Absolute Zero, Segment 5

One-Stop Planner CD-ROM **with Test Generator**

Chapter 2 • States of Matter

CLASSROOM WORKSHEETS, TRANSPARENCIES, AND RESOURCES	SCIENCE INTEGRATION AND CONNECTIONS	REVIEW AND ASSESSMENT
Directed Reading Worksheet **Science Puzzlers, Twisters & Teasers**		
Directed Reading Worksheet, Section 1 **Transparency 206,** Models of a Solid, a Liquid, and a Gas **Transparency 207,** Boyle's Law **Transparency 207,** Charles's Law **Reinforcement Worksheet,** Make a State-ment	**Connect to Life Science,** p. 33 in ATE **Math and More,** p. 35 in ATE **MathBreak,** Gas Law Graphs, p. 36 **Apply,** p. 37	**Self-Check,** p. 34 **Section Review,** p. 34 **Homework,** p. 36 in ATE **Section Review,** p. 37 **Quiz,** p. 37 in ATE **Alternative Assessment,** p. 37 in ATE
Directed Reading Worksheet, Section 2 **Math Skills for Science Worksheet,** Checking Division with Multiplication **Transparency 208,** Summarizing the Changes of State **Critical Thinking Worksheet,** What a State! **Transparency 141,** The Water Cycle **Transparency 209,** Changing the State of Water	**Multicultural Connection,** p. 39 in ATE **Meteorology Connection,** p. 41 **Math and More,** p. 41 in ATE **Science, Technology, and Society:** Guiding Lightning, p. 50 **Eureka!** Full Steam Ahead! p. 51	**Self-Check,** p. 40 **Homework,** p. 40 in ATE **Section Review,** p. 43 **Quiz,** p. 43 in ATE **Alternative Assessment,** p. 43 in ATE

internet connect

go.hrw.com
Holt, Rinehart and Winston On-line Resources
go.hrw.com

For worksheets and other teaching aids related to this chapter, visit the HRW Web site and type in the keyword: **HSTSTA**

*sci*LINKS NSTA
National Science Teachers Association
www.scilinks.org

Encourage students to use the *sci*LINKS numbers listed in the internet connect boxes to access information and resources on the **NSTA** Web site.

END-OF-CHAPTER REVIEW AND ASSESSMENT

Chapter Review in Study Guide
Vocabulary and Notes in Study Guide
Chapter Tests with Performance-Based Assessment, Chapter 2 Test
Chapter Tests with Performance-Based Assessment, Performance-Based Assessment 2
Concept Mapping Transparency 3

Chapter Resources & Worksheets

Visual Resources

TEACHING TRANSPARENCIES

#206 — Models of a Solid, a Liquid, and a Gas — *Holt Science and Technology* — 206

#207 — Boyle's Law / Charles's Law — *Holt Science and Technology* — Teaching Transparency 207

#208 — Summarizing the Changes of State — *Holt Science and Technology* — 208

#209 — Changing the State of Water — *Holt Science and Technology* — 209

TEACHING TRANSPARENCIES

#141 — The Water Cycle — *Holt Science and Technology* — Teaching Transparency 141

LINK TO EARTH SCIENCE

CONCEPT MAPPING TRANSPARENCY

#3 — States of Matter — *Holt Science and Technology* — Concept Mapping Transparency 3

Use the following terms to complete the concept map below: changes of state, plasma, melting, vaporization, liquid, condensation, states of matter, solid

Meeting Individual Needs

DIRECTED READING

#2 — DIRECTED READING WORKSHEET — States of Matter

REINFORCEMENT & VOCABULARY REVIEW

#2 — REINFORCEMENT WORKSHEET — Make a State-ment

#2 — VOCABULARY REVIEW WORKSHEET — Know Your States

SCIENCE PUZZLERS, TWISTERS & TEASERS

#2 — SCIENCE PUZZLERS, TWISTERS & TEASERS — States of Matter — Mystery Jars

Review & Assessment

STUDY GUIDE

#2 VOCABULARY & NOTES WORKSHEET
States of Matter

By studying the Vocabulary and Notes listed for each section below, you can gain a better understanding of this chapter.

SECTION 1
Vocabulary

In your own words, write a definition for each of the following terms in the space provided.

1. states of matter

2. solid

3. liquid

4. gas

5. pressure

6. Boyle's law

7. Charles's law

8. plasma

#2 CHAPTER REVIEW WORKSHEET
States of Matter

USING VOCABULARY

For each pair of terms, explain the difference in their meanings.

1. solid/liquid

2. Boyle's Law/Charles's Law

3. evaporation/boiling

4. melting/freezing

UNDERSTANDING CONCEPTS
Multiple Choice

5. Which of the following best describes the particles of a liquid?
   a. The particles are far apart and moving fast.
   b. The particles are close together but moving past each other.
   c. The particles are far apart and moving slowly.
   d. The particles are closely packed and vibrate in place.

6. Boiling points and freezing points are examples of
   a. chemical properties.   c. energy.
   b. physical properties.   d. matter.

7. During which change of state do atoms or molecules become more ordered?
   a. boiling   c. melting
   b. condensation   d. sublimation

CHAPTER TESTS WITH PERFORMANCE-BASED ASSESSMENT

#2 STATES OF MATTER
Chapter 2 Test

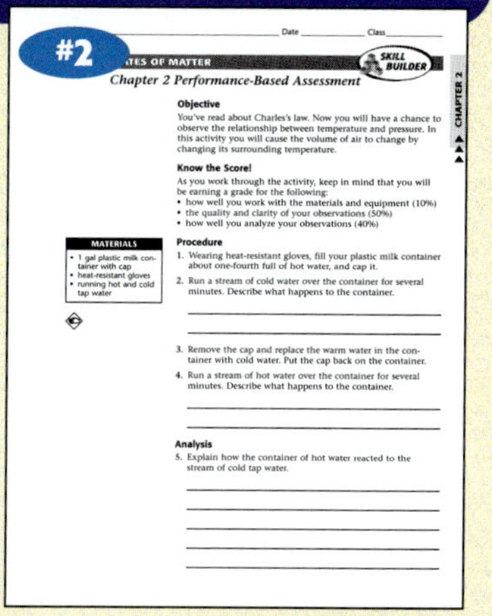

USING VOCABULARY

To complete the following sentences, choose the correct term from each pair of terms listed, and write the term in the blank.

1. The drops of water that appear on the outside of a glass of cold juice on a warm day are an example of _____. (condensation or evaporation)

2. The way a balloon decreases in volume when the temperature is decreased illustrates _____. (Boyle's law or Charles's law)

3. _____ is the change of state from a liquid to a gas. (Vaporization or Condensation)

4. Sublimation is a change of state from a solid directly to a _____. (liquid or gas)

5. In an _____ change, energy is added to a substance. (endothermic or exothermic)

6. Magnetic fields are used to contain _____. (gases or plasmas)

UNDERSTANDING CONCEPTS
Multiple Choice
Circle the correct answer.

7. Boyle's law explains the relationship between volume and pressure for a fixed amount of a
   a. solid.   c. gas.
   b. liquid.   d. plasma.

8. Which of the following examples involves an exothermic change?
   a. ice melting on a warm day
   b. water boiling in a tea kettle
   c. gaseous water particles coming together to form fog
   d. air in a bicycle tire gaining pressure after a long ride

9. Which of these factors could affect the temperature at which water boils?
   a. the volume of water in the pot
   b. the atmospheric pressure at which the water is heated
   c. the amount of energy added to the water
   d. the type of fuel used to heat the water

10. The particles of water that evaporate from an open container have _____ than the particles that remain.
   a. more speed   c. higher energy
   b. greater order   d. more speed and higher energy

#2 STATES OF MATTER
Chapter 2 Performance-Based Assessment

Objective
You've read about Charles's law. Now you will have a chance to observe the relationship between temperature and pressure. In this activity you will cause the volume of air to change by changing its surrounding temperature.

Know the Score!
As you work through the activity, keep in mind that you will be earning a grade for the following:
• how well you work with materials and equipment (10%)
• the quality and clarity of your observations (50%)
• how well you analyze your observations (40%)

Procedure
1. Wearing heat-resistant gloves, fill your plastic milk container about one-fourth full of hot water, and cap it.
2. Run a stream of cold water over the container for several minutes. Describe what happens to the container.

3. Remove the cap and replace the warm water in the container with cold water. Put the cap back on the container.
4. Run a stream of hot water over the container for several minutes. Describe what happens to the container.

Analysis
5. Explain how the container of hot water reacted to the stream of cold tap water.

Lab Worksheets

LABS YOU CAN EAT

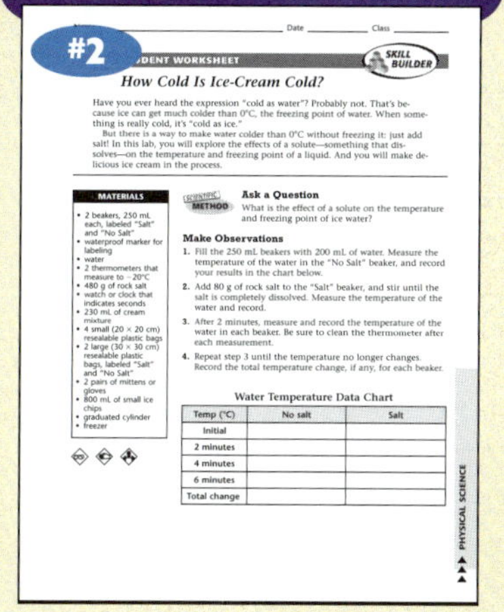

#2 STUDENT WORKSHEET
How Cold Is Ice-Cream Cold?

Have you ever heard the expression "cold as water"? Probably not. That's because ice can get much colder than 0°C, the freezing point of water. When something is really cold, it's "cold as ice."
But there is a way to make water colder than 0°C without freezing it: just add salt! In this lab, you will explore the effects of a solute—something that dissolves—on the temperature and freezing point of a liquid. And you will make delicious ice cream in the process.

Ask a Question
What is the effect of a solute on the temperature and freezing point of ice water?

Make Observations
1. Fill the 250 mL beakers with 200 mL of water. Measure the temperature of the water in the "No Salt" beaker, and record your results in the chart below.
2. Add 80 g of rock salt to the "Salt" beaker, and stir until the salt is completely dissolved. Measure the temperature of the water and record.
3. After 2 minutes, measure and record the temperature of the water in each beaker. Be sure to clean the thermometer after each measurement.
4. Repeat step 3 until the temperature no longer changes. Record the total temperature change, if any, for each beaker.

Water Temperature Data Chart

Temp (°C)	No salt	Salt
Initial		
2 minutes		
4 minutes		
6 minutes		
Total change		

What to Do
1. Explain that you have two batches of popcorn kernels. One was soaked in

WHIZ-BANG DEMONSTRATIONS

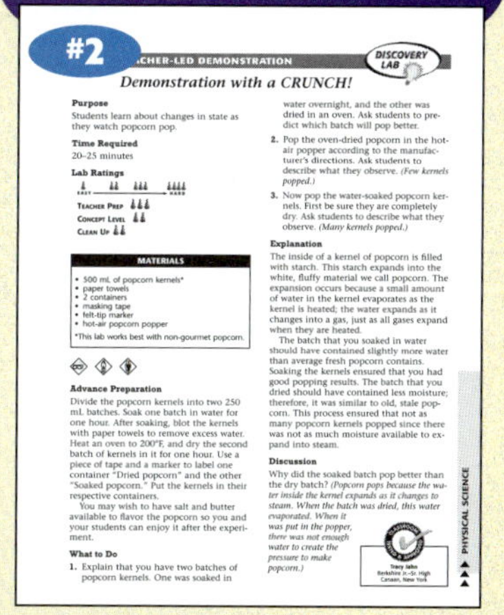

#2 TEACHER-LED DEMONSTRATION
Demonstration with a CRUNCH!

Purpose
Students learn about changes in state as they watch popcorn pop.

Time Required
20–25 minutes

Lab Ratings

TEACHER PREP
CONCEPT LEVEL
CLEAN UP

MATERIALS
• 500 mL of popcorn kernels*
• paper towels
• 2 containers
• masking tape
• hot-tip marker
• hot-air popcorn popper
*This lab works best with non-gourmet popcorn.

Advance Preparation
Divide the popcorn kernels into two 250 mL batches. Soak one batch in water for one hour. After soaking, blot the kernels with paper towels to remove excess water. Heat an oven to 200°F, and dry the second batch of kernels in it for one hour. Use a piece of tape and a marker to label one container "Dried popcorn" and the other "Soaked popcorn." Put the kernels in their respective containers.

What to Do
1. Explain that you have two batches of popcorn kernels. One was soaked in

water overnight, and the other was dried in an oven. Ask students to predict which batch will pop better.
2. Pop the oven-dried popcorn in the hot-air popper according to the manufacturer's directions. Ask students to describe what they observe. *(Few kernels popped.)*
3. Now pop the water-soaked popcorn kernels. First be sure they are completely dry. Ask students to describe what they observe. *(Many kernels popped.)*

Explanation
The inside of a kernel of popcorn is filled with starch. This starch expands into the white, fluffy material we call popcorn. The expansion occurs because a small amount of water in the kernel evaporates as the kernel is heated; the water expands as it changes into a gas, just as all gases expand when they are heated.
The batch that you soaked in water should have contained slightly more water than average fresh popcorn contains. Soaking the kernels ensured that you had good popping results. The batch that you dried should have contained less moisture, therefore, it was similar to old, stale popcorn. This process ensured that not as many popcorn kernels popped since there was not as much moisture available to expand into steam.

Discussion
Why did the soaked batch pop better than the dry batch? *(Popcorn pops because the water inside the kernel expands as it changes to steam. When the batch was dried, this water evaporated. When it was put in the popper, there was not enough water to create the pressure to make the popcorn.)*

LONG-TERM PROJECTS & RESEARCH IDEAS

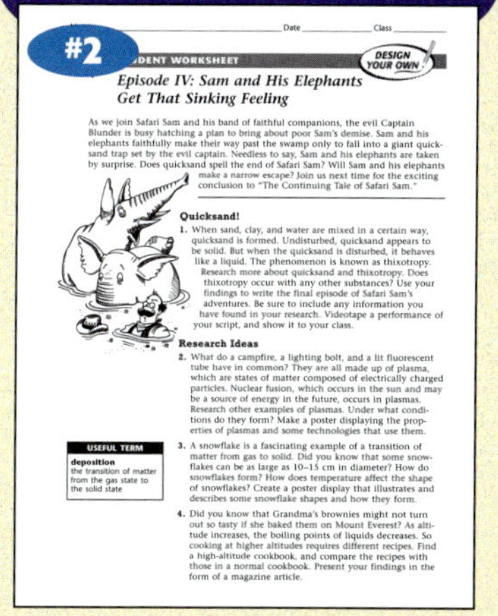

#2 STUDENT WORKSHEET
Episode IV: Sam and His Elephants Get That Sinking Feeling

As we join Safari Sam and his band of faithful companions, the evil Captain Blunder is busy hatching a plan to bring about poor Sam's demise. Sam and his elephants faithfully make their way past the swamp only to fall into a giant quicksand trap set by the evil captain. Needless to say, Sam and his elephants are taken by surprise. Does quicksand really exist? Will Sam and his elephants make a narrow escape? Join us next time for the exciting conclusion to "The Continuing Tale of Safari Sam."

Quicksand!
When sand, clay, and water are mixed in a certain way, quicksand is formed. Undisturbed, quicksand appears to be solid. But when the quicksand is disturbed, it behaves like a liquid. The phenomenon is known as thixotropy. Research more about quicksand and thixotropy. Does thixotropy occur with any other substances? Use your findings to write the final episode of Safari Sam's adventures. Be sure to include any information you have found in your research. Videotape a performance of your script, and show it to your class.

Research Ideas
2. What do a campfire, a lighting bolt, and a lit fluorescent tube have in common? They are all made up of plasma, which are states of matter composed of electrically charged particles. Nuclear fusion, which occurs in the sun and may be a source of energy in the future, occurs in plasmas. Research other examples of plasmas. Under what conditions do they form? Make a poster displaying the properties of plasmas and some technologies that use them.

3. A snowflake is a fascinating example of a transition of matter from gas to solid. Did you know that some snowflakes can be as large as 10–15 cm in diameter? How do snowflakes form? How does temperature affect the shape of snowflakes? Create a poster display that illustrates and describes some snowflake shapes and how they form.

USEFUL TERM
deposition
the transition of matter from the gas state to the solid state

4. Did you know that Grandma's brownies might not turn out so tasty if she baked them on Mount Everest? As altitude increases, the boiling points of liquids decreases. So cooking at higher altitudes requires different recipes. Find a high-altitude cookbook, and compare the recipes with those in a normal cookbook. Present your findings in the form of a magazine article.

DATASHEETS FOR LABBOOK

Name ___ Date ___ Class ___

#2 Full of Hot Air!

#2 Can Crusher

#2 A Hot and Cool Lab

3. Make a data table like the one below to list your observations. Make as many observations as you can about the potatoes in Group A, Group B, and Group C.

Observations	
Group A:	
Group B:	

Form a Hypothesis
4. You have identified a problem and made your observations. Now you can make a hypothesis. Write a clear hypothesis about what you think will be the outcome of your tests.

Applications & Extensions

CRITICAL THINKING & PROBLEM SOLVING

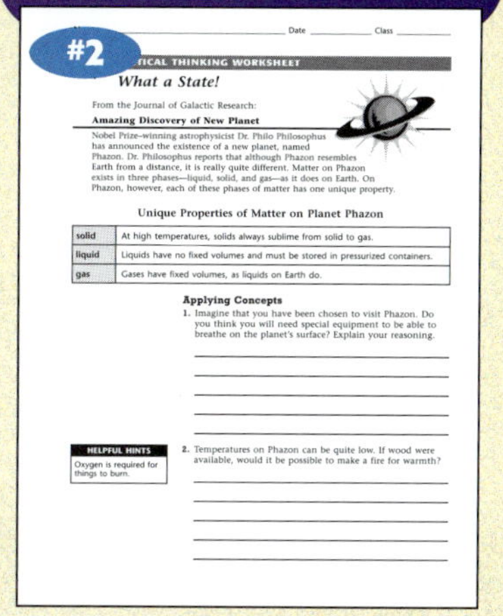

#2 CRITICAL THINKING WORKSHEET
What a State!

From the Journal of Galactic Research:
Amazing Discovery of New Planet
Nobel Prize-winning astrophysicist Dr. Philo Philosophus has announced the existence of a new planet, named Phazon. Dr. Philosophus reports that although Phazon resembles Earth from a distance, it is really quite different. Matter on Phazon exists in three phases—liquid, solid, and gas—as it does on Earth. On Phazon, however, each of these phases of matter has one unique property.

Unique Properties of Matter on Planet Phazon

solid	At high temperatures, solids always sublime from solid to gas.
liquid	Liquids have no fixed volumes and must be stored in pressurized containers.
gas	Gases have fixed volumes, as liquids on Earth do.

Applying Concepts
1. Imagine that you have been chosen to visit Phazon. Do you think you will need special equipment to be able to breathe on the planet's surface? Explain your reasoning.

HELPFUL HINTS
Oxygen is required for things to burn.

2. Temperatures on Phazon can be quite low. If wood were available, would it be possible to make a fire for warmth?

SCIENTISTS IN ACTION

Name ___ Date ___ Class ___

#5 Science in the News: Critical Thinking Worksheets

In Search of Absolute Zero

1. What do scientists believe will happen to a gas as its temperature gets very close to absolute zero?

2. How is the motion of atoms related to temperature?

3. Why is absolute zero considered the limit to how cold a substance can become? Why is it considered an impossible goal?

4. How do you think scientists get a gas very cold?

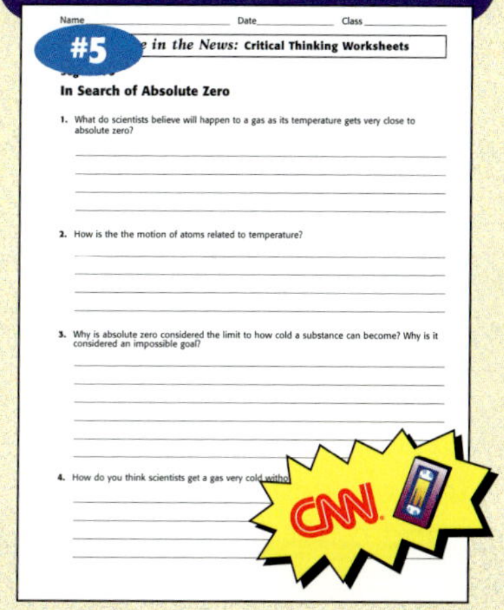

Chapter Background

SECTION 1

Four States of Matter

▶ Solids

In solids, particles vibrate about fixed points. If the particles are arranged in a regular, repeating pattern, the solid is defined as a crystalline solid. If a crystalline solid is melted and cooled down quickly, it usually forms an amorphous solid. Amorphous solid particles are not arranged in regular, repeating patterns.

▶ Liquids

The properties of liquids are caused by *cohesion*, the attraction between atoms and molecules of the liquid, and *adhesion*, the attraction between atoms and molecules of the liquid and other atoms and molecules. Because the surface of a liquid has no liquid particles above it, the particles at the surface cohere to the liquid below, and the surface exhibits surface tension.

▶ Gases

The defining property of gases is the ability to expand indefinitely. Gases are extremely compressible. Gases are also miscible with other gases in all proportions.

▶ Plasma

Plasma is ionized gas. Matter changes into the plasma state when gaseous particles collide with such intensity that electrons are torn away from the atoms, producing an ionized gas that conducts electric current and is affected by magnetic fields. Slightly ionized plasmas, like those found in plasma balls and fluorescent lights, can also conduct an electric current. Fluorescent lights and plasma balls are cool to the touch because electrical energy, instead of thermal energy, is used to break apart the particles.

- A great deal of scientific research is being done to find useful applications for plasmas. For example, plasmas are being used to destroy chemical weapons. Plasmas are also necessary in fusion research.

- In 1920, American chemist Irving Langmuir coined the name *plasma*. Langmuir and other scientists

discovered that many substances reach a unique state at temperatures above 3,000°C in which their particles have properties like a gas (no definite shape or volume) but are electrically charged.

▶ Robert Boyle and Boyle's Law

Robert Boyle (1627–1691) was born in Ireland and educated at Eton, Geneva, and Oxford. In 1662, Boyle was experimenting with mercury in a closed J-shaped tube. He discovered the inverse relationship between the volume of a confined gas and its pressure. Through experimentation, Boyle discovered that if the volume of a gas at a constant temperature is doubled, the pressure is reduced by half. And for any decrease in volume, there is a proportional increase in pressure.

▶ Jacques Charles and Charles's Law

Jacques Alexander Charles (1746–1823) was a professor of physics at the University of Paris and a friend of Benjamin Franklin. Charles was an avid balloonist. From his work with balloons and gases, he realized that hydrogen would be ideal for balloon flight. He built a balloon and used hydrogen to fill it. He made several flights with his hydrogen balloon and once flew to a height of over 1.7 km.

- Charles's law states that if an ideal gas is held at a constant pressure, its volume will increase as temperature increases and decrease as temperature decreases. Charles's research with gases was used by Lord Kelvin to formulate the absolute, or Kelvin, temperature scale.

IS THAT A FACT!

- ◄ Many purists argue that amorphous materials are not true solids. When defining amorphous substances, these scientists prefer to call such substances supercooled liquids.

- ◄ Crystalline solids can be classified into seven crystal systems: cubic, tetragonal, hexagonal, rhombohedral, orthorhombic, monoclinic, and triclinic. A few examples of common substances with their corresponding crystal system are: table salt—cubic system; most metals—cubic or hexagonal system; diamond—cubic system (but graphite, which, like diamond, is a form of carbon, has a layered hexagonal system).

◆ All stars, fluorescent lights, fire, and lightning are examples of matter in the plasma state.

Changes of State

▶ **Change of State**
For a solid substance to melt, it must gain sufficient energy to overcome intermolecular attraction. As a substance such as ice absorbs energy, the individual molecules of the ice vibrate faster and faster, breaking the bonds that hold the molecules together. This allows the molecules to begin sliding past one another. If sufficient energy is added, the liquid begins to boil.

▶ **Temperature**
Temperature is a measure of the average speed of a substance's particles. Temperature differences indicate in which direction thermal energy will move.

▶ **Fahrenheit Scale**
Gabriel Daniel Fahrenheit (1686–1736), a German physicist, developed the Fahrenheit temperature scale.

Fahrenheit's zero point was the freezing temperature of icy brine water. He chose icy brine water because in the late 1600s, it was the lowest known temperature. Pure water froze at 32° on Fahrenheit's scale. The scale has 180 divisions between the freezing point and boiling point of pure water.

▶ **Celsius Scale**
Swedish astronomer Anders Celsius (1704–1744) developed his thermometer so scientists could have a common scale and standard by which to compare experiments. The Celsius scale, also known as the centigrade scale, has 100 divisions between the freezing and boiling points of pure water at 1 atm. Celsius assigned the freezing point of water as 100° and the boiling point as 0°. This was later changed to the scale we use today.

▶ **William Thomson, Lord Kelvin**
William Thomson, Lord Kelvin, (1824–1907) was born in Ireland. Thomson was considered a child prodigy—he began his studies at the University of Glasgow when he was only 11 years old. Thomson had many interests (for instance, he investigated the age of Earth), but he is probably best known for his work with absolute temperature.

▶ **Kelvin Scale**
The Kelvin scale's divisions are based on the centigrade scale, but the scale does not have any negative numbers. Pure water boils at 373 K and freezes at 273 K.

IS THAT A FACT!

◆ There is no apparent limit to how hot a substance can become, but there is a limit to how cold something can become. Lord Kelvin stated that temperature is related to volume and energy, and at absolute zero a substance's volume and energy would achieve their lowest values.

For background information about teaching strategies and issues, refer to the _Professional Reference for Teachers_.

CHAPTER

2

States of Matter

Sections

Pre-Reading Questions

1. What are the four most familiar states of matter?

2. Compare the motion of particles in a solid, a liquid, and a gas.

3. Name three ways matter changes from one state to another.

28

It Takes Mettle to Melt Metal

If you wanted to make a flavored ice pop, you would pour juice into a mold and freeze it. You are able to make the ice pop into the desired shape because, unlike solids, liquids will take the shape of their container. Metal workers apply this important property of liquids when they create metal parts that have complicated shapes. They melt the metal at extremely high temperatures and then pour it into a mold. In this chapter, you will find out more about the properties of different states of matter.

START-UP Activity

VANISHING ACT

In this activity, you will use rubbing alcohol to investigate a change of state.

Procedure

1. Pour **rubbing alcohol** into a **small plastic cup** until the alcohol just covers the bottom of the cup.

2. Moisten the tip of a **cotton swab** by dipping it into the alcohol in the cup.

3. Rub the cotton swab on the palm of your hand.

4. Record your observations in your ScienceLog.

5. Wash your hands thoroughly.

Analysis

6. Explain what happened to the alcohol.

7. Did you feel a sensation of hot or cold? If so, how do you explain what you observed?

8. Record your answers in your ScienceLog.

START-UP Activity

VANISHING ACT

MATERIALS
FOR EACH GROUP: • rubbing alcohol   • small plastic cup   • cotton swab

Safety Caution

Remind students to review all safety cautions and icons before beginning this activity. Students should wear safety goggles and aprons during this activity.

Teacher's Notes

Only a small amount of alcohol is needed for this activity. Demonstrate how little alcohol is needed by pouring an amount sufficient for this activity into your cup.

Answers to START-UP Activity

6. The alcohol disappeared by evaporating.

7. Students should feel a cooling sensation. As the alcohol evaporates, it absorbs energy from the student's hand.

29

Focus

Four States of Matter

This section introduces four states of matter, and students explore the similarities and differences among these four states. Students also examine the effect of temperature and pressure on gases.

The terms *phase* and *state* are sometimes used interchangeably, but they are not the same. *State* refers to states of matter (solid, liquid, gas, plasma). *Phase* refers to a region of material with distinct boundaries and uniform properties for that region. For example, both corn oil and water are in the liquid state. A mixture of corn oil and water has two phases, the corn oil phase and the water phase.

🔔 Bellringer

Pose the following question to your students:

In the kitchen, you might find three different forms of water. What are these three forms of water, and where exactly in the kitchen would you find them?

1 Motivate

DEMONSTRATION

From one corner of the room, or from the very front, spray room deodorizer into the air. Ask students to raise their hand when they smell the deodorizer. Discuss a possible model that would explain why different students smelled the deodorizer at different times. Sheltered English

---

Terms to Learn

states of matter	pressure
solid	Boyle's law
liquid	Charles's law
gas	plasma

What You'll Do

- Describe the properties shared by particles of all matter.
- Describe the four states of matter discussed here.
- Describe the differences between the states of matter.
- Predict how a change in pressure or temperature will affect the volume of a gas.

Four States of Matter

Figure 1 shows a model of the earliest known steam engine, invented about A.D. 60 by Hero, a scientist who lived in Alexandria, Egypt. This model also demonstrates the four most familiar states of matter: solid, liquid, gas, and plasma. The **states of matter** are the physical forms in which a substance can exist. For example, water commonly exists in three different states of matter: solid (ice), liquid (water), and gas (steam).

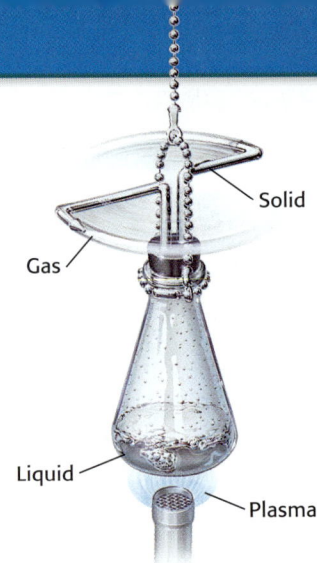

Figure 1 *This model of Hero's steam engine spins as steam escapes through the nozzles.*

Solid
Gas
Liquid
Plasma

Moving Particles Make Up All Matter

Matter consists of tiny particles called atoms and molecules (MAHL i KYOOLZ) that are too small to see without an amazingly powerful microscope. These atoms and molecules are always in motion and are constantly bumping into one another. The state of matter of a substance is determined by how fast the particles move and how strongly the particles are attracted to one another. **Figure 2** illustrates three of the states of matter—solid, liquid, and gas—in terms of the speed and attraction of the particles.

Figure 2   Models of a Solid, a Liquid, and a Gas

Particles of a solid do not move fast enough to overcome the strong attraction between them, so they are held tightly in place. The particles vibrate in place.

Particles of a liquid move fast enough to overcome some of the attraction between them. The particles are able to slide past one another.

Particles of a gas move fast enough to overcome nearly all of the attraction between them. The particles move independently of one another.

Teaching Transparency 206 "Models of a Solid, a Liquid, and a Gas"

Directed Reading Worksheet Section 1

Solids Have Definite Shape and Volume

Look at the ship in **Figure 3.** Even in a bottle, it keeps its original shape and volume. If you moved the ship to a larger bottle, the ship's shape and volume would not change. Scientifically, the state in which matter has a definite shape and volume is **solid.** The particles of a substance in a solid are very close together. The attraction between them is stronger than the attraction between the particles of the same substance in the liquid or gaseous state. The atoms or molecules in a solid move, but not fast enough to overcome the attraction between them. Each particle vibrates in place because it is locked in position by the particles around it.

Figure 3 *Because this ship is a solid, it does not take the shape of the bottle.*

Two Types of Solids Solids are often divided into two categories—*crystalline* and *amorphous* (uh MOHR fuhs). Crystalline solids have a very orderly, three-dimensional arrangement of atoms or molecules. That is, the particles are arranged in a repeating pattern of rows. Examples of crystalline solids include iron, diamond, and ice. Amorphous solids are composed of atoms or molecules that are in no particular order. That is, each particle is in a particular spot, but the particles are in no organized pattern. Examples of amorphous solids include rubber and wax. **Figure 4** illustrates the differences in the arrangement of particles in these two solids.

Activity

Imagine that you are a particle in a solid. Your position in the solid is your chair. In your ScienceLog, describe the different types of motion that are possible even though you cannot leave your chair.

Figure 4 *Differing arrangements of particles in crystalline solids and amorphous solids lead to different properties. Imagine trying to hit a home run with a rubber bat!*

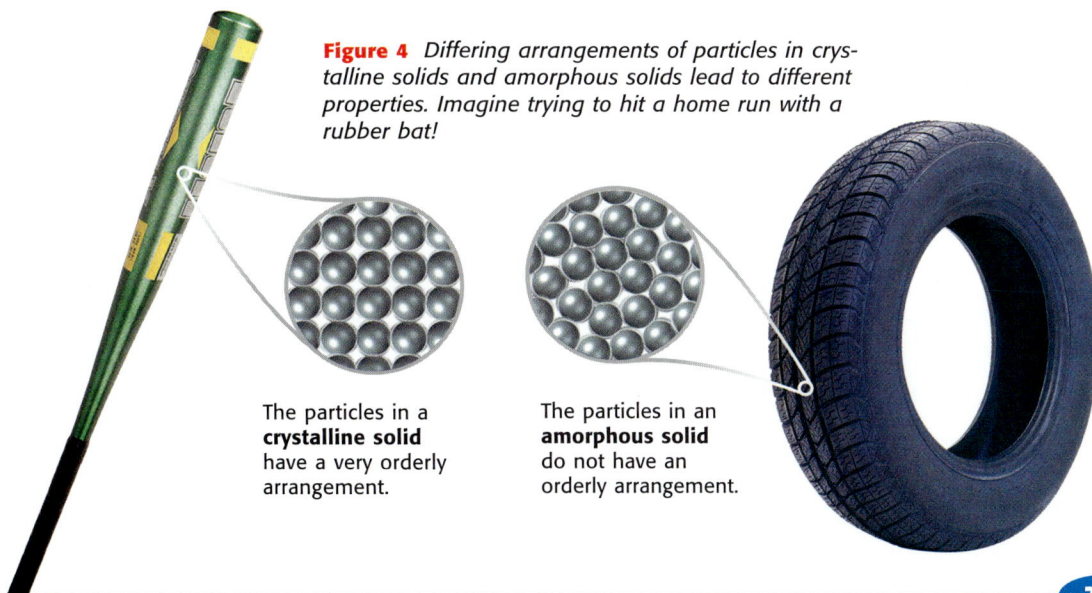

The particles in a **crystalline solid** have a very orderly arrangement.

The particles in an **amorphous solid** do not have an orderly arrangement.

31

WEIRD SCIENCE

Even the atoms or molecules that make up a solid are constantly in motion. However, as matter is cooled to extremely cold temperatures, its particles move more slowly. Theoretically, matter can be cooled enough for all particle motion to stop. This temperature is known as absolute zero, or 0 K (–273°C). A temperature of absolute zero has never been achieved by scientists in a laboratory.

DISCUSSION

The images in this chapter depict particles in matter as gray spheres. Although these particles can be atoms or molecules, the general term *particle* is frequently used to help students better grasp the concepts of particle arrangement and behavior in each of the states of matter. Discuss with students the movement of particles (atoms and molecules) in each of the states of matter. Have them predict what might happen to the particles if the temperature or pressure on the matter in the jar is changed.

Answer to Activity

Students should see that they can still bounce and move back and forth in their chair, simulating the vibrating motion of particles in a solid. Students might also bend or turn their body. These motions are possible for molecules made up of several atoms bonded together, such as a water molecule. Molecules will be discussed in a later chapter. For now, all particles will be represented as single spheres to avoid confusion.

MISCONCEPTION ALERT

A common misconception is that crystalline solids hold their shape while amorphous solids do not. It is true that the particles in an amorphous solid are not arranged in a definite pattern, but each particle remains in position relative to surrounding particles.

Learners Having Difficulty

Provide students with a hand lens and samples of salt, flour, sugar, margarine or butter, and a rubber band. Give them time to look at and compare all the samples. Encourage them to investigate and describe the visual differences between an amorphous solid and a crystalline solid.

DISCUSSION

Lead a discussion about the differences between solids and liquids. Ask students to explain the differences in terms of molecular attraction, motion, and distances. Ask students to explain the behavior of salt, sugar, dust, and so on. These substances pour easily but are considered solids. Explaining this pouring behavior of small solid particles can lead to a discussion of how a liquid's molecules slide past one another while still remaining in contact with one another.

MISCONCEPTION ALERT

Students often believe erroneously that the distances between liquid molecules are much greater than distances between molecules of a solid. Point out that while most materials are more dense in the solid state than in the liquid, the densities of solids and liquids are very similar. In fact, water molecules are closer in the liquid state than they are in the solid state.

Liquids Change Shape but Not Volume

A liquid will take the shape of whatever container it is put in. You are reminded of this every time you pour yourself a glass of juice. The state in which matter takes the shape of its container and has a definite volume is **liquid.** The atoms or molecules in liquids move fast enough to overcome some of the attractions between them. The particles slide past each other until the liquid takes the shape of its container. **Figure 5** shows how the particles in juice might look if they were large enough to see.

Even though liquids change shape, they do not readily change volume. You know that a can of soda contains a certain volume of liquid regardless of whether you pour it into a large container or a small one. **Figure 6** illustrates this point using a beaker and a graduated cylinder.

Figure 5 Particles in a liquid slide past one another until the liquid conforms to the shape of its container.

Figure 6 Even when liquids change shape, they don't change volume.

BRAIN FOOD

The Boeing 767 Freighter, a type of commercial airliner, has 187 km (116 mi) of hydraulic tubing.

The Squeeze Is On   Because the particles in liquids are close to one another, it is difficult to push them closer together. This makes liquids ideal for use in hydraulic (hie DRAW lik) systems. For example, brake fluid is the liquid used in the brake systems of cars. Stepping on the brake pedal applies a force to the liquid. The particles in the liquid move away rather than squeezing closer together. As a result, the fluid pushes the brake pads outward against the wheels, which slows the car.

IS THAT A FACT!

A gel is a liquid that has tiny particles of a solid suspended in it. Gels are best known for their elasticity or ability to bounce. In a gel, the solid particles remain suspended, unaffected by gravity. These suspended solids give gels their limited firmness. Examples of gels are flavored gelatin and some kinds of toothpaste.

A Drop in the Bucket Two other important properties of liquids are *surface tension* and *viscosity* (vis KAHS uh tee). Surface tension is the force acting on the particles at the surface of a liquid that causes the liquid to form spherical drops, as shown in **Figure 7.** Different liquids have different surface tensions. For example, rubbing alcohol has a lower surface tension than water, but mercury has a higher surface tension than water.

Viscosity is a liquid's resistance to flow. In general, the stronger the attractions between a liquid's particles are, the more viscous the liquid is. Think of the difference between pouring honey and pouring water. Honey flows more slowly than water because it has a higher viscosity than water.

Figure 7 *Liquids form spherical drops as a result of surface tension.*

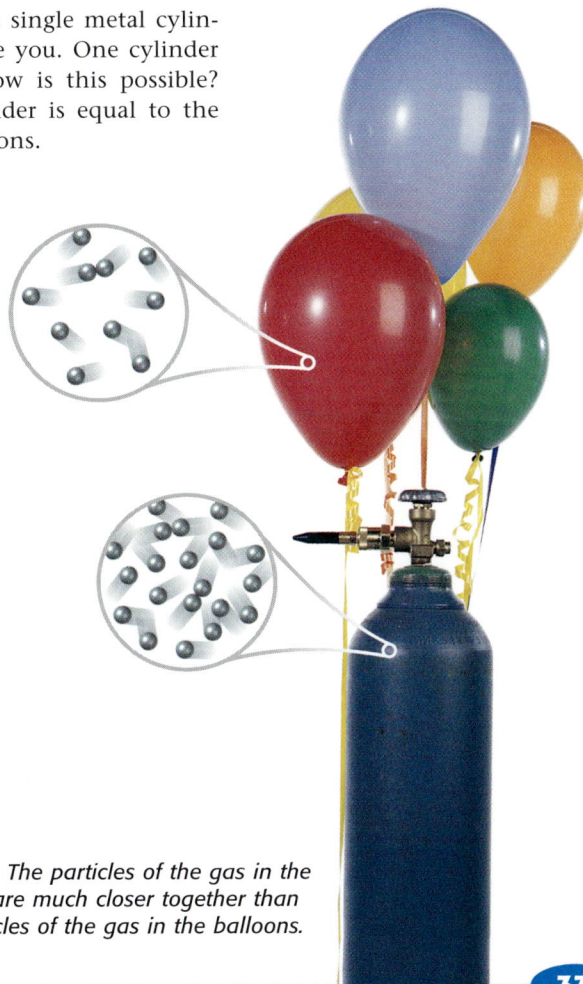

Gases Change Both Shape and Volume

How many balloons can be filled from a single metal cylinder of helium? The number may surprise you. One cylinder can fill approximately 700 balloons. How is this possible? After all, the volume of the metal cylinder is equal to the volume of only about five inflated balloons.

It's a Gas! Helium is a gas. **Gas** is the state in which matter changes in both shape and volume. The atoms or molecules in a gas move fast enough to break away completely from one another. Therefore, the particles of a substance in the gaseous state have less attraction between them than particles of the same substance in the solid or liquid state. In a gas, there is empty space between particles.

The amount of empty space in a gas can change. For example, the helium in the metal cylinder consists of atoms that have been forced very close together, as shown in **Figure 8.** As the helium fills the balloon, the atoms spread out, and the amount of empty space in the gas increases. As you continue reading, you will learn how this empty space is related to pressure.

Figure 8 *The particles of the gas in the cylinder are much closer together than the particles of the gas in the balloons.*

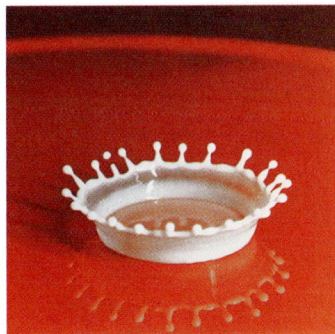

33

CONNECT TO LIFE SCIENCE

The tallest living organism is a tree known as the giant sequoia. These trees, which grow primarily in California, can grow 100 m tall. Surface tension, at least in part, helps water reach the top of these giant trees. As water evaporates from the leaves, a thin column of water molecules travels up through the tissues of the trunk and limbs. The surface tension of the water molecules, a result of cohesion, causes the molecules to stick together and, along with adhesion, allows them to be pulled all the way to the top of the tree.

ACTIVITY

Safety Caution: Caution students to wear safety goggles and an apron when performing this activity.

Provide groups with flour, paper plates, food coloring (with eye-dropper), one toothpick or pin dipped in liquid dish soap, and one clean toothpick or pin. Have students drop one or two drops of food coloring onto the flour. Ask what happened. (Surface tension causes beading of the water.)

Have students insert the toothpick with soap gently into one of the spheres and note any changes. (Spheres disappear and leave a wet spot on the powder.)

Have students insert the clean toothpick in a different sphere. Discuss any differences. (The clean toothpick should produce no changes in the sphere.)
Sheltered English

Prediction Guide Before students read this page, ask:

Assuming that a beach ball is the same size and volume as a basketball, which do you think contains more particles of air? Explain your answer.

Have students evaluate their answers after they read about pressure.

Answer to Self-Check

The pressure would increase.

DEMONSTRATION

Be sure to wear safety goggles and an apron when doing this demonstration. Obtain two small plastic containers of different sizes with push-on lids. Place two teaspoonfuls of baking soda and two tablespoonfuls of vinegar into the smaller container. Quickly snap the lid in place. Shake the container once, and then leave it on the desk. The lid should pop off within seconds. Ask why. (Pressure from the gas caused the lid to pop off.)

Repeat the demonstration using the larger container, but use the same amount of reactants. It should take longer for the lid to pop off. Discuss why. (More gas molecules are needed to create the same pressure in a larger volume.)

MISCONCEPTION ///ALERT\\\

A common misconception is that gases do not have mass. Gases do have mass. A cubic kilometer of air at sea level has a mass of about 1×10^9 kg.

Gas Under Pressure

Pressure is the amount of force exerted on a given area. You can think of this as the number of collisions of particles against the inside of the container. Compare the basketball with the beach ball in **Figure 9.** The balls have the same volume and contain particles of gas (air) that constantly collide with one another and with the inside surface of the balls. Notice, however, that there are more particles in the basketball than in the beach ball. As a result, more particles collide with the inside surface of the basketball than with the inside surface of the beach ball. When the number of collisions increases, the force on the inside surface of the ball increases. This increased force leads to increased pressure.

> ### ✔ Self-Check
>
> How would an increase in the speed of the particles affect the pressure of gas in a metal cylinder? *(See page 168 to check your answer.)*

Figure 9 *Both balls shown here are full of air, but the pressure in the basketball is higher than the pressure in the beach ball.*

The basketball has a higher pressure than the beach ball because the greater number of particles of gas are closer together. Therefore, they collide with the inside of the ball at a faster rate.

The beach ball has a lower pressure than the basketball because the lesser number of particles of gas are farther apart. Therefore, they collide with the inside of the ball at a slower rate.

internet connect

SCiLINKS
NSTA

TOPIC: Solids, Liquids, and Gases
GO TO: www.scilinks.org
*sci*LINKS NUMBER: HSTP060

SECTION REVIEW

1. List two properties that all particles of matter have in common.

2. Describe solids, liquids, and gases in terms of shape and volume.

3. Why can the volume of a gas change?

4. **Applying Concepts** Explain what happens inside the ball when you pump up a flat basketball.

▼ **Answers to Section Review**

1. They move constantly and are attracted to one another.

2. Solids have definite shape and volume. Liquids take the shape of their container but have a definite volume. Gases take the shape and volume of their container.

3. The volume of a gas can change because the atoms or molecules of gas can move closer together due to the large amount of space between the particles.

4. Pumping up a flat basketball increases the number of atoms and molecules of air in the ball. The greater number of particles would cause the pressure in the ball to increase.

Laws Describe Gas Behavior

Earlier in this chapter, you learned about the atoms and molecules in both solids and liquids. You learned that compared with gas particles, the particles of solids and liquids are closely packed together. As a result, solids and liquids do not change volume very much. Gases, on the other hand, behave differently; their volume can change by a large amount.

It is easy to measure the volume of a solid or liquid, but how do you measure the volume of a gas? Isn't the volume of a gas the same as the volume of its container? The answer is yes, but there are other factors, such as pressure, to consider.

Boyle's Law Imagine a diver at a depth of 10 m blowing a bubble of air. As the bubble rises, its volume increases. By the time the bubble reaches the surface, its original volume will have doubled due to the decrease in pressure. The relationship between the volume and pressure of a gas is known as Boyle's law because it was first described by Robert Boyle, a seventeenth-century Irish chemist. **Boyle's law** states that for a fixed amount of gas at a constant temperature, the volume of a gas increases as its pressure decreases. Likewise, the volume of a gas decreases as its pressure increases. Boyle's law is illustrated by the model in **Figure 10.**

Figure 10 **Boyle's Law**
Each illustration shows the same piston and the same amount of gas at the same temperature.

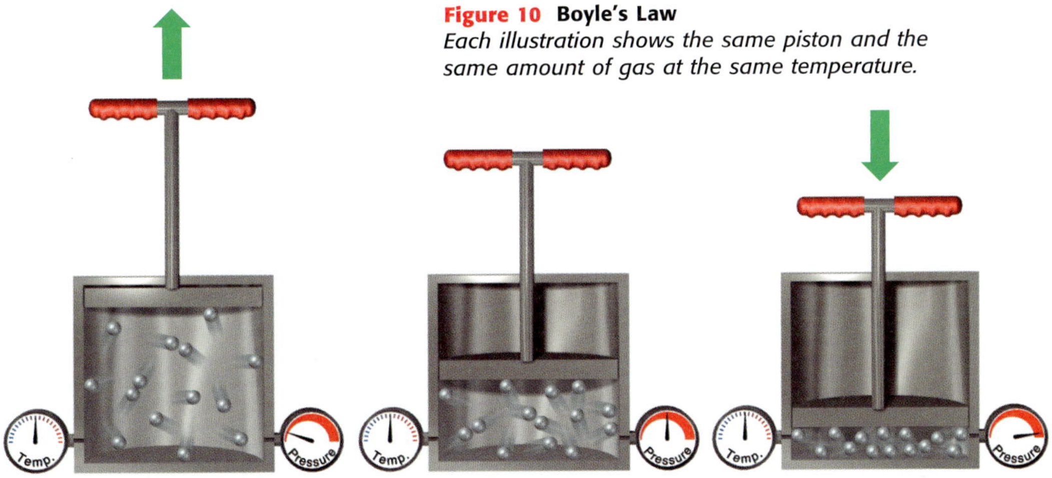

Lifting the plunger decreases the pressure of the gas. The particles of gas spread farther apart. The volume of the gas increases as the pressure decreases.

Releasing the plunger allows the gas to change to an intermediate volume and pressure.

Pushing the plunger increases the pressure of the gas. The particles of gas are forced closer together. The volume of the gas decreases as the pressure increases.

35

READING STRATEGY

Prediction Guide Before students read the passage about the behavior of gases, ask whether the following statements are true or false:

1. The atoms or molecules in a gas are closely packed. (false)
2. With an increase in pressure, the volume of the gas will decrease. (true)
3. When the pressure of a gas decreases, the volume increases. (true)

USING THE FIGURE

The pressure and volume of a gas are inversely proportional. For example, doubling the pressure reduces the volume by half, and doubling the volume reduces the pressure by half. Using **Figure 10,** discuss how changing the pressure applied to a gas will change its volume. Make sure students notice that the temperature remains constant. Ask students to explain what would happen if they applied pressure to a solid or liquid.

MATH and MORE

Use mathematics to discuss Boyle's law and inverse relationships. According to Boyle's law, if the pressure is doubled, the volume of the gas would decrease by one-half. Ask students what would happen to the volume if the pressure decreased by one-half. (The volume would double.)

Teaching Transparency 207
"Boyle's Law"

SCIENTISTS AT ODDS

Many early scientists believed that all matter was made of four elements: earth, air, fire, and water. In 1661, Robert Boyle wrote a book called the *Sceptical Chymist* in which he disagreed with this belief. Boyle proposed that matter was made of primitive and simple bodies called corpuscles. He believed that corpuscles had different shapes and sizes that mixed together to give elements their unique properties.

USING THE FIGURE

The temperature and volume of a gas are proportional, and they are directly proportional when the temperature is measured in kelvins. When the Kelvin temperature is doubled, the volume of the gas is also doubled. In **Figure 11,** what happens to the gas in the piston when the temperature is changed? Have students predict what might happen to solids or liquids in a container when the temperature is changed.

LabBook **PG 138**

Full of Hot Air!

Answers to MATHBREAK

1. In Graph A, the missing variable increases as the volume increases. In Graph B, the missing variable decreases as the volume increases.

2. Graph A represents Charles's law. Graph B represents Boyle's law.

3. In Graph A, the *x*-axis should be labeled "Temperature." In Graph B, the label for the *x*-axis should be "Pressure."

4. Graph A is linear and shows that both variables increase together. Graph B is nonlinear and shows that as one variable increases, the other variable decreases.

LabBook

See Charles's law in action for yourself using a balloon on page 138 of the LabBook.

$$\div \; 5 \; \div \; \frac{\Omega}{\infty} \; \leq \; \infty \; + \Omega^{\sqrt{}} \; 9 \; _{\infty} \; \overset{\leq}{\Sigma} \; 2$$

MATH BREAK

Gas Law Graphs

Each graph below illustrates a gas law. However, the variable on one axis of each graph is not labeled. Answer the following questions for each graph:

1. As the volume increases, what happens to the missing variable?

2. Which gas law is shown?

3. What label belongs on the axis?

4. Is the graph linear or nonlinear? What does this tell you?

Graph A

Graph B

Weather balloons demonstrate a practical use of Boyle's law. A weather balloon carries equipment into the atmosphere to collect information used to predict the weather. This balloon is filled with only a small amount of gas because the pressure of the gas decreases and the volume increases as the balloon rises. If the balloon were filled with too much gas, it would pop as the volume of the gas increased.

Charles's Law An inflated balloon will also pop when it gets too hot, demonstrating another gas law—Charles's law. **Charles's law** states that for a fixed amount of gas at a constant pressure, the volume of the gas increases as its temperature increases. Likewise, the volume of the gas decreases as its temperature decreases. Charles's law is illustrated by the model in **Figure 11.** You can see Charles's law in action by putting an inflated balloon in the freezer. Wait about 10 minutes, and see what happens!

Figure 11 Charles's Law
Each illustration shows the same piston and the same amount of gas at the same pressure.

Lowering the temperature of the gas causes the particles to move more slowly. They hit the sides of the piston less often and with less force. As a result, the volume of the gas decreases.

Raising the temperature of the gas causes the particles to move more quickly. They hit the sides of the piston more often and with greater force. As a result, the volume of the gas increases.

Teaching Transparency 207 "Charles's Law"

internet connect

SCiLINKS **NSTA**
TOPIC: Natural and Artificial Plasma
GO TO: www.scilinks.org
sciLINKS NUMBER: HSTP065

Homework

Plasmas (page 37) may be the key to an almost unlimited energy resource—nuclear fusion. Have students research nuclear fusion and plasma containment. Have them make a poster showing the powerful magnets and the tokamak reactor. Remind them to explain the possible benefits of controlling nuclear fusion on Earth.

IS THAT A FACT!

Scuba is an acronym for *s*elf-*c*ontained *u*nderwater *b*reathing *a*pparatus. Credit for the invention of scuba in 1943 is usually given to Jacques-Yves Cousteau and Emile Gagnan.

APPLY

Charles's Law and Bicycle Tires

One of your friends overinflated the tires on her bicycle. Use Charles's law to explain why she should let out some of the air before going for a ride on a hot day.

Plasmas

Scientists estimate that more than 99 percent of the known matter in the universe, including the sun and other stars, is made of a state of matter called plasma. **Plasma** is the state of matter that does not have a definite shape or volume and whose particles have broken apart.

Plasmas have some properties that are quite different from the properties of gases. Plasmas conduct electric current, while gases do not. Electric and magnetic fields affect plasmas but do not affect gases. In fact, strong magnetic fields are used to contain very hot plasmas that would destroy any other container.

Natural plasmas are found in lightning, fire, and the incredible light show in **Figure 12,** called the aurora borealis (ah ROHR uh BOHR ee AL is). Artificial plasmas, found in fluorescent lights and plasma balls, are created by passing electric charges through gases.

Figure 12 *Auroras, like the aurora borealis seen here, form when high-energy plasma collides with gas particles in the upper atmosphere.*

Quiz

1. What are the two categories of solids? (crystalline and amorphous)
2. How are they different? (The particles in crystalline solids have a set pattern; particles in amorphous solids have no pattern.)
3. How are gases different from solids and liquids? (Gases have the greatest energy. Liquids have less energy, and solids have the least. Gas takes the shape and volume of its container. Liquids take the shape of their container but do not change volume. Solids do not change either their shape or their volume.)
4. What behavior of a gas does Boyle's law describe? (The volume of a gas changes with pressure. The volume increases with a decrease in pressure and decreases with an increase in pressure.)

SECTION REVIEW

1. When scientists record the volume of a gas, why do they also record the temperature and the pressure?
2. List two differences between gases and plasmas.
3. **Applying Concepts** What happens to the volume of a balloon left on a sunny windowsill? Explain.

internet**connect**

*SCi*LINKS
NSTA

TOPIC: Natural and Artificial Plasma
GO TO: www.scilinks.org
*sci*LINKS NUMBER: HSTP065

ALTERNATIVE ASSESSMENT

Concept Mapping Have students create a concept map using the vocabulary terms for this section. Concept maps should include examples of each state of matter.

Reinforcement Worksheet
"Make a State-ment"

▼ **Answers to Section Review**
1. The volume of a gas can be changed by changing the temperature and the pressure.
2. Plasmas conduct electric current and are affected by electric and magnetic fields.
3. The volume of the balloon will increase. Leaving the balloon on a sunny windowsill will cause the temperature of the gas in the balloon to increase. According to Charles's law, the volume will increase as the temperature increases.

Focus

Changes of State

This section examines how matter changes from state to state. Changes in state are explained in terms of matter gaining or losing energy. Certain physical properties are explained in terms of molecular motion and molecular attraction.

🔔 Bellringer

Have students describe what must be done to liquid water to change it to ice or to change it to steam. Then have students use these explanations to predict what must happen, in general, to cause matter to change state. Have them write their explanations and predictions in their ScienceLog.

1 Motivate

DEMONSTRATION

Ask four student volunteers to stand close together to form a square. Wrap masking tape around the students several times. Tell the class that the four students represent the particles in a solid. Have the students demonstrate that they can still move a bit without breaking the tape, just as particles in a solid move a bit but generally stay close together. Then have the four students move around more and more until they break the tape. Discuss with the class that when particles in a solid move faster, they move apart, and the solid changes to the liquid state.

Terms to Learn

change of state — boiling
melting — evaporation
freezing — condensation
vaporization — sublimation

What You'll Do

- Describe how substances change from state to state.
- Explain the difference between an exothermic change and an endothermic change.
- Compare the changes of state.

Want to learn how to get power from changes of state? Steam ahead to page 51.

Changes of State

A **change of state** is the conversion of a substance from one physical form to another. All changes of state are physical changes. In a physical change, the identity of a substance does not change. In **Figure 13,** the ice, liquid water, and steam are all the same substance—water. In this section, you will learn about the four changes of state illustrated in Figure 13 as well as a fifth change of state called *sublimation* (SUHB li MAY shuhn).

Figure 13 *The terms in the arrows are changes of state. Water commonly goes through the changes of state shown here.*

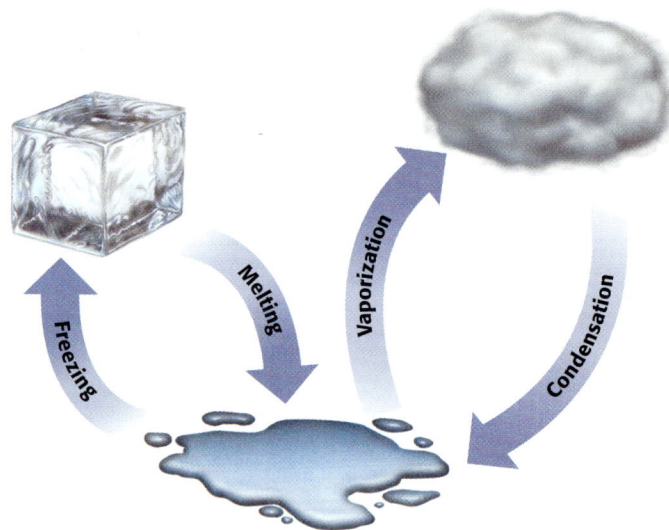

Energy and Changes of State

During a change of state, the energy of a substance changes. The *energy* of a substance is related to the motion of its particles. The molecules in the liquid water in Figure 13 move faster than the molecules in the ice. Therefore, the liquid water has more energy than the ice.

If energy is added to a substance, its particles move faster. If energy is removed, its particles move slower. The *temperature* of a substance is a measure of the speed of its particles and therefore is a measure of its energy. For example, steam has a higher temperature than liquid water, so particles in steam have more energy than particles in liquid water. A transfer of energy, known as *heat,* causes the temperature of a substance to change, which can lead to a change of state.

IS THAT A FACT!

Water is the only substance that can be found as a solid, a liquid, and a gas at normal surface temperatures on Earth.

Melting: Solids to Liquids

Melting is the change of state from a solid to a liquid. This is what happens when an ice cube melts. **Figure 14** shows a metal called gallium melting. What is unusual about this metal is that it melts at around 30°C. Because your normal body temperature is about 37°C, gallium will melt right in your hand!

The *melting point* of a substance is the temperature at which the substance changes from a solid to a liquid. Melting points of substances vary widely. The melting point of gallium is 30°C. Common table salt, however, has a melting point of 801°C.

Most substances have a unique melting point that can be used with other data to identify them. Because the melting point does not change with different amounts of the substance, melting point is a *characteristic property* of a substance.

Absorbing Energy For a solid to melt, particles must overcome some of their attractions to each other. When a solid is at its melting point, any energy it absorbs increases the motion of its atoms or molecules until they overcome the attractions that hold them in place. Melting is an *endothermic* change because energy is absorbed by the substance as it changes state.

Freezing: Liquids to Solids

Freezing is the change of state from a liquid to a solid. The temperature at which a liquid changes into a solid is its *freezing point*. Freezing is the reverse process of melting, so freezing and melting occur at the same temperature, as shown in **Figure 15**.

Removing Energy For a liquid to freeze, the motion of its atoms or molecules must slow to the point where attractions between them overcome their motion. If a liquid is at its freezing point, removing more energy causes the particles to begin locking into place. Freezing is an *exothermic* change because energy is removed from, or taken out of, the substance as it changes state.

Figure 14 *Even though gallium is a metal, it would not be very useful as jewelry!*

Figure 15 *Liquid water freezes at the same temperature that ice melts—0°C.*

If energy is added at 0°C, the ice will melt.

If energy is removed at 0°C, the liquid water will freeze.

39

WEIRD SCIENCE

To change the freezing point of water, just add salt. People rely on this phenomenon when making homemade ice cream. To achieve temperatures cold enough to help freeze the ice cream, rock salt is added to the ice, thus lowering the freezing point of the ice-salt mixture.

USING THE FIGURE

Have students look at **Figure 13.** During which changes are the substances gaining energy? (melting and vaporization)

How can students tell? Help them make a concept map to describe and compare the particle motion before and after the change in state.

MISCONCEPTION ALERT

Most students associate the term *freezing* with cold temperatures. However, be sure to point out that the term applies to any change of state from liquid to solid, regardless of temperature. Freezing can occur at high or low temperatures. For example, ammonia freezes at –77.7°C, while magnesium freezes at 650°C.

Multicultural CONNECTION

Frederick McKinley Jones (1892–1961) was an African-American inventor with more than 60 patents to his name. Perhaps his most important invention was his compact, shockproof, automatic-refrigeration unit for trucks hauling meat or produce to market. This invention, which was later adapted for trains, is still in use around the world today.

Have students find out more about Jones's inventions, especially the one that gave us "talking movies."

Directed Reading Worksheet Section 2

READING STRATEGY

Prediction Guide Before students read this page, ask them if they agree with the following three statements.

- Evaporation can occur at any temperature. (true)
- Boiling occurs only at the surface of a liquid. (false)
- Vaporization is simply a liquid changing to a gas. (true)

Homework

Traditional clothing varies from culture to culture. People who live in warm climates wear lightweight and loose-fitting clothes that allow air to circulate near the body. This causes perspiration to evaporate, which cools the body. Examples of cultures that wear such lightweight clothing are those of the Middle East and Northern Africa. Ask students to research other examples of how people dress to fit the climate where they live. Have them make posters showing the examples they found.

Answer to Self-Check

endothermic

Vaporization: Liquids to Gases

One way to experience vaporization (VAY puhr i ZAY shuhn) is to iron a shirt—carefully!—using a steam iron. You will notice steam coming up from the iron as the wrinkles are eliminated. This steam results from the vaporization of liquid water by the iron. **Vaporization** is simply the change of state from a liquid to a gas.

Boiling is vaporization that occurs throughout a liquid. The temperature at which a liquid boils is called its *boiling point*. Like the melting point, the boiling point is a characteristic property of a substance. The boiling point of water is 100°C, whereas the boiling point of liquid mercury is 357°C. **Figure 16** illustrates the process of boiling and a second form of vaporization—evaporation (ee VAP uh RAY shuhn).

Evaporation is vaporization that occurs at the surface of a liquid below its boiling point, as shown in Figure 16. When you perspire, your body is cooled through the process of evaporation. Perspiration is mostly water. Water absorbs energy from your skin as it evaporates. You feel cooler because your body transfers energy to the water. Evaporation also explains why water in a glass on a table disappears after several days.

✓ Self-Check

Is vaporization an endothermic or exothermic change? (See page 168 to check your answer.)

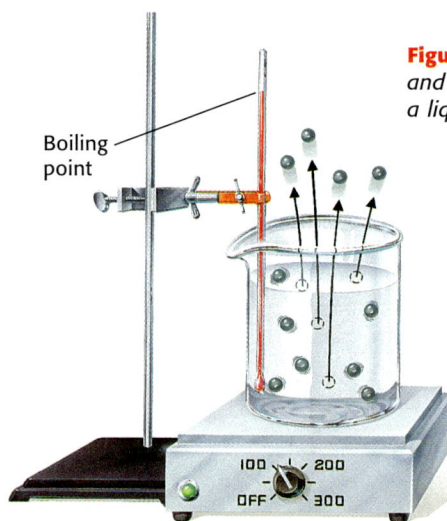 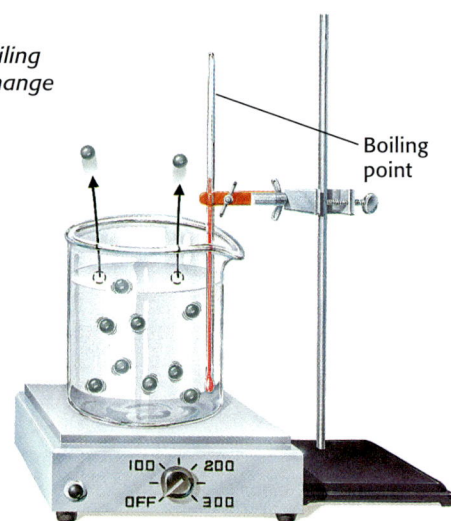

Figure 16 *Both boiling and evaporation change a liquid to a gas.*

Boiling point

Boiling point

Boiling occurs in a liquid at its boiling point. As energy is added to the liquid, particles throughout the liquid move fast enough to break away from the particles around them and become a gas.

Evaporation occurs in a liquid below its boiling point. Some particles at the surface of the liquid move fast enough to break away from the particles around them and become a gas.

40

Science Bloopers

Most people do not want to get too sweaty, so they wear an antiperspirant. In addition to covering unpleasant odors, antiperspirants contain compounds that clog pores in the skin. This prevents sweat from being excreted by the body.

However, sweat is the body's natural air conditioner. When sweat evaporates, it cools the skin. By wearing antiperspirant, people cause their bodies to cool less efficiently.

Pressure Affects Boiling Point Earlier you learned that water boils at 100°C. In fact, water only boils at 100°C at sea level because of atmospheric pressure. Atmospheric pressure is caused by the weight of the gases that make up the atmosphere. Atmospheric pressure varies depending on where you are in relation to sea level. Atmospheric pressure is lower at higher elevations. The higher you go above sea level, the less air there is above you, and the lower the atmospheric pressure is. If you were to boil water at the top of a mountain, the boiling point would be lower than 100°C. For example, Denver, Colorado, is 1.6 km (1 mi) above sea level and water boils there at about 95°C. You can make water boil at an even lower temperature by doing the QuickLab at right.

Condensation: Gases to Liquids

Look at the cool glass of lemonade in **Figure 17.** Notice the beads of water on the outside of the glass. These form as a result of condensation. **Condensation** is the change of state from a gas to a liquid. The *condensation point* of a substance is the temperature at which the gas becomes a liquid and is the same temperature as the boiling point at a given pressure. Thus, at sea level, steam condenses to form water at 100°C—the same temperature at which water boils.

For a gas to become a liquid, large numbers of atoms or molecules must clump together. Particles clump together when the attraction between them overcomes their motion. For this to occur, energy must be removed from the gas to slow the particles down. Therefore, condensation is an exothermic change.

Figure 17 *Gaseous water in the air will become liquid when it contacts a cool surface.*

Meteorology
CONNECTION

The amount of gaseous water that air can hold decreases as the temperature of the air decreases. As the air cools, some of the gaseous water condenses to form small drops of liquid water. These drops form clouds in the sky and fog near the ground.

41

QuickLab
MATERIALS

FOR EACH STUDENT:
• syringe
• warm water

Safety Caution: Remind all students to review all safety cautions and icons before beginning this activity. Students should wear safety goggles and aprons during this activity.

Answers to QuickLab

Since the temperature of the water required depends on the size of the syringes used, it would be best to determine the necessary temperature for the syringes your students will be using.

5. Bubbles form in the water as the plunger is pulled out.
6. The boiling water is not 100°C. The lower pressure causes the water to boil at a much lower temperature.

MATH and MORE

The average atmospheric pressure on top of Mount Everest, at an elevation of 8,850 m, is only about 33,000 Pa. The average atmospheric pressure at sea level is 101,000 Pa. Assuming that the boiling point of water is approximately 1°C lower for every decrease of 2,230 Pa, what would be the approximate boiling point of water on the top of Mount Everest? (The pressure drops approximately 68,000 Pa, so the boiling point drops approximately 30° to 70°C.)

Math Skills Worksheet "Checking Division with Multiplication"

IS THAT A FACT!
If all of the water vapor in Earth's atmosphere were to suddenly condense, the falling water would cover the United States to a depth of 8 m.

LabBook PG 139
Can Crusher

GUIDED PRACTICE

Draw a sequence of state changes. For example, a solid is melting or a gas is condensing. Ask students how the next drawing will look. After completing the drawings, ask the students to create a concept map showing changes of state and separating endothermic and exothermic changes. (This assignment can also be used as an Alternative Assessment.)

Teaching Transparency 208 "Summarizing the Changes of State"

Critical Thinking Worksheet "What a State!"

internetconnect

SCILINKS
NSTA

TOPIC: Changes of State
GO TO: www.scilinks.org
sciLINKS NUMBER: HSTP070

Sublimation: Solids Directly to Gases

Look at the solids shown in **Figure 18.** The solid on the left is ice. Notice the drops of liquid collecting as it melts. On the right, you see carbon dioxide in the solid state, also called dry ice. It is called dry ice because instead of melting into a liquid, it goes through a change of state called sublimation. **Sublimation** is the change of state from a solid directly into a gas. Dry ice is colder than ice, and it doesn't melt into a puddle of liquid. It is often used to keep food, medicine, and other materials cold without getting them wet.

For a solid to change directly into a gas, the atoms or molecules must move from being very tightly packed to being very spread apart. The attractions between the particles must be completely overcome. Because this requires the addition of energy, sublimation is an endothermic change.

Figure 18 Ice melts, but dry ice, on the right, turns directly into a gas.

Comparing Changes of State

As you learned in Section 1 of this chapter, the state of a substance depends on how fast its atoms or molecules move and how strongly they are attracted to each other. A substance may undergo a physical change from one state to another by an endothermic change (if energy is added) or an exothermic change (if energy is removed). The table below shows the differences between the changes of state discussed in this section.

Summarizing the Changes of State			
Change of state	Direction	Endothermic or exothermic?	Example
Melting	solid ⟶ liquid	endothermic	Ice melts into liquid water at 0°C.
Freezing	liquid ⟶ solid	exothermic	Liquid water freezes into ice at 0°C.
Vaporization	liquid ⟶ gas	endothermic	Liquid water vaporizes into steam at 100°C.
Condensation	gas ⟶ liquid	exothermic	Steam condenses into liquid water at 100°C.
Sublimation	solid ⟶ gas	endothermic	Solid dry ice sublimes into a gas at −78°C.

42

IS THAT A FACT!

Helium, an unreactive gas, has one of the lowest boiling points. Helium boils at 4.2 K, a little above absolute zero.

WEIRD SCIENCE

Ice and snow sometimes sublime directly to water vapor. If the air is dry after a snowstorm, a thin layer of ice on a driveway or street may simply disappear in a matter of hours, even though the temperature never rises above freezing.

Temperature Change Versus Change of State

When most substances lose or absorb energy, one of two things happens to the substance: its temperature changes or its state changes. Earlier in the chapter, you learned that the temperature of a substance is a measure of the speed of the particles. This means that when the temperature of a substance changes, the speed of the particles also changes. But while a substance changes state, its temperature does not change until the change of state is complete, as shown in **Figure 19.**

Figure 19 Changing the State of Water

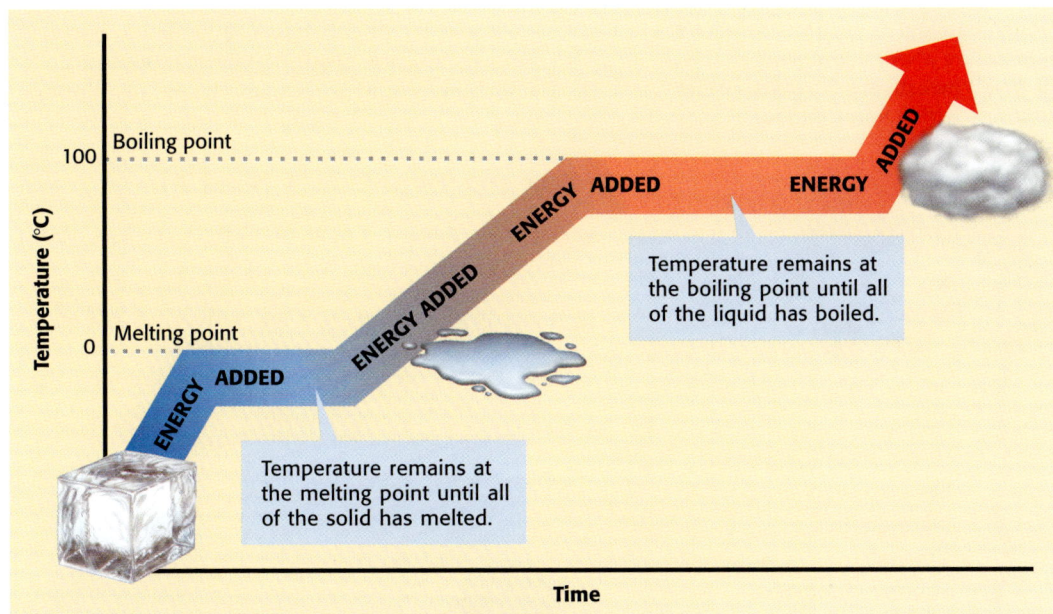

Boiling point

100

Melting point

0

Temperature (°C)

ENERGY ADDED

ENERGY ADDED

ENERGY ADDED

ENERGY ADDED

Temperature remains at the melting point until all of the solid has melted.

Temperature remains at the boiling point until all of the liquid has boiled.

Time

SECTION REVIEW

1. Compare endothermic and exothermic changes.

2. Classify each change of state (melting, freezing, vaporization, condensation, and sublimation) as endothermic or exothermic.

3. Describe how the motion and arrangement of particles change as a substance freezes.

4. **Comparing Concepts** How are evaporation and boiling different? How are they similar?

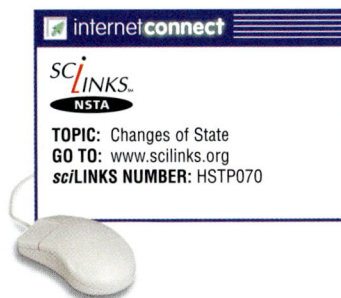

internet**connect**

SCiLINKS
NSTA

TOPIC: Changes of State
GO TO: www.scilinks.org
sciLINKS NUMBER: HSTP070

▼ *Answers to Section Review*

1. During endothermic changes, energy is absorbed. During exothermic changes, energy is released.

2. exothermic: freezing, condensation; endothermic: melting, vaporization, sublimation

3. As a substance freezes, its particles lose some of their freedom of motion and become more orderly.

4. Evaporation occurs only at the surface of a liquid while boiling occurs throughout a liquid. Both evaporation and boiling are endothermic processes that change a liquid to a gas.

Quiz

Pose the following situation to students:

A scientist and her assistant have an unmarked thermometer on which they want to mark the temperatures of 0°C and 100°C. They heat a beaker of water and take several cubes of ice out of the freezer. Just as they are about to mark the thermometer, the phone rings. The scientist spends 15 minutes on the phone. When she is finished, she again places the thermometer on the ice to mark 0°C. The scientist's assistant stops her and says that after 15 minutes the ice has begun to melt and must be warmer than 0°C. Also, since the water has been boiling for 15 minutes, it must be hotter than 100°C. Explain why the assistant's assumptions are not correct.

ALTERNATIVE ASSESSMENT

Use "The Water Cycle" teaching transparency to help students understand the changes of state that water undergoes as it cycles through the environment. Have students explain in their own words how changing the state of water (**Figure 19**) relates to the water cycle.

Teaching Transparency 141
"The Water Cycle"

LINK TO
EARTH
SCIENCE

Teaching Transparency 209
"Changing the State of Water"

A Hot and Cool Lab
Teacher's Notes

Time Required

One or two 45-minute class periods

Lab Ratings

EASY ————————→ HARD

TEACHER PREP   🧪🧪🧪
STUDENT SET-UP   🧪🧪
CONCEPT LEVEL   🧪🧪🧪
CLEAN UP   🧪🧪

MATERIALS

The materials listed are for a group of 3–4 students.

Safety Caution

Remind students to review all safety cautions and icons before beginning this lab activity. Hot plates should have flat metal surfaces and not metal coils. To prevent spills, caution students to keep all power cords away from the beakers and pans of hot water.

Answer

1. Accept all reasonable predictions.

CLASSROOM TESTED & APPROVED

C. John Graves
Monforton Middle School
Bozeman, Montana

Discovery Lab

USING SCIENTIFIC METHODS

A Hot and Cool Lab

When you add energy to a substance through heating, does the substance's temperature always go up? When you remove energy from a substance through cooling, does the substance's temperature always go down? In this lab, you'll investigate these important questions with a very common substance—water.

MATERIALS

- 250 or 400 mL beaker
- water
- heat-resistant gloves
- hot plate
- thermometer
- stopwatch
- 100 mL graduated cylinder
- large coffee can
- crushed ice
- rock salt
- wire-loop stirring device
- graph paper

Form a Hypothesis

1. In your ScienceLog, answer the following questions: What happens to the temperature of boiling water when you continue to add energy through heating? (Part A) What happens to the temperature of freezing water when you continue to remove energy through cooling? (Part B)

Test the Hypothesis (Part A)

2. Make a table like the one on the next page. Label the table "Heating Water."

3. Fill the beaker one-half full with water. Put on heat-resistant gloves. Turn on the hot plate. Put the beaker on the burner. Place the thermometer in the beaker.
Caution: Do not touch the burner.

Preparation Notes

To construct the wire-loop stirring device, make a loop slightly smaller than the inside of the graduated cylinder at one end of a straightened 25-cm piece of copper wire. The loop should easily fit into the graduated cylinder with the thermometer in place. Angle the loop so that it is perpendicular to the rest of the wire. At the other end of the wire, make a handle that extends in the opposite direction of the loop. Students will use this device to stir the contents of the graduated cylinder by placing the loop around the thermometer and using the handle to move the device up and down.

Time (s)	30	60	90	120	150	180	210	etc.
Temperature (°C)		DO NOT WRITE IN BOOK						

4 Record the temperature of the water every 30 seconds. Continue taking readings until about one-fourth of the water boils away. Note the first temperature reading at which the water steadily boils.

5 Turn off the hot plate. While the beaker is cooling, make a graph of temperature (*y*-axis) versus time (*x*-axis). Draw an arrow pointing to the first temperature reading at which the water was steadily boiling.

6 After you finish the graph, follow your teacher's instructions for cleanup and disposal.

Test the Hypothesis (Part B)

7 Put 20 mL of water in the graduated cylinder. Place the graduated cylinder in the coffee can, and fill the can with crushed ice. Pour rock salt on the ice. Place the thermometer and the wire-loop stirring device in the graduated cylinder.

8 Label a new table "Cooling Water." Record the temperature of the water in the graduated cylinder every 30 seconds. Add ice and rock salt to the can as needed. Stir the water with the stirring device.
Caution: Do not stir with the thermometer.

9 Once the water begins to freeze, stop stirring. Do not try to pull the thermometer out of the solid ice in the cylinder.

10 Record the temperature when you first see ice crystals forming in the water. Continue taking readings until the water is completely frozen.

11 Make a graph of temperature (*y-axis*) versus time (*x-axis*). Draw an arrow to the temperature reading at which the first ice crystals formed in the water.

Analyze the Results (Parts A and B)

12 What does the slope of each graph represent?

13 How does the slope when the water is boiling compare with the slope before the water boils? Explain why the slopes differ.

14 How does the slope when the water is freezing compare with the slope before the water freezes? Explain why the slopes differ.

Draw Conclusions (Parts A and B)

15 Adding or removing energy leads to changes in the movement of particles that make up solids, liquids, and gases. Use this idea to explain why the temperature graphs of the two experiments look the way they do.

45

Datasheets for LabBook

Science Skills Worksheet
"Grasping Graphing"

Discovery Lab

Answers

12. The slope of each graph represents the rate of temperature change.

13. The slope is less steep (line should be horizontal) when the water starts to boil. The slope is different because the energy added to the water through heating is making steam rather than increasing the temperature.

14. The slope is less steep (the line should be horizontal) when the water starts to freeze. The slope is different because the removal of energy from the water allows crystal structures (ice) to form rather than decreasing the temperature.

15. Adding energy to liquid water makes the particles speed up, thereby increasing the temperature. When the particles speed up enough, water can become gas (steam), which has more energy at the same temperature. Even though energy is being added the whole time, the temperature stops rising when the liquid starts changing into a gas. This explains the part of the graph that levels off. When energy is removed, the temperature stops falling, and the liquid turns to solid. At this point, the particles have less energy, but the temperature of the water stays the same. This explains the part of the graph that levels off.

Chapter Highlights

Chapter Highlights

VOCABULARY DEFINITIONS

SECTION 1

states of matter the physical forms in which a substance can exist; states include solid, liquid, gas, and plasma

solid the state in which matter has a definite shape and volume

liquid the state in which matter takes the shape of its container and has a definite volume

gas the state in which matter changes in both shape and volume

pressure the amount of force exerted on a given area

Boyle's law the law that states that for a fixed amount of gas at a constant temperature, the volume of a gas increases as its pressure decreases

Charles's law the law that states that for a fixed amount of gas at a constant pressure, the volume of a gas increases as its temperature increases

plasma the state of matter that does not have a definite shape or volume and whose particles have broken apart

---

SECTION 1

Vocabulary

states of matter (p. 30)
solid (p. 31)
liquid (p. 32)
gas (p. 33)
pressure (p. 34)
Boyle's law (p. 35)
Charles's law (p. 36)
plasma (p. 37)

Section Notes

- The states of matter are the physical forms in which a substance can exist. The four most familiar states are solid, liquid, gas, and plasma.

- All matter is made of tiny particles called atoms and molecules that attract each other and move constantly.

- A solid has a definite shape and volume.

- A liquid has a definite volume but not a definite shape.

- A gas does not have a definite shape or volume. A gas takes the shape and volume of its container.

- Pressure is a force per unit area. Gas pressure increases as the number of collisions of gas particles increases.

- Boyle's law states that the volume of a gas increases as the pressure decreases if the temperature does not change.

- Charles's law states that the volume of a gas increases as the temperature increases if the pressure does not change.

- Plasmas are composed of particles that have broken apart. Plasmas do not have a definite shape or volume.

Labs

Full of Hot Air! (p. 138)

✓ Skills Check

Math Concepts

GRAPHING DATA The relationship between measured values can be seen by plotting the data on a graph. The top graph shows the linear relationship described by Charles's law—as the temperature of a gas increases, its volume increases. The bottom graph shows the nonlinear relationship described by Boyle's law—as the pressure of a gas increases, its volume decreases.

Visual Understanding

PARTICLE ARRANGEMENT Many of the properties of solids, liquids, and gases are due to the arrangement of the atoms or molecules of the substance. Review the models in Figure 2 on page 30 to study the differences in particle arrangement between the solid, liquid, and gaseous states.

SUMMARY OF THE CHANGES OF STATE Review the table on page 42 to study the direction of each change of state and whether energy is absorbed or removed during each change.

46

Lab and Activity Highlights

A Hot and Cool Lab   PG 44

Full of Hot Air!   PG 138

Can Crusher   PG 139

Datasheets for LabBook
(blackline masters for these labs)

SECTION 2

SECTION 2

Vocabulary

change of state *(p. 38)*

melting *(p. 39)*

freezing *(p. 39)*

vaporization *(p. 40)*

boiling *(p. 40)*

evaporation *(p. 40)*

condensation *(p. 41)*

sublimation *(p. 42)*

Section Notes

- A change of state is the conversion of a substance from one physical form to another. All changes of state are physical changes.

- Exothermic changes release energy. Endothermic changes absorb energy.

- Melting changes a solid to a liquid. Freezing changes a liquid to a solid. The freezing point and melting point of a substance are the same temperature.

- Vaporization changes a liquid to a gas. There are two kinds of vaporization: boiling and evaporation.

- Boiling occurs throughout a liquid at the boiling point.

- Evaporation occurs at the surface of a liquid, at a temperature below the boiling point.

- Condensation changes a gas to a liquid.

- Sublimation changes a solid directly to a gas.

- Temperature does not change during a change of state.

Labs

Can Crusher *(p. 139)*

change of state the conversion of a substance from one physical form to another

melting the change of state from a solid to a liquid

freezing the change of state from a liquid to a solid

vaporization the change of state from a liquid to a gas; includes boiling and evaporation

boiling vaporization that occurs throughout a liquid

evaporation vaporization that occurs at the surface of a liquid below its boiling point

condensation the change of state from a gas to a liquid

sublimation the change of state from a solid directly into a gas

Vocabulary Review Worksheet

Blackline masters of these Chapter Highlights can be found in the **Study Guide.**

---

internetconnect

GO TO: go.hrw.com

Visit the **HRW** Web site for a variety of learning tools related to this chapter. Just type in the keyword:

KEYWORD: HSTSTA

SC/LINKSSM
N S T A

GO TO: www.scilinks.org

Visit the **National Science Teachers Association** on-line Web site for Internet resources related to this chapter. Just type in the *sci*LINKS number for more information about the topic:

TOPIC: Forms and Uses of Glass	*sci*LINKS NUMBER: HSTP055
TOPIC: Solids, Liquids, and Gases	*sci*LINKS NUMBER: HSTP060
TOPIC: Natural and Artificial Plasma	*sci*LINKS NUMBER: HSTP065
TOPIC: Changes of State	*sci*LINKS NUMBER: HSTP070
TOPIC: The Steam Engine	*sci*LINKS NUMBER: HSTP075

47

Lab and Activity Highlights

LabBank

Whiz-Bang Demonstrations, Demonstration with a CRUNCH!

Calculator-Based Lab, A Hot and Cool Lab

Labs You Can Eat, How Cold Is Ice-Cream Cold?

Long-Term Projects & Research Ideas, Episode IV: Sam and His Elephants Get That Sinking Feeling

Chapter Review
Answers

USING VOCABULARY

1. Solid is the state of matter in which the substance has a definite shape and volume; liquid is the state in which the substance takes the shape of its container but has a definite volume.
2. Boyle's law states that when the pressure of a gas increases, its volume decreases. Charles's law states that when the temperature of a gas increases, its volume increases.
3. Evaporation is the change of a liquid to a gas at the surface of a liquid. Boiling is the change of a liquid to a gas throughout a liquid.
4. Melting changes a solid to a liquid. Freezing changes a liquid to a solid.

UNDERSTANDING CONCEPTS

Multiple Choice

5. b
6. b
7. b
8. b
9. a
10. c
11. a
12. d
13. c

Short Answer

14. The particles of liquid water can move past one another and take the shape of a container. Particles in an ice cube are locked in place and cannot move past one another. An ice cube holds its shape no matter what container you put it in.
15. gases, liquids, solids
16. Iron in the solid state is more dense than iron in the liquid or gaseous state. The density of gaseous iron is lower than the density of solid or liquid iron.

Chapter Review

USING VOCABULARY

For each pair of terms, explain the difference in meaning.

1. solid/liquid

2. Boyle's law/Charles's law

3. evaporation/boiling

4. melting/freezing

UNDERSTANDING CONCEPTS

Multiple Choice

5. Which of the following best describes the particles of a liquid?
   a. The particles are far apart and moving fast.
   b. The particles are close together but moving past each other.
   c. The particles are far apart and moving slowly.
   d. The particles are closely packed and vibrate in place.

6. Boiling points and freezing points are examples of
   a. chemical properties.   c. energy.
   b. physical properties.   d. matter.

7. During which change of state do atoms or molecules become more ordered?
   a. boiling   c. melting
   b. condensation   d. sublimation

8. Which of the following describes what happens as the temperature of a gas in a balloon increases?
   a. The speed of the particles decreases.
   b. The volume of the gas increases and the speed of the particles increases.
   c. The volume decreases.
   d. The pressure decreases.

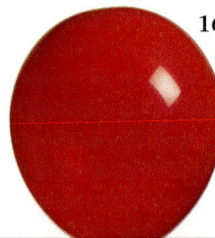

9. Dew collects on a spider web in the early morning. This is an example of
   a. condensation.   c. sublimation.
   b. evaporation.   d. melting.

10. Which of the following changes of state is exothermic?
    a. evaporation   c. freezing
    b. sublimation   d. melting

11. What happens to the volume of a gas inside a piston if the temperature does not change but the pressure is reduced?
    a. increases
    b. stays the same
    c. decreases
    d. not enough information

12. The atoms and molecules in matter
    a. are attracted to one another.
    b. are constantly moving.
    c. move faster at higher temperatures.
    d. All of the above

13. Which of the following contains plasma?
    a. dry ice   c. a fire
    b. steam   d. a hot iron

Short Answer

14. Explain why liquid water takes the shape of its container but an ice cube does not.

15. Rank solids, liquids, and gases in order of decreasing particle speed.

16. Compare the density of iron in the solid, liquid, and gaseous states.

48

Concept Mapping

17. An answer to this exercise can be found at the front of this book.

Concept Mapping Transparency 3

Concept Mapping

17. Use the following terms to create a concept map: states of matter, solid, liquid, gas, plasma, changes of state, freezing, vaporization, condensation, melting.

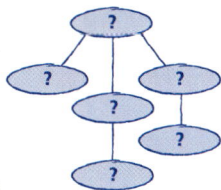

CRITICAL THINKING AND PROBLEM SOLVING

18. After taking a shower, you notice that small droplets of water cover the mirror. Explain how this happens. Be sure to describe where the water comes from and the changes it goes through.

19. In the photo below, water is being split to form two new substances, hydrogen and oxygen. Is this a change of state? Explain your answer.

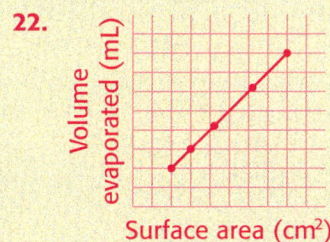

20. To protect their crops during freezing temperatures, orange growers spray water onto the trees and allow it to freeze. In terms of energy lost and energy gained, explain why this practice protects the oranges from damage.

21. At sea level, water boils at 100°C, while methane boils at –161°C. Which of these substances has a stronger force of attraction between its particles? Explain your reasoning.

MATH IN SCIENCE

22. Kate placed 100 mL of water in five different pans, placed the pans on a windowsill for a week, and measured how much water evaporated. Draw a graph of her data, shown below, with surface area on the *x*-axis. Is the graph linear or nonlinear? What does this tell you?

Pan number	1	2	3	4	5
Surface area (cm²)	44	82	20	30	65
Volume evaporated (mL)	42	79	19	29	62

23. Examine the graph below, and answer the following questions:
  a. What is the boiling point of the substance? What is the melting point?
  b. Which state is present at 30°C?
  c. How will the substance change if energy is added to the liquid at 20°C?

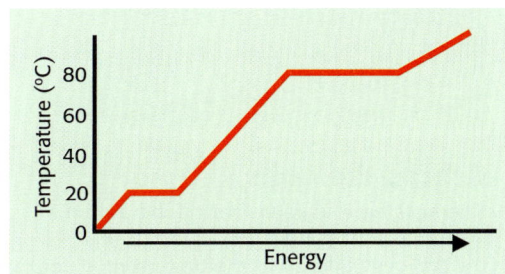

Reading Check-up Take a minute to review your answers to the Pre-Reading Questions found at the bottom of page 28. Have your answers changed? If necessary, revise your answers based on what you have learned since you began this chapter.

CRITICAL THINKING AND PROBLEM SOLVING

18. As you take a shower, some of the liquid water evaporates and becomes a gas. When the gaseous water touches the mirror, the water releases energy to the mirror and condenses into drops of liquid water.

19. The splitting of water into hydrogen and oxygen is not a change of state because the substance (water) does not keep its identity during the change. The water is changed into two new substances, hydrogen and oxygen.

20. Freezing is an exothermic change. As the water freezes, it releases energy. The oranges absorb some of this energy and warm up. (The ice also helps to insulate the oranges from the cold air.)

21. Water has a stronger force of attraction between its particles. A higher temperature, and therefore more energy, is required to separate the water particles from one another than is needed to separate the methane particles from one another.

MATH IN SCIENCE

22.

The graph is linear, which tells you that both variables (surface area and volume evaporated) increase together.

23. a. 80°C; 20°C
  b. liquid
  c. The temperature of the liquid will rise.

Blackline masters of this Chapter Review can be found in the **Study Guide.**

Teaching Strategy

In order to help students connect this feature with the topic of the unit, review the term *plasma* with students.

Research Activity You may wish to have a group of students do research on different types of lightning that exist planetwide. Examples include ribbon lightning, sheet lightning, and bead lightning. Ribbon lightning looks like a broad stream of fire; sheet lightning causes a cloud to glow; and bead lightning breaks up, creating a beadlike chain across the sky.

Students may also find information about "sprites" (red and blue balls of lightning generated in some thunderstorms). This type of lightning is not well understood and may be related to particulate matter in the atmosphere. There was a great increase in the number of sprites following the forest fires in Mexico in 1998.

Science, Technology, and Society

Guiding Lightning

By the time you finish reading this sentence, lightning will have flashed more than 500 times around the world. This common phenomenon can have devastating results. Each year in the United States alone, lightning kills almost a hundred people and causes several hundred million dollars in damage. While controlling this awesome outburst of Mother Nature may seem impossible, scientists around the world are searching for ways to reduce the destruction caused by lightning.

Behind the Bolts

Scientists have learned that during a normal lightning strike several events occur. First, electric charges build up at the bottom of a cloud. The cloud then emits a line of negatively charged air particles that zigzags toward the Earth. The attraction between these negatively charged air particles and positively charged particles from objects on the ground forms a *plasma channel.* This channel is the pathway for a lightning bolt. As soon as the plasma channel is complete, BLAM!—between 3 and 20 lightning bolts separated by thousandths of a second travel along it.

A Stroke of Genius

Armed with this information, scientists have begun thinking of ways to redirect these naturally occurring plasma channels. One idea is to use laser beams. In theory, a laser beam directed into a thundercloud can charge the air particles in its path, causing a plasma channel to develop and forcing lightning to strike.

By creating the plasma channels themselves, scientists can, in a way, catch a bolt of lightning before it strikes and direct it to a safe area of the ground. So scientists simply use lasers to direct naturally occurring lightning to strike where they want it to.

A Bright Future?

Laser technology is not without its problems, however. The machines that generate laser beams are large and expensive, and they can themselves be struck by misguided lightning bolts. Also, it is not clear whether creating these plasma channels will be enough to prevent the devastating effects of lightning.

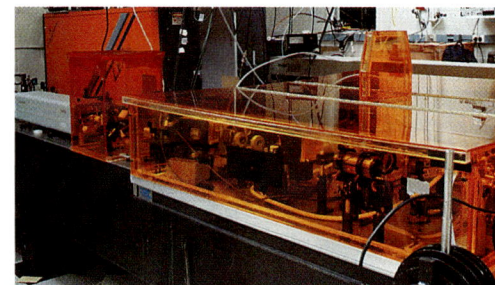

▲ *Sometime in the future, a laser like this might be used to guide lightning away from sensitive areas.*

Find Out for Yourself

▶ Use the Internet or an electronic database to find out how rockets have been used in lightning research. Share your findings with the class.

Sample Answer to Find Out for Yourself

Researchers from NASA have sent up test rockets with trailing wires to try to determine the atmospheric conditions necessary for aircraft and rockets to trigger lightning strikes. Scientists have found that a rocket is more likely to trigger lightning when a thunderstorm is relatively inactive.

The Electric Power Research Institute uses small rockets to provoke and direct lightning strikes in order to study the impact of lightning on electric utility equipment, such as power lines and transformers.

Eureka!

Full Steam Ahead!

It was huge. It was 40 m long, about 5 m high, and it weighed 245 metric tons. It could pull a 3.28 million kilogram train at 100 km/h. It was a 4-8-8-4 locomotive, called a Big Boy, delivered in 1941 to the Union Pacific Railroad in Omaha, Nebraska. It was also one of the final steps in a 2,000-year search to harness steam power.

A Simple Observation

For thousands of years, people used wind, water, gravity, dogs, horses, and cattle to replace manual labor. But until about 300 years ago, they had limited success. Then in 1690, Denis Papin, a French mathematician and physicist, observed that steam expanding in a cylinder pushed a piston up. As the steam then cooled and contracted, the piston fell. Watching the motion of the piston, Papin had an idea: attach a water-pump handle to the piston. As the pump handle rose and fell with the piston, water was pumped.

More Uplifting Ideas

Eight years later, an English naval captain named Thomas Savery made Papin's device more efficient by using water to cool and condense the steam. Savery's improved pump was used in British coal mines. As good as Savery's pump was, the development of steam power didn't stop there!

In 1712, an English blacksmith named Thomas Newcomen improved Savery's device by adding a second piston and a horizontal beam that acted like a seesaw. One end of the beam was attached to the piston in the steam cylinder. The other end of the beam was attached to the pump piston. As the steam piston moved up and down, it created a vacuum in the pump cylinder and sucked water up from the mine. Newcomen's engine was the most widely used steam engine for more than 50 years.

Watt a Great Idea!

In 1764, James Watt, a Scottish technician, was repairing a Newcomen engine. He realized that heating the cylinder, letting it cool, then heating it again wasted an enormous amount of energy. Watt added a separate chamber where the steam could cool and condense. The two chambers were connected by a valve that let the steam escape from the boiler. This improved the engine's efficiency—the boiler could stay hot all the time!

A few years later, Watt turned the whole apparatus on its side so that the piston was moving horizontally. He added a slide valve that admitted steam first to one end of the chamber (pushing the piston in one direction) and then to the other end (pushing the piston back). This changed the steam pump into a true steam engine that could drive a locomotive the size of Big Boy!

Explore Other Inventions

▶ Watt's engine helped trigger the Industrial Revolution as many new uses for steam power were found. Find out more about the many other inventors, from tinkerers to engineers, who harnessed the power of steam.

Background

In A.D. 60, an Egyptian writer named Hero wrote about a machine made of a metal sphere supported by two hollow pipes over a kettle of water. When a fire was lit under the kettle, the water turned into steam, which traveled through the pipes into the sphere. Two L-shaped nozzles in the sphere allowed the steam to escape, causing the sphere to rotate. This interesting device was invented just for fun, but it is an early example of how steam can be used to set objects in motion.

James Watt is also well known for his patented design of the double-acting piston. In this engine, the steam entered the cylinder on one side of the piston, pushing the piston to one end of the cylinder, and then on the other side of the piston, pushing the piston back to the other end of the cylinder. This allowed the steam engine to operate continuously. The addition of a flywheel made rotary motion possible.

internet**connect**

SCLINKS
NSTA

TOPIC: The Steam Engine
GO TO: www.scilinks.org
*sci*LINKS NUMBER: HSTP075

51

Sample Answer to Explore Other Inventions

In the early nineteenth century, Oliver Evans, an American inventor, built a stationary high-pressure steam engine. His engine was originally used to drive a rotary crusher to produce pulverized limestone for agricultural use. Later, he designed a high-pressure steam engine that was used in driving sawmills, sowing grain, and powering a dredge to clear the Philadelphia waterfront. Evans's steam engines were also used to process paper, cotton, and tobacco.

Richard Trevithick, an English mechanical engineer and inventor, used the high-pressure steam engine to construct the world's first steam railway locomotive in 1803. By 1805, Trevithick had adapted his high-pressure steam engine to propel a barge using paddle wheels and to drive an iron-rolling mill. Trevithick's engines also powered one of the first steam dredges in 1806 and drove a threshing machine on a farm in 1812.

Chapter Organizer

CHAPTER ORGANIZATION	TIME MINUTES	OBJECTIVES	LABS, INVESTIGATIONS, AND DEMONSTRATIONS
Chapter Opener pp. 52–53	45	National Standards: UCP 1, 2, 3, SAI 1, 2, ST 1, SPSP 3, 4, 5, HNS 1, PS 1a, 1b	**Start-Up Activity,** Mystery Mixture, p. 53
Section 1 Elements	90	▶ Describe pure substances. ▶ Describe the characteristics of elements, and give examples. ▶ Explain how elements can be identified. ▶ Classify elements according to their properties. UCP 1, 2, SAI 1, PS 1a, 1b, 1c	**Demonstration,** p. 54 in ATE **Interactive Explorations CD-ROM,** What's the Matter? A **Worksheet** is also available in the **Interactive Explorations Teacher's Edition.** **Labs You Can Eat,** An Iron-ic Cereal Experience
Section 2 Compounds	90	▶ Describe the properties of compounds. ▶ Identify the differences between an element and a compound. ▶ Give examples of common compounds. UCP 1, 2, SAI 2, SPSP 5, PS 1b, 1c; Labs UCP 1, 2, SPSP 5	**Demonstration,** p. 58 in ATE **QuickLab,** Compound Confusion, p. 59 **Discovery Lab,** Flame Tests, p. 70 **Datasheets for LabBook,** Flame Tests
Section 3 Mixtures	90	▶ Describe the properties of mixtures. ▶ Describe methods of separating the components of a mixture. ▶ Analyze a solution in terms of its solute, solvent, and concentration. ▶ Compare the properties of solutions, suspensions, and colloids. UCP 1, 2, ST 2, SPSP 1, 3, PS 1a; Labs UCP 1, 2, SAI 1, 2, ST 2, SPSP 5	**Demonstration,** Separation, p. 63 in ATE **Discovery Lab,** A Sugar Cube Race! p. 140 **Datasheets for LabBook,** A Sugar Cube Race! **Skill Builder,** Making Butter, p. 141 **Datasheets for LabBook,** Making Butter **Making Models,** Unpolluting Water, p. 142 **Datasheets for LabBook,** Unpolluting Water **Inquiry Labs,** Separation Anxiety **Whiz-Bang Demonstrations,** Dense Suspense **EcoLabs & Field Activities,** Ozone News Zone **Long-Term Projects & Research Ideas,** A Coin-cidence?

See page **T23** *for a complete correlation of this book with the*

NATIONAL SCIENCE EDUCATION STANDARDS.

TECHNOLOGY RESOURCES

Guided Reading Audio CD
English or Spanish, Chapter 3

Science Discovery Videodiscs
Image and Activity Bank with Lesson Plans: It Takes Concentration
Science Sleuths: Fortune or Fraud?

CNN. **Eye on the Environment,** Treating Toxic Waste, Segment 24

Interactive Explorations CD-ROM
CD 1, Exploration 4, What's the Matter?

One-Stop Planner CD-ROM **with Test Generator**

hapter 3 • Elements, Compounds, and Mixtures

CLASSROOM WORKSHEETS, TRANSPARENCIES, AND RESOURCES	SCIENCE INTEGRATION AND CONNECTIONS	REVIEW AND ASSESSMENT
Directed Reading Worksheet **Science Puzzlers, Twisters & Teasers**		
Directed Reading Worksheet, Section 1 **Math Skills for Science Worksheet,** Percentages, Fractions, and Decimals **Science Skills Worksheet,** Introduction to Graphs **Transparency 210,** The Three Major Categories of Elements **Problem Solving Worksheet,** Jet Smart	**Math and More,** p. 55 in ATE **Multicultural Connection,** p. 55 in ATE **Real-World Connection,** p. 57 in ATE	**Section Review,** p. 57 **Quiz,** p. 57 in ATE **Alternative Assessment,** p. 57 in ATE
Directed Reading Worksheet, Section 2	**Physics Connection,** p. 60 **Holt Anthology of Science Fiction,** *The Strange Case of Dr. Jekyll and Mr. Hyde*	**Self-Check,** p. 59 **Section Review,** p. 61 **Quiz,** p. 61 in ATE **Alternative Assessment,** p. 61 in ATE
Transparency 211, Separation of a Mixture **Directed Reading Worksheet,** Section 3 **Math Skills for Science Worksheet,** Parts of 100: Calculating Percentages **Math Skills for Science Worksheet,** Using Proportions and Cross-Multiplication **Transparency 212,** Solubility Graph **Transparency 134,** Three Types of Volcanoes **Reinforcement Worksheet,** It's All Mixed Up	**Cross-Disciplinary Focus,** p. 64 in ATE **Math and More,** p. 65 in ATE **Cross-Disciplinary Focus,** p. 65 in ATE **MathBreak,** Calculating Concentration, p. 66 **Math and More,** p. 66 in ATE **Connect to Earth Science,** p. 66 in ATE **Connect to Life Science,** p. 67 in ATE **Biology Connection,** p. 68 **Apply,** p. 68 **Connect to Earth Science,** p. 68 in ATE **Science, Technology, and Society:** Perfume, p. 76	**Section Review,** p. 64 **Homework,** pp. 64, 67 in ATE **Self-Check,** p. 65 **Section Review,** p. 69 **Quiz,** p. 69 in ATE **Alternative Assessment,** p. 69 in ATE

internet connect

go.hrw.com

Holt, Rinehart and Winston On-line Resources

go.hrw.com

For worksheets and other teaching aids related to this chapter, visit the HRW Web site and type in the keyword: **HSTMIX**

*sci*LINKS NSTA

National Science Teachers Association

www.scilinks.org

Encourage students to use the *sci*LINKS numbers listed in the internet connect boxes to access information and resources on the **NSTA** Web site.

END-OF-CHAPTER REVIEW AND ASSESSMENT

Chapter Review in Study Guide
Vocabulary and Notes in Study Guide
Chapter Tests with Performance-Based Assessment, Chapter 3 Test
Chapter Tests with Performance-Based Assessment, Performance-Based Assessment 3
Concept Mapping Transparency 4

Chapter Resources & Worksheets

Visual Resources

TEACHING TRANSPARENCIES

#210
Holt Science and Technology — Teaching Transparency 210
The Three Major Categories of Elements

Metals
Metals are elements that are shiny and are good conductors of thermal energy and electric current. They are malleable (they can be hammered into thin sheets) and ductile (they can be drawn into thin wires). Iron has many uses in building and automobile construction. Copper is used in wires and coins.

Lead
Copper
Tin

Nonmetals
Nonmetals are elements that are dull (not shiny) and that are poor conductors of thermal energy and electric current. Solid nonmetals tend to be brittle and unmalleable. Few familiar objects are made of only nonmetals. The neon used in lights is a nonmetal, as is the graphite (carbon) used in pencils.

Sulfur
Bromine
Neon

Metalloids
Metalloids, also called semiconductors, are elements that have properties of both metals and nonmetals. Some metalloids are shiny, while others are dull. Metalloids are somewhat malleable and ductile. Some metalloids conduct thermal energy and electric current well. Silicon is used to make computer chips. However, other elements must be added to silicon to make a working chip.

Antimony
Silicon
Boron

#211
Holt Science and Technology
Separation of a Mixture

A mixture of the compound sodium chloride (table salt) with the element sulfur requires more than one separation step.

The **first step** is to mix them with another compound—water. Salt dissolves in water, but sulfur does not.

In the **second step**, the mixture is poured through a filter. The filter traps the solid sulfur.

In the **third step**, the sodium chloride is separated from the water by simply evaporating the water.

#212
Holt Science and Technology
Solubility Graph

The solubility of most solids increases as the temperature gets higher. Thus, more solute can dissolve at higher temperatures. However, some solids, such as cerium sulfate, are less soluble at higher temperatures.

Solubility (g/100 mL of water)
Temperature (°C)
Cerium sulfate
Sodium chloride
Potassium nitrate
Sodium chlorate

TEACHING TRANSPARENCIES

#134
Holt Science and Technology — Teaching Transparency 134
Three Types of Volcanoes

Shield volcano
Cinder cone volcano
Composite volcano

LINK TO EARTH SCIENCE

CONCEPT MAPPING TRANSPARENCY

#4
Holt Science and Technology — Concept Mapping Transparency 4
Elements, Compounds, and Mixtures
Use the following terms to complete the concept map below:
mixture, colloid, filter, element, suspension, solution, compound

Matter
can be classified as a(n)
which may be categorized as a
which is separated by a

Meeting Individual Needs

DIRECTED READING

#3
Date _____ Class _____
DIRECTED READING WORKSHEET
Elements, Compounds, and Mixtures

Chapter Introduction
As you begin this chapter, answer the following.
1. Read the title of the chapter. List three things that you already know about this subject.

2. Write two questions about this subject that you would like answered by the time you finish this chapter.

Start-Up Activity (p. 00)
3. What do you think will happen to the ink of the black marker in this activity?

REINFORCEMENT & VOCABULARY REVIEW

#3
Date _____ Class _____
REINFORCEMENT WORKSHEET
It's All Mixed Up

Complete this worksheet after you finish reading Chapter 3, Section 3.
Label each figure below with the type of substance it BEST models: colloid, compound, element, solution, or suspension.

6. Why did you label the figures on the previous page as you did?

Professor Jumble's confusion
In her lab, Professor Jumble has four shelves labeled Suspensions, Solutions, Compounds, and Colloids, respectively. Last night, the Professor set one beaker of clear liquid on each of the four shelves. When the professor walked into her lab this morning, all four beakers were on the same shelf and she didn't know which was which. She tested each beaker and the results are below.
Use the test results to help Professor Jumble unjumble the beakers, and write the identity of each mixture in the blanks.

Beaker A:	Beaker B:
Light passes right through. Particles do not separate in a centrifuge.	After heating, a different-looking liquid remains. Does not separate in a centrifuge.

Beaker C:	Beaker D:
Centrifuged into two different-colored layers. Scatters light. Solute was left behind in the filter.	Scatters light. Passes through a filter without leaving a residue.

#3
Date _____ Class _____
VOCABULARY REVIEW WORKSHEET
An ELEMENTary Word Puzzle

Give this puzzle a try after you read Chapter 4.
Identify each term described by the clues. Then find and circle each term in the puzzle on the next page. Words may appear forward or backward, horizontally, vertically, or diagonally.

1. _____ Measure of amount of solute needed to make a saturated solution
2. _____ Dispersed particles are too light to settle out
3. _____ Substance in which another is dissolved
4. _____ Amount of solute in a given amount of solvent
5. _____ Pure substance made up of only one type of particle
6. _____ Two or more substances that are combined; not a compound
7. _____ Pure substance made up of at least two elements
8. _____ Characteristic property measured in grams per cubic centimeter
9. _____ Mixed with another element can become a good conductor
10. _____ Solution of a metal or nonmetal dissolved in a metal
11. _____ Dissolved substance
12. _____ Shiny element; good conductor of heat and electricity
13. _____ Mixture in which particles of one substance are large enough to settle out of another substance
14. _____ Mixture composed of evenly-distributed particles of two or more substances
15. _____ Poor conductor and a brittle solid

SCIENCE PUZZLERS, TWISTERS & TEASERS

#3
Date _____ Class _____
SCIENCE PUZZLERS, TWISTERS & TEASERS
Elements, Compounds, and Mixtures

I Am Not a Metalloid
1. Professor Medeno, the mad scientist, wants to create an army of metalloids to take over the world. He has made a number of prototypes and named them after his friends. Professor Peña must now test the prototypes. Based on each test result, help Professor Peña identify (in the spaces below) his creations as metal, nonmetal, or metalloid.

a. Carrie is difficult to shape, and when struck with an anvil she shatters.

b. Tom is shiny and easy to shape, but when dropped into water he rusts.

c. Lydia is shiny and may be hammered into a thin sheet or drawn into a small wire, but when thrown in the fire she gets hot too quickly.

d. Pete is dull in appearance, but when mixed with silicon he becomes a working computer chip.

e. Molly fills whatever container she occupies, but when thrown in the fire she is slow to react.

f. Chia has a shiny surface, and when mixed with others she becomes a good conductor.

Review & Assessment

STUDY GUIDE

#3 VOCABULARY & NOTES WORKSHEET
Elements, Compounds, and Mixtures

#3 CHAPTER REVIEW WORKSHEET
Elements, Compounds, and Mixtures

CHAPTER TESTS WITH PERFORMANCE-BASED ASSESSMENT

#3 ELEMENTS, COMPOUNDS AND MIXTURES
Chapter 3 Test

#3 ELEMENTS, COMPOUNDS, AND MIXTURES
Chapter 3 Performance-Based Assessment

Lab Worksheets

INQUIRY LABS

#3 STUDENT WORKSHEET — DESIGN YOUR OWN
Separation Anxiety

LABS YOU CAN EAT

#3 STUDENT WORKSHEET — DISCOVERY LAB
An Iron-ic Cereal Experience

WHIZ-BANG DEMONSTRATIONS

#3 TEACHER-LED DEMONSTRATION — DISCOVERY LAB
Dense Suspense

ECOLABS & FIELD ACTIVITIES

#3 STUDENT WORKSHEET — DESIGN YOUR OWN
Ozone News Zone

LONG-TERM PROJECTS & RESEARCH IDEAS

#3 STUDENT WORKSHEET — DESIGN YOUR OWN
A Coin-cidence?

DATASHEETS FOR LABBOOK

#3 Flame Tests
#3 A Sugar Cube Race!
#3 Making Butter
#3 Unpolluting Water

Applications & Extensions

CRITICAL THINKING & PROBLEM SOLVING

#3 PROBLEM SOLVING WORKSHEET
Jet Smart

EYE ON THE ENVIRONMENT

#24 in the News: Critical Thinking Worksheets
Treating Toxic Waste

INTERACTIVE EXPLORATIONS

#1-4 Exploration 4 Worksheet
What's the Matter?

Chapter Background

SECTION 1

Elements

▶ **Electrolysis**
Sir Humphry Davy (1778–1829) was a professor at England's Royal Institution. After the creation of the voltaic pile in 1800, Davy built a battery made of voltaic cells and applied the electric current to potash and to soda. The compounds decomposed to form two previously unknown elements—potassium and sodium. He later used electrolysis to isolate the elements magnesium, calcium, strontium, barium, boron, and silicon.

▶ **Gold—a Metal**
Gold is often found in its elemental state because it is not a chemically reactive element. The ancient Egyptians hammered gold into a sheet so thin that it took 367,000 leaves to make a pile 2.5 cm high. Gold is often used with other metals in jewelry because it can be reshaped easily.

▶ **Sulfur—a Nonmetal**
Sulfur is found both uncombined and combined in nature. Sulfur is found in deposits on the slopes of volcanoes and in crystals at the mouth of volcanic vents. A series of chemical reactions causes uncombined sulfur to form and fall onto the volcanoes' slopes.

IS THAT A FACT!
 ☛ Jupiter's moon Io appears yellow due to large deposits of sulfur from volcanic activity.

▶ **Elements in the Body**
Sodium and potassium compounds are vital to the human body in blood, muscle tissue, and nerve tissue. These are only two of many elements present in compounds that keep the human body functioning properly.

SECTION 2

Compounds

▶ **Sodium Compounds**
Sodium compounds are very important commercially. For example, several million tons of sodium hydroxide are used each year to make paper, other chemicals, and petroleum products. Sodium sulfate is used in the manufacture of paper, glass, and detergents, and sodium silicate is used in the manufacture of soaps and detergents.

▶ **Silicon Compounds**
In nature, most silicon is combined with oxygen to form silicon dioxide. Silicon dioxide makes up sand, flint, quartz, and opal. Silicon is also found combined with iron, aluminum, magnesium, and other metals. One reason for the large number of silicon compounds is that silicon forms long chains of atoms, often in the form of silicates (compounds of silicon, a metal, and oxygen).

 • When silicon dioxide contains small amounts of manganese or iron, it forms a type of quartz called amethyst, which is purple in color and is used as a gemstone. Many silicates, such as emerald, jade, aquamarine, garnet, opal, onyx, and moonstone, are valued as gemstones.

IS THAT A FACT!
 ☛ The greenish color of the Statue of Liberty is caused by the compound copper(II) carbonate, which formed when the copper in the statue reacted with carbon dioxide and water.

▶ **Carbon Compounds**
The science of carbon compounds, organic chemistry, has produced plastics, fuels, medicines, fibers, and armor. Carbon compounds are everywhere, and our

lives would be vastly different without plastic containers, gasoline, penicillin, nylon, and Teflon®, all of which are made from carbon compounds.

- In 1945, Dorothy Hodgkin used X-ray diffraction to determine the structure of penicillin. Once its structure was known, penicillin could be synthesized and mass produced. Hodgkin went on to determine the structures of insulin and vitamin B_{12}.

SECTION 3

Mixtures

▶ Describing Solutions

Students are shown three methods of describing how much solute is dissolved in a solvent. The most general method uses the terms *concentrated* and *dilute*. Describing a solution in terms of its *saturation* is a little more specific because it relates the amount of solute to a specific number value. Calculating the *concentration* is the most specific method because it gives a number value for the solution.

▶ Supersaturated Solutions

A solution that contains more dissolved solute than is normally possible is *supersaturated*. Supersaturated

solutions will become saturated solutions when solute comes out of solution. This can occur if more solute is added or if the solution is shaken. To make a supersaturated solution, make a saturated solution at a higher temperature; then allow it to cool undisturbed. Several varieties of reusable hand warmers make use of supersaturated solutions.

▶ Colloids

Colloids resist separation for several reasons. The small particle size makes gravitational force less effective in causing their separation. Another factor is the constant, random motion of the colloid particles, called Brownian motion. Third, most colloid particles are electrically charged. In any colloid, all the particles have the same charge and should repel each other. This keeps them from clumping together and from settling out.

▶ Colloids You Know

Gels are colloids in which solid particles are spread out in a liquid. Aerosols are colloids made with solid or liquid particles that are suspended in a gas. Emulsions are colloids made of two liquids. Smog is a colloid of dust and other solid particles in the air.

For background information about teaching strategies and issues, refer to the *Professional Reference for Teachers.*

CHAPTER 3

Elements, Compounds, and Mixtures

Pre-Reading Questions

Students may not know the answers to these questions before reading the chapter, so accept any reasonable response.

Suggested Answers

1. An element is a pure substance that cannot be separated into simpler substances by physical or chemical means.
2. A compound is a pure substance composed of two or more elements that are chemically combined. A mixture can be physically separated into its components, but a compound can be separated into the elements that compose it only by chemical means.
3. The two components of a solution are the solute and the solvent.

CHAPTER 3

Sections

Pre-Reading Questions

1. What is an element?
2. What is a compound? How are compounds and mixtures different?
3. What are the components of a solution called?

Elements, Compounds, and Mixtures

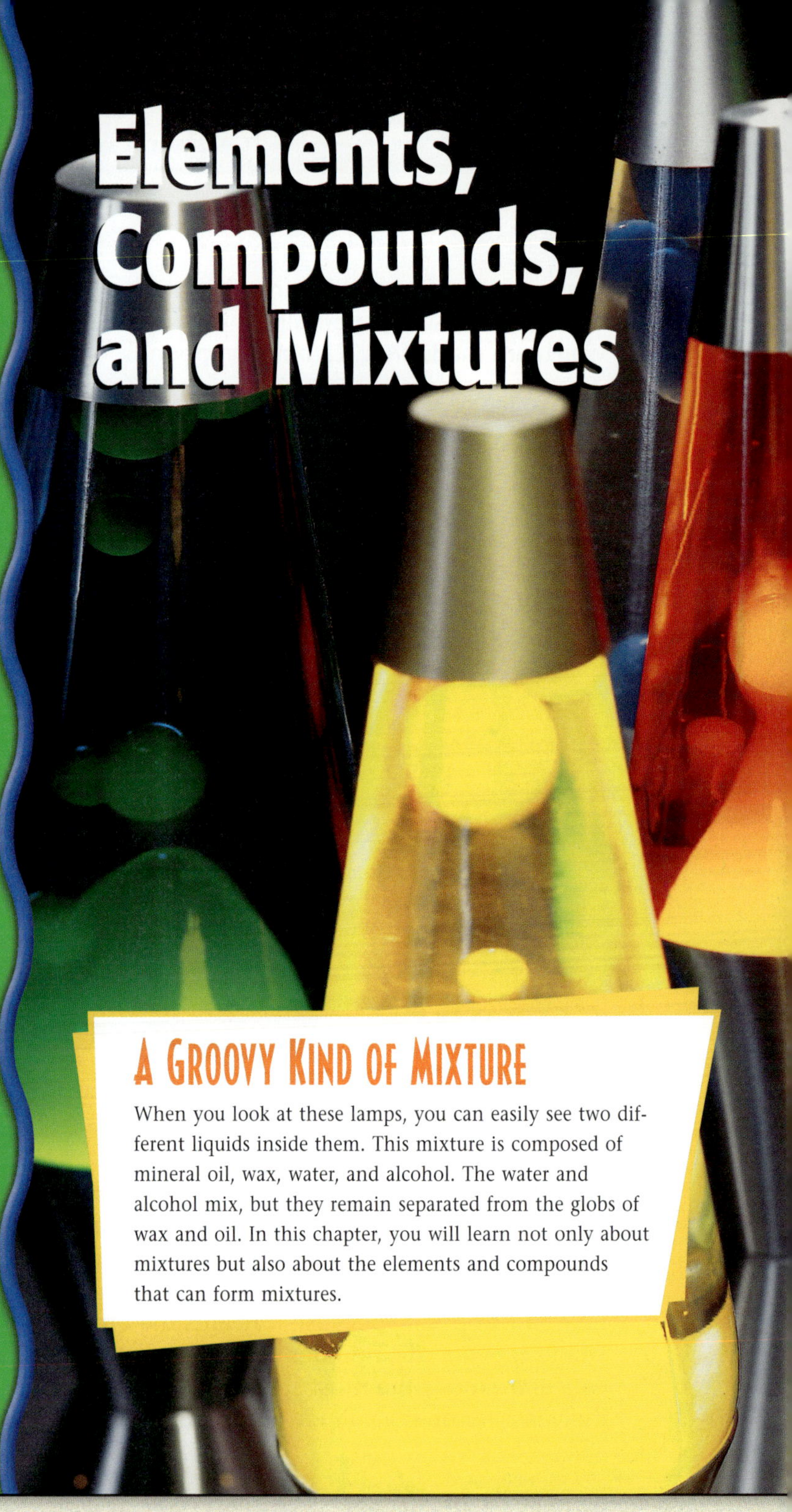

A Groovy Kind of Mixture

When you look at these lamps, you can easily see two different liquids inside them. This mixture is composed of mineral oil, wax, water, and alcohol. The water and alcohol mix, but they remain separated from the globs of wax and oil. In this chapter, you will learn not only about mixtures but also about the elements and compounds that can form mixtures.

52

internet connect

HRW On-line Resources

go.hrw.com
For worksheets and other teaching aids, visit the HRW Web site and type in the keyword: **HSTMIX**

www.scilinks.com
Use the *sci*LINKS numbers at the end of each chapter for additional resources on the **NSTA** Web site.

www.si.edu/hrw
Visit the Smithsonian Institution Web site for related on-line resources.

www.cnnfyi.com
Visit the CNN Web site for current events coverage and classroom resources.

MYSTERY MIXTURE

In this activity, you will separate the different dyes found in an ink mixture.

Procedure

1. Tear a strip of paper (about 3 cm × 15 cm) from a **coffee filter.** Wrap one end of the strip around a **pencil** so that the other end will just touch the bottom of a **clear plastic cup.** Use **tape** to attach the paper to the pencil.

2. Take the paper out of the cup. Using a **water-soluble black marker,** make a small dot in the center of the strip about 2 cm from the bottom.

3. Pour **water** in the cup to a depth of 1 cm. Carefully lower the paper into the cup. Be sure the dot is not under water.

4. Remove the paper when the water is 1 cm from the top. Record your observations in your ScienceLog.

Analysis

5. Infer what happened as the filter paper soaked up the water.

6. Which colors were mixed to make your black ink?

7. Compare your results with those of your classmates. Record your observations.

8. Infer whether the process used to make the ink involved a physical or chemical change. Explain.

53

MYSTERY MIXTURE

MATERIALS

FOR EACH GROUP:
- 3 x 15 cm strip of paper from a coffee filter or filter paper
- pencil
- clear plastic cup or beaker
- piece of tape
- water-soluble, black felt-tip marker or pen (not permanent marker)
- water

Teacher's Notes

You might wish to cut strips from the coffee filters ahead of time. The size of the strips can be adjusted to better fit the cups that students will be using, thereby reducing waste. You might want students to wear lab aprons during this activity.

This activity works best if students are given a variety of brands of markers. Brands that are known to work include Mr. Sketch®, Vis-à-vis®, Crayola Washable®, and Flair®. Test the markers for suitability. Demonstrate how to use this procedure (called chromatography) to identify a sample: Students can determine the type of marker you used by comparing the pattern of colors on your paper with the pattern on theirs.

Answers to START-UP Activity

5. The ink in the dot separated into several colors and moved up the paper. (The colors will vary with the brand of marker used.)

6. Answers will vary. If several varieties of markers are used, differences in the order and colors of ink can be seen.

8. The process involved is a physical change. The colors of ink are separated without changing their chemical makeup.

Focus

Elements

This section explains the characteristics of elements and gives examples of these characteristics. It also explains how to identify and classify elements as metals, nonmetals, and metalloids based on their properties.

🔔 Bellringer

Pose the following questions to your students:

What do gold, iron, and aluminum have in common? What do oxygen, neon, and sulfur have in common? How is silicon different from aluminum or oxygen?

1 ▶ Motivate

DEMONSTRATION

Hold up a large piece of heavy-duty aluminum foil, and ask students to identify it. Fold the foil into a strip. Demonstrate that it will conduct electric current by touching both electrodes of a conductivity apparatus to the strip. (Do not touch the metal strip while it is attached to the apparatus.)

Next bunch the foil into a ball, then flatten it by pounding it with a hammer. Ask students to list as many properties of aluminum as they can.

📄 **Directed Reading
Worksheet** Section 1

Terms to Learn

element	nonmetals
pure substance	metalloids
metals	

What You'll Do

- Describe pure substances.
- Describe the characteristics of elements, and give examples.
- Explain how elements can be identified.
- Classify elements according to their properties.

Elements

Imagine you are working as a lab technician for the Break-It-Down Corporation. Your job is to break down materials into the simplest substances you can obtain. One day a material seems particularly difficult to break down. You crush and grind it. You notice that the resulting pieces are smaller, but they are still the same material. You try other physical changes, including melting, boiling, and filtering it, but the material does not change into anything simpler.

Next you try some chemical changes. You pass an electric current through the material but it still does not become any simpler. After recording your observations, you analyze the results of your tests. You then draw a conclusion: the substance must be an element. An **element** is a pure substance that cannot be separated into simpler substances by physical or chemical means, as shown in **Figure 1.**

Figure 1 *No matter what kind of physical or chemical change you attempt, an element cannot be changed into a simpler substance!*

Figure 2 *The atoms of the element iron are alike whether they are in a meteorite or in a common iron skillet.*

An Element Has Only One Type of Particle

A **pure substance** is a substance in which there is only one type of particle. Because elements are pure substances, each element contains only one type of particle. For example, every particle (atom) in a 5 g nugget of the element gold is like every other particle of gold. The particles of a pure substance are alike no matter where that substance is found, as shown in **Figure 2.** Although a meteorite might travel more than 400 million kilometers (about 248 million miles) to reach Earth, the particles of iron in a meteorite are identical to the particles of iron in objects around your home!

54

BRAIN FOOD

Three types of particles are used when discussing pure substances. The term *atom* is used for most elements. The term *molecule* is used for diatomic elements. Molecule is also used for covalent compounds. The term *formula unit* is used for ionic compounds, although the term *molecule* is also widely used and accepted.

Every Element Has a Unique Set of Properties

Each element has a unique set of properties that allows you to identify it. For example, each element has its own *characteristic properties*. These properties do not depend on the amount of material present in a sample of the element. Characteristic properties include some physical properties, such as boiling point, melting point, and density, as well as chemical properties, such as reactivity with acid. The elements helium and krypton are unreactive gases. However, the density (mass per unit volume) of helium is less than the density of air. Therefore, a helium-filled balloon will float up if it is released. Krypton is more dense than air, so a krypton-filled balloon will sink to the ground if it is released.

Identifying Elements by Their Properties Look at the elements cobalt, iron, and nickel, shown in **Figure 3.** Even though these three elements have some similar properties, each can be identified by its unique set of properties.

Notice that the physical properties for the elements in Figure 3 include melting point and density. Other physical properties, such as color, hardness, and texture, could be added to the list. Also, depending on the elements being identified, other chemical properties might be useful. For example, some elements, such as hydrogen and carbon, are flammable. Other elements, such as sodium, react immediately with oxygen. Still other elements, such as zinc, are reactive with acid.

Figure 3 *Like all other elements, cobalt, iron, and nickel can be identified by their unique combination of properties.*

Melting point is 1,495°C.
Density is 8.9 g/cm³.
Conducts electric current and thermal energy.
Unreactive with oxygen in the air.

Cobalt

Melting point is 1,535°C.
Density is 7.9 g/cm³.
Conducts electric current and thermal energy.
Combines slowly with oxygen in the air to form rust.

Iron

Melting point is 1,455°C.
Density is 8.9 g/cm³.
Conducts electric current and thermal energy.
Unreactive with oxygen in the air.

Nickel

2 Teach

USING THE FIGURE

Have students look at the elements shown in **Figure 3.** Ask them if they could use density to tell the three elements apart. (Students could identify iron but could not tell nickel and cobalt apart because these two elements have the same density.)

Ask students if they could use conductivity or reactivity with oxygen to tell them apart. (No; all three conduct thermal energy and electric current, and only iron reacts with oxygen.)

Ask if melting point could be used to identify the elements. (Yes; each has a different melting point.)

Emphasize that it is the *set* of properties of an element, and not any one single property, that identifies an element.

MATH and MORE

The percentages by mass of the elements composing the compounds that make up the human body are:

oxygen, 64.6 percent; carbon, 18.0 percent; hydrogen, 10.0 percent; nitrogen, 3.1 percent; calcium, 1.9 percent; phosphorus, 1.1 percent; other elements, 1.3 percent.

Have students prepare a pie chart or bar graph to illustrate this information.

Math Skills Worksheet
"Percentages, Fractions, and Decimals"

Science Skills Worksheet
"Introduction to Graphs"

IS THAT A FACT!

People in ancient times are thought to have used iron from meteorites before they learned to mine and process ore. A dagger found in the tomb of the Egyptian pharaoh Tutankhamen is thought to have been made from an iron meteorite.

Multicultural CONNECTION

People have long enjoyed the special characteristics and beauty of gold. Gold has also been vital to commerce. Have students research the importance and use of gold in different cultures, such as the ancient Egyptians and Minoans. Students can present their findings on a poster or in a story.

Learners Having Difficulty
Allow students to observe samples of elements, such as aluminum, iron, sulfur, iodine, and silicon, in closed plastic containers. Ask students to describe the properties of the elements that they can see. Then ask them to classify the elements as metals, nonmetals, or metalloids.

MISCONCEPTION ///ALERT///

It is useful to group objects with similar properties into categories based on those properties. However, students should realize that exceptions exist in many categories. For example, all metals except mercury are solids at room temperature. Students should be prepared for exceptions and "gray areas."

3 Extend

GOING FURTHER

Have students find out which elements are nutrients needed by the human body for proper functioning and which foods are good sources of these elements. Have them make a poster or a concept map that shows the major and minor elements and their food sources.

RESEARCH

Ask each student to choose an element to gather information about. Students should research the properties of the element, the history of its discovery, and how it is used. Have them create a poster or a booklet showing their results. Sheltered English

Elements Are Classified by Their Properties

Consider how many different breeds of dogs there are. Consider also how you tell one breed from another. Most often you can tell just by their appearance, or what might be called physical properties. **Figure 4** shows several breeds of dogs, which all happen to be terriers. Many terriers are fairly small in size and have short hair. Although not all terriers are exactly alike, they share enough common properties to be classified in the same group.

Figure 4 Even though these dogs are different breeds, they have enough in common to be classified as terriers.

Elements Are Grouped into Categories

Elements are classified into groups according to their shared properties. Recall the elements iron, nickel, and cobalt. All three are shiny, and all three conduct thermal energy and electric current. Using these shared properties, scientists have grouped these three elements, along with other similar elements, into one large group called metals. Metals are not all exactly alike, but they do have some properties in common.

If You Know the Category, You Know the Properties If you have ever browsed at a music store, you know that the CDs are categorized by type of music. If you like rock-and-roll, you would go to the rock-and-roll section. You might not recognize a particular CD, but you know that it must have the characteristics of rock-and-roll for it to be in this section.

Likewise, you can predict some of the properties of an unfamiliar element by knowing the category to which it belongs. As shown in the concept map in **Figure 5**, elements are classified into three categories—metals, nonmetals, and metalloids. Cobalt, iron, and nickel are classified as metals. If you know that a particular element is a metal, you know that it shares certain properties with iron, nickel, and cobalt. The chart on the next page shows examples of each category and describes the properties that identify elements in each category.

Figure 5 Elements are divided into three categories: metals, nonmetals, and metalloids.

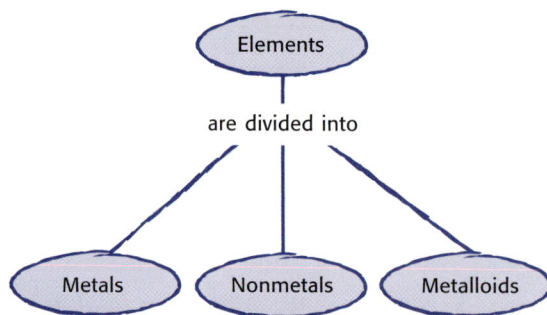

56

internet**connect**

SCILINKS
NSTA

TOPIC: Elements
GO TO: www.scilinks.org
*sci*LINKS NUMBER: HSTP085

WEIRD SCIENCE

Nitrogen forms about four-fifths of Earth's atmosphere but is so unreactive with other elements that it makes up only a tiny percentage of Earth's crust.

The Three Major Categories of Elements

Metals

Metals are elements that are shiny and are good conductors of thermal energy and electric current. They are *malleable* (they can be hammered into thin sheets) and *ductile* (they can be drawn into thin wires). Iron has many uses in building and automobile construction. Copper is used in wires and coins.

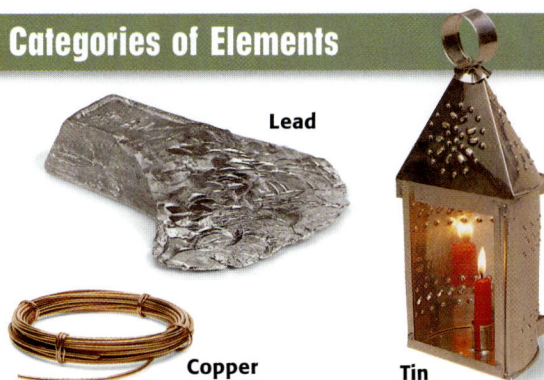

Lead

Copper

Tin

Nonmetals

Sulfur

Nonmetals are elements that are dull (not shiny) and that are poor conductors of thermal energy and electric current. Solid nonmetals tend to be brittle and unmalleable. Few familiar objects are made of only nonmetals. The neon used in lights is a nonmetal, as is the graphite (carbon) used in pencils.

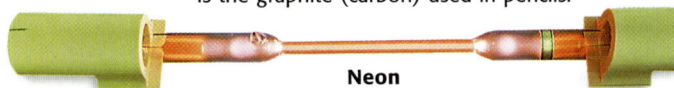

Bromine

Neon

Metalloids

Metalloids, also called semiconductors, are elements that have properties of both metals and nonmetals. Some metalloids are shiny, while others are dull. Metalloids are somewhat malleable and ductile. Some metalloids conduct thermal energy and electric current well. Silicon is used to make computer chips. However, other elements must be added to silicon to make a working chip.

Antimony

Silicon

Boron

SECTION REVIEW

1. What is a pure substance?

2. List three properties that can be used to classify elements.

3. **Applying Concepts**   Which category of element would be the least appropriate choice for making a container that can be dropped without shattering? Explain your reasoning.

internetconnect

SC**LINKS.**
NSTA

TOPIC: Elements
GO TO: www.scilinks.org
*sci***LINKS NUMBER:** HSTP085

▼ *Answers to Section Review*

1. A pure substance is a substance in which there is only one type of particle.

2. Accept all reasonable responses. Answers may include melting point, boiling point, density, reactivity with acid, color, hardness, texture, and flammability.

3. Sample answer: The least appropriate choice for this container would be a nonmetal. Nonmetals tend to be brittle and would most likely crack when dropped.

---

4) Close

Quiz

Ask students to identify the group or groups of elements that have each of the following properties:

1. good conductors of electric current (metals, some metalloids)

2. brittle and nonmalleable (nonmetals)

3. shiny (metals, some metalloids)

4. poor conductors of thermal energy (nonmetals, some metalloids)

ALTERNATIVE ASSESSMENT

Concept Mapping Have students make a concept map that compares the properties of metals, nonmetals, and metalloids.
Sheltered English

REAL-WORLD CONNECTION

For hundreds of years, lead and lead compounds have been used in and around homes. But lead, although common, is toxic to humans (especially to children). It disrupts enzymes that are important to the function of brain cells.

Teaching Transparency 210 "Three Major Categories of Elements"

Problem Solving Worksheet "Jet Smart"

Interactive Explorations CD-ROM "What's the Matter?"

Focus

Compounds

This section describes the properties of compounds and explains the differences between compounds and elements. Students also learn about the properties and importance of common compounds.

Bellringer

Point out to students that the word *compound* refers to something that consists of two or more parts. Ask students to write in their ScienceLog how they might make a compound using elements. Have them list any compounds they know about.

1) Motivate

DEMONSTRATION

Safety Caution: Follow and model proper safety procedures.

The reaction between magnesium and oxygen is easily demonstrated for your students. The light produced in the chemical change is VERY bright. Looking directly at the flame may cause damage to the retina. Inform students of this, and remind them to look away as the magnesium begins to burn. In a darkened classroom, this demonstration is even more dramatic. This reaction was used in old flashbulbs.

Directed Reading Worksheet Section 2

Terms to Learn

compound

What You'll Do

- Describe the properties of compounds.
- Identify the differences between an element and a compound.
- Give examples of common compounds.

Familiar Compounds

- **table salt—** sodium and chlorine
- **water—** hydrogen and oxygen
- **sugar—** carbon, hydrogen, and oxygen
- **carbon dioxide—** carbon and oxygen
- **baking soda—** sodium, hydrogen, carbon, and oxygen

Compounds

Most elements take part in chemical changes fairly easily, so few elements are found alone in nature. Instead, most elements are found combined with other elements as compounds.

A **compound** is a pure substance composed of two or more elements that are chemically combined. In a compound, a particle is formed when atoms of two or more elements join together. In order for elements to combine, they must *react,* or undergo a chemical change, with one another. In **Figure 6,** you see magnesium reacting with oxygen to form a compound called magnesium oxide. The compound is a new pure substance that is different from the elements that reacted to form it. Most substances you encounter every day are compounds. The table at left lists some familiar examples.

Figure 6 *As magnesium burns, it reacts with oxygen and forms the compound magnesium oxide.*

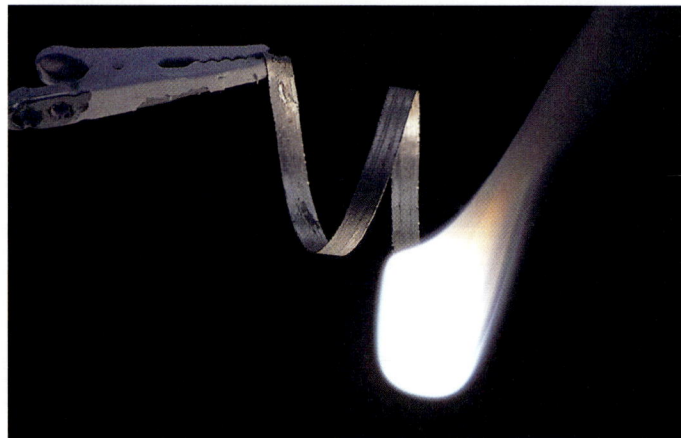

Elements Combine in a Definite Ratio to Form a Compound

Compounds are not random combinations of elements. When a compound forms, the elements join in a specific ratio according to their masses. For example, the ratio of the mass of hydrogen to the mass of oxygen in water is always the same—1 g of hydrogen to 8 g of oxygen. This mass ratio can be written as 1:8 or as the fraction 1/8. Every sample of water has this 1:8 mass ratio of hydrogen to oxygen. If a sample of a compound has a different mass ratio of hydrogen to oxygen, the compound cannot be water.

IS THAT A FACT!

Rubber is a compound of hydrogen and carbon, with the basic formula of $(C_5H_8)n$, where n is about 3,000. The basic unit in the molecule is isoprene. The composition of rubber was proved in 1860 when rubber was heated and broken down into isoprene; its composition was confirmed in 1884 when rubber was produced by the accidental polymerization of isoprene that had been left in a bottle.

Every Compound Has a Unique Set of Properties

Each compound has a unique set of properties that allows you to distinguish it from other compounds. Like elements, each compound has its own physical properties, such as boiling point, melting point, density, and color. Compounds can also be identified by their different chemical properties. Some compounds, such as the calcium carbonate found in chalk, react with acid. Others, such as hydrogen peroxide, react when exposed to light. You can see how chemical properties can be used to identify compounds in the QuickLab at right.

A compound has different properties from the elements that form it. Did you know that ordinary table salt is a compound made from two very dangerous elements? Table salt—sodium chloride—consists of sodium (which reacts violently with water) and chlorine (which is poisonous). Together, however, these elements form a harmless compound with unique properties. Take a look at **Figure 7.** Because a compound has different properties from the elements that react to form it, sodium chloride is safe to eat and dissolves (without exploding!) in water.

Figure 7 *Table salt is formed when the elements sodium and chlorine join. The properties of salt are different from the properties of sodium and chlorine.*

Sodium is a soft, silvery white metal that reacts violently with water.

Chlorine is a poisonous, greenish yellow gas.

Sodium chloride, or table salt, is a white solid that dissolves easily in water and is safe to eat.

✓ Self-Check

Do the properties of pure water from a glacier and from a desert oasis differ? *(See page 168 to check your answer.)*

QuickLab

Compound Confusion

1. Measure 4 g (1 tsp) of **compound A,** and place it in a **clear plastic cup.**

2. Measure 4 g (1 tsp) of **compound B,** and place it in a **second clear plastic cup.**

3. Observe the color and texture of each compound. Record your observations.

4. Add 5 mL (1 tsp) of **vinegar** to each cup. Record your observations.

5. Baking soda reacts with vinegar, while powdered sugar does not. Which of these compounds is compound A, and which is compound B?

59

BRAIN FOOD

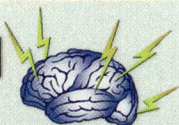

In ancient times, salt was a precious commodity. It was even traded for an equal weight of gold. Soldiers in ancient Rome, as part of their pay, often received a *salarium,* a special ration of salt. (The Latin word for salt is *sal.*) This term eventually evolved into the English word *salary,* a payment for work. Have students research how salt is produced, and have them present their results to the class in some sort of demonstration, a report, or a poster.

---

2) Teach

MEETING INDIVIDUAL NEEDS

Learners Having Difficulty

Help students make a table or chart to describe and compare the properties of sodium, chlorine, and sodium chloride. Then tell students that oxygen is a colorless, odorless gas that supports burning and that hydrogen is a colorless, odorless gas that is highly explosive. Have students compare the properties of these elements with the properties of a compound made from them—water.

QuickLab

MATERIALS
FOR EACH STUDENT:
• baking soda
• powdered sugar
• vinegar
• teaspoon
• 2 clear plastic cups

Safety Caution: Caution students to wear an apron, safety goggles, and gloves while doing this activity.

Label the unknown compounds A and B. The two unknown substances are baking soda and powdered sugar. The materials can be disposed of in a sink with running water.

Answers to QuickLab

5. One compound (baking soda) reacts with vinegar by bubbling; the other (powdered sugar) does not react.

Answer to Self-Check

No, the properties of pure water are the same no matter what its source is.

Prediction Guide Before students read this page, ask them whether the following statements are true or false:

1. Compounds cannot be broken down by any means. (false)
2. Compounds can be broken down only by chemical changes. (true)
3. Heating can break down some compounds. (true)

RETEACHING

Write the alphabet on the board, and get a large dictionary. Discuss with students the millions of words and other combinations of letters the alphabet can make. Then discuss that elements, like letters, can combine in many different ways to form millions of compounds. Point out that of the more than 4 million compounds that exist, carbon is found in 94 percent of them. Sheltered English

3 Extend

GOING FURTHER

Ask students to find out how electrolysis is used in industry and to prepare a poster showing one of these uses.

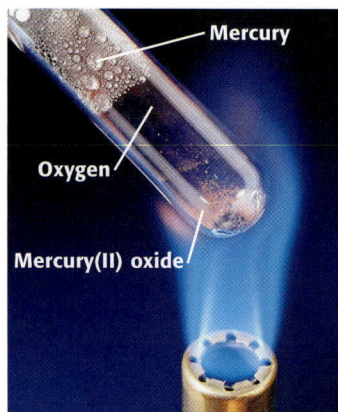

Compounds Can Be Broken Down into Simpler Substances

Some compounds can be broken down into elements through chemical changes. Look at **Figure 8.** When the compound mercury(II) oxide is heated, it breaks down into the elements mercury and oxygen. Likewise, if an electric current is passed through melted table salt, the elements sodium and chlorine are produced.

Other compounds undergo chemical changes to form simpler compounds. These compounds can be broken down into elements through additional chemical changes. For example, carbonic acid is a compound that helps to give carbonated beverages their "fizz," as shown in **Figure 9.** The carbon dioxide and water that are formed can be further broken down into the elements carbon, oxygen, and hydrogen through additional chemical changes.

Figure 8 *Heating mercury(II) oxide causes a chemical change that separates it into the elements mercury and oxygen.*

Figure 9 *Opening a carbonated drink can be messy as carbonic acid breaks down into two simpler compounds—carbon dioxide and water.*

Physics CONNECTION

The process of using electric current to break compounds into simpler compounds and elements is known as electrolysis. Electrolysis can be used to separate water into hydrogen and oxygen. The elements aluminum and copper and the compound hydrogen peroxide are important industrial products obtained through electrolysis.

Compounds Cannot Be Broken Down by Physical Changes

The only way to break down a compound is through a chemical change. If you pour water through a filter, the water will pass through the filter unchanged. Filtration is a physical change, so it cannot be used to break down a compound. Likewise, a compound cannot be broken down by being ground into a powder or by any other physical process.

60

IS THAT A FACT!

Some metals can have more than one charge when they form compounds. To identify the charge used in a particular compound, a Roman numeral is used in the name. Thus, the mercury in mercury(II) oxide has a 2+ charge.

WEIRD SCIENCE

In 1772, Joseph Priestly discovered nitrous oxide. This gas was nontoxic, but it produced unusual effects when inhaled. People would often sing, fight, or laugh. This led to the popular name for nitrous oxide—"laughing gas." The gas is still used in dental surgery.

Compounds in Your World

You are always surrounded by compounds. Compounds make up the food you eat, the school supplies you use, the clothes you wear—even you!

Compounds in Nature Proteins are compounds found in all living things. The element nitrogen is needed to make proteins. **Figure 10** shows how some plants get the nitrogen they need. Other plants use nitrogen compounds that are in the soil. Animals get the nitrogen they need by eating plants or by eating animals that have eaten plants. As an animal digests food, the proteins in the food are broken down into smaller compounds that the animal's cells can use.

Another compound that plays an important role in life is carbon dioxide. You exhale carbon dioxide that was made in your body. Plants take in carbon dioxide and use it to make other compounds, including sugar.

Compounds in Industry The element nitrogen is combined with the element hydrogen to form a compound called ammonia. Ammonia is manufactured for use in fertilizers. Plants can use ammonia as a source of nitrogen for their proteins. Other manufactured compounds are used in medicines, food preservatives, and synthetic fabrics.

The compounds found in nature are usually not the raw materials needed by industry. Often, these compounds must be broken down to provide elements used as raw material. For example, the element aluminum, used in cans, airplanes, and building materials, is not found alone in nature. It is produced by breaking down the compound aluminum oxide.

Figure 10 *The bumps on the roots of this pea plant are home to bacteria that form compounds from atmospheric nitrogen. The pea plant makes proteins from these compounds.*

SECTION REVIEW

1. What is a compound?
2. What type of change is needed to break down a compound?
3. **Analyzing Ideas** A jar contains samples of the elements carbon and oxygen. Does the jar contain a compound? Explain.

internetconnect

SCLINKS
NSTA

TOPIC: Compounds
GO TO: www.scilinks.org
*sci*LINKS NUMBER: HSTP090

61

---

4) Close

Quiz

1. How are compounds and elements alike? How do they differ? (Both are pure substances, but elements cannot be broken down into simpler substances; compounds can be broken down into simpler substances.)
2. What are two ways to break down a compound? (heating and electrolysis)
3. How do the properties of sodium chloride compare with the properties of sodium and of chlorine? (Sodium chloride is a white solid that dissolves in water and is safe to eat. Sodium is a soft, silvery white metal that reacts violently with water, and chlorine is a poisonous, greenish yellow gas.)

ALTERNATIVE ASSESSMENT

Have each student pick a compound and conduct research to find out more about the compound. (You may want to post a list of interesting compounds.) Then have students create a poster or other presentation about their compound. The presentation should contain written information as well as pictures or other visual aids. Display the posters in the classroom.

▼ **Answers to Section Review**

1. A compound is a pure substance composed of two or more elements that are chemically combined.
2. a chemical change
3. A jar containing samples of carbon and oxygen does not contain a compound because the two elements are not chemically combined.

Focus

Mixtures

This section explains the properties of mixtures. Students learn how mixtures can be separated. The concepts of solutes, concentration, and solvents are covered. Finally, students compare solutions, suspensions, and colloids.

🔔 Bellringer

Have students respond to the following situation:

When you add sugar to coffee, tea, iced tea, or lemonade, the sugar disappears. What do you think happens to the sugar? Write your answer in your ScienceLog.

① Motivate

ACTIVITY

MATERIALS

FOR EACH GROUP:
- two beakers with water
- tablespoon of ground coffee (not instant)
- two sugar cubes or sugar packets
- funnel
- coffee filters or filter paper
- paper towels
- stirring rod

Divide students into groups of three or four. Have students put the coffee into a beaker of water and stir it vigorously. Their task is to recover as much of the coffee as possible. Give them time. Have them record their steps, observations, and results in their ScienceLog.

Repeat the activity using the sugar. Have students record everything in their ScienceLog. Discuss the results.

Terms to Learn

mixture	concentration
solution	solubility
solute	suspension
solvent	colloid

What You'll Do

- Describe the properties of mixtures.
- Describe methods of separating the components of a mixture.
- Analyze a solution in terms of its solute, solvent, and concentration.
- Compare the properties of solutions, suspensions, and colloids.

Mixtures

Have you ever made your own pizza? You roll out the dough, add a layer of tomato sauce, then add toppings like green peppers, mushrooms, and olives—maybe even some pepperoni! Sprinkle cheese on top, and you're ready to bake. You have just created not only a pizza but also a mixture—and a delicious one at that!

Properties of Mixtures

All mixtures—even pizza—share certain properties. A **mixture** is a combination of two or more substances that are not chemically combined. Two or more materials together form a mixture if they do not react to form a compound. For example, cheese and tomato sauce do not react when they are used to make a pizza.

Figure 11 *Colorless quartz, pink feldspar, and black mica make up the mixture granite.*

Substances in a Mixture Retain Their Identity Because no chemical change occurs, each substance in a mixture has the same chemical makeup it had before the mixture formed. That is, each substance in a mixture keeps its identity. In some mixtures, such as the pizza above or the piece of granite shown in **Figure 11,** you can even see the individual components. In other mixtures, such as salt water, you cannot see all the components.

Mixtures Can Be Physically Separated If you don't like mushrooms on your pizza, you can pick them off. This is a physical change of the mixture. The identities of the substances did not change. In contrast, compounds can be broken down only through chemical changes.

Not all mixtures are as easy to separate as a pizza. You cannot simply pick salt out of a saltwater mixture, but you can separate the salt from the water by heating the mixture. When the water changes from a liquid to a gas, the salt remains behind. Several common techniques for separating mixtures are shown on the following page.

62

IS THAT A FACT!

The Liberty Bell is a mixture of 70 percent copper, 25 percent tin, and small amounts of lead, zinc, arsenic, gold, and silver.

Common Techniques for Separating Mixtures

Distillation is a process that separates a mixture based on the boiling points of the components. Here you see pure water being distilled from a saltwater mixture. In addition to water purification, distillation is used to separate crude oil into its components, such as gasoline and kerosene.

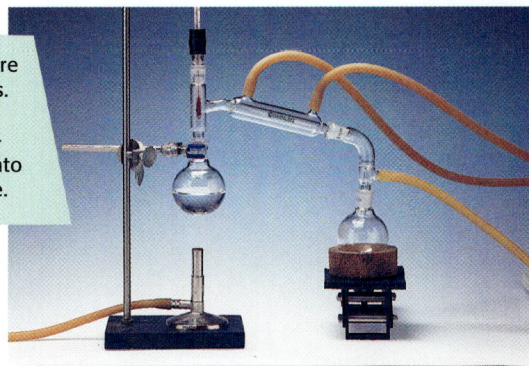

A **magnet** can be used to separate a mixture of the elements iron and aluminum. Iron is attracted to the magnet, but aluminum is not.

The components that make up blood are separated using a machine called a **centrifuge.** This machine separates mixtures according to the densities of the components.

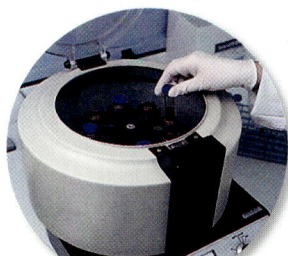

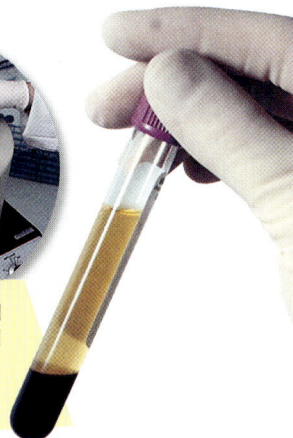

A mixture of the compound sodium chloride (table salt) with the element sulfur requires more than one separation step.

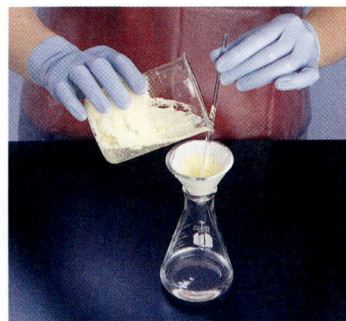

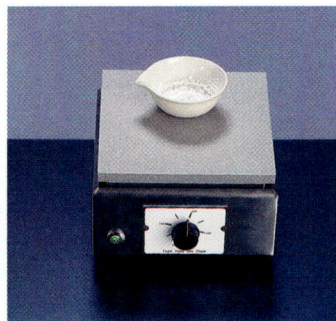

1 The **first step** is to mix them with another compound—water. Salt dissolves in water, but sulfur does not.

2 In the **second step,** the mixture is poured through a filter. The filter traps the solid sulfur.

3 In the **third step,** the sodium chloride is separated from the water by simply evaporating the water.

63

BRAIN FOOD

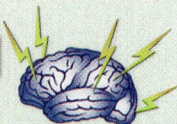

The property of density is used to separate a mixture of ripe and unripe cranberries. Ripe cranberries float in water, but unripe cranberries sink. During harvesting, cranberry bogs are flooded, and the floating cranberries are skimmed from the water.

---

2 Teach

DISCUSSION
Discuss with students the idea that pizza is a mixture. Ask students to write a recipe for their favorite pizza. Have volunteers write their pizza recipes on the board. Have the class compare recipes and discuss how they vary.

MEETING INDIVIDUAL NEEDS
Advanced Learners Have students prepare posters that illustrate how the components of crude oil are separated by distillation. Encourage students to be creative as they share their work with the class. Sheltered English

DEMONSTRATION
Separation Demonstrate the separation of salt (sodium chloride) and sand (silicon dioxide) using the method shown in the text. Ask students which compound dissolves in water and which does not.

GROUP ACTIVITY
Have students work in small groups to figure out a way to separate a mixture of sand, sawdust, and gravel into its components. Let each group come up with a method and present the results to the class.

Teaching Transparency 211 "Separation of a Mixture"

Directed Reading Worksheet Section 3

RETEACHING

Show students a bottle of an oil-and-vinegar salad dressing. Read aloud the ingredients label. Allow the bottle to sit undisturbed so that the ingredients separate. Discuss how the ingredients in the dressing retain their identity and how they could be mixed in any proportions.

CROSS-DISCIPLINARY FOCUS

Art Give students samples of red, blue, and yellow paint. Allow students to discover how many different colors they can make by mixing various amounts of two or three of the colors. Let students make paintings using their mixtures.

Homework

Remind students that many garments are mixtures of fibers. Ask students to examine clothing labels and to list the mixtures and percentages of each material in the mixture. What fabrics are most common? Which ones are not used very often?

Mixtures vs. Compounds	
Mixtures	**Compounds**
Components are elements, compounds, or both	Components are elements
Components keep their original properties	Components lose their original properties
Separated by physical means	Separated by chemical means
Formed using any ratio of components	Formed using a set mass ratio of components

BRAIN FOOD

Many substances are soluble in water, including salt, sugar, alcohol, and oxygen. Water does not dissolve everything, but it dissolves so many different solutes that it is often called the universal solvent.

The Components of a Mixture Do Not Have a Definite Ratio Recall that a compound has a specific mass ratio of the elements that form it. Unlike compounds, the components of a mixture do not need to be combined in a definite ratio. For example, granite that has a greater amount of feldspar than mica or quartz appears to have a pink color. Granite that has a greater amount of mica than feldspar or quartz appears black. Regardless of which ratio is present, this combination of materials is always a mixture—and it is always called granite.

Air is a mixture composed mostly of nitrogen and oxygen, with smaller amounts of other gases, such as carbon dioxide and water vapor. Some days the air has more water vapor, or is more humid, than on other days. But regardless of the ratio of the components, air is still a mixture. The chart at left summarizes the differences between mixtures and compounds.

SECTION REVIEW

1. What is a mixture?

2. Is a mixture separated by physical or chemical changes?

3. **Applying Concepts** Suggest a procedure to separate iron filings from sawdust. Explain why this procedure works.

Solutions

A **solution** is a mixture that appears to be a single substance but is composed of particles of two or more substances that are distributed evenly amongst each other. Solutions are often described as *homogeneous mixtures* because they have the same appearance and properties throughout the mixture.

The process in which particles of substances separate and spread evenly throughout a mixture is known as *dissolving*. In solutions, the **solute** is the substance that is dissolved, and the **solvent** is the substance in which the solute is dissolved. A solute is *soluble*, or able to dissolve, in the solvent. A substance that is *insoluble*, or unable to dissolve, forms a mixture that is not homogeneous and therefore is not a solution.

Salt water is a solution. Salt is soluble in water, meaning that salt dissolves in water. Therefore, salt is the solute and water is the solvent. When two liquids or two gases form a solution, the substance with the greater volume is the solvent.

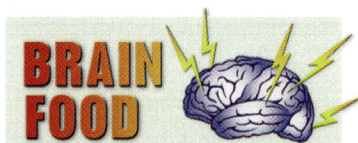

64

1. A mixture is a combination of two or more substances that are not chemically combined.

2. physical

3. Sample answer: Use a magnet to attract the iron filings away from the sawdust. Iron is attracted to a magnet, while sawdust is not.

You may think of solutions as being liquids. And, in fact, tap water, soft drinks, gasoline, and many cleaning supplies are liquid solutions. However, solutions may also be gases, such as air, and solids, such as steel. *Alloys* are solid solutions of metals or nonmetals dissolved in metals. Brass is an alloy of the metal zinc dissolved in copper. Steel is an alloy made of the nonmetal carbon and other elements dissolved in iron. Look at the chart below for examples of the different states of matter used as solutes and solvents in solutions.

✓ **Self-Check**

Yellow gold is an alloy made from equal parts copper and silver combined with a greater amount of gold. Identify each component of yellow gold as a solute or solvent.
(See page 168 to check your answer.)

Examples of Different States in Solutions

Gas in gas	Dry air (oxygen in nitrogen)
Gas in liquid	Soft drinks (carbon dioxide in water)
Liquid in liquid	Antifreeze (alcohol in water)
Solid in liquid	Salt water (salt in water)
Solid in solid	Brass (zinc in copper)

Particles in Solutions Are Extremely Small The particles in solutions are so small that they never settle out, nor can they be filtered out of these mixtures. In fact, the particles are so small, they don't even scatter light. Look at **Figure 12** and see for yourself. The jar on the left contains a solution of sodium chloride in water. The jar on the right contains a mixture of gelatin in water.

Figure 12 *Both of these jars contain mixtures. The mixture in the jar on the left, however, is a solution. The particles in solutions are so small they don't scatter light. Therefore, you can't see the path of light through it.*

65

READING 📖 STRATEGY

Prediction Guide Before students read this page, ask them: Which of these are solutions:

air, soft drinks, ocean water, antifreeze, and brass (all)

Discuss students' answers.

USING THE CHART

Discuss with students that although they may be most familiar with solutions of solids dissolved in liquids, there are many other kinds of solutions. Use the chart **Examples of Different States in Solutions** to start a discussion of the different kinds of solutions.

Answer to Self-Check

Copper and silver are solutes. Gold is the solvent.

MATH and MORE

Pure gold is said to be 24 karat. A 12-karat gold item is 50 percent gold. The common alloy used in jewelry is 14 karat. Have students calculate the percentage of gold in 14-karat gold. (58 percent) Tell them that gold coins are 22 karat. Ask what percentage of gold that is. (92 percent)

Math Skills Worksheet "Parts of 100: Calculating Percentages"

SCIENCE HUMOR

Q: What did the compound say to the solution?

A: You're all mixed up!

CROSS-DISCIPLINARY FOCUS

Music The most commonly used brass instruments in a symphony orchestra are the trumpet, the French horn, the trombone, and the tuba. Brass instruments are based on ancient valveless horns that were used for signaling or hunting.

📶 internetconnect

SCiLINKS
NSTA
TOPIC: Mixtures
GO TO: www.scilinks.org
*sci*LINKS NUMBER: HSTP095

ACTIVITY

MATERIALS

FOR EACH PAIR:
• food coloring
• water
• clear plastic cups

Safety Caution: Caution students to wear an apron while doing this activity.

Students work in pairs. One member of each pair makes a concentrated solution of food coloring and water. The other partner makes a dilute solution of the same.

Place all the cups containing concentrated solutions in front of a white sheet of paper so that students can compare the solutions' colors. Do the same thing with all the dilute solutions. Mention that not all of the concentrated solutions have the same color, nor do all the dilute solutions. This will reinforce that the terms *concentrated* and *dilute* are relative and do not specify amounts.

MATH and MORE

Extend the MathBreak:

Suppose you have 45 g of sodium chloride (salt) dissolved in 150 mL of water, and you need 250 mL more of the same solution. How much sodium chloride do you need to make the additional solution? (75 g)

What is the concentration of the solution? (0.3 g/mL)

Math Skills Worksheet "Using Proportions and Cross-Multiplication"

MATH BREAK

Calculating Concentration

Many solutions are colorless. Therefore, you cannot always compare the concentrations of solutions by looking at the color—you have to compare the actual calculated concentrations. One way to calculate the concentration of a liquid solution is to divide the grams of solute by the milliliters of solvent. For example, the concentration of a solution in which 35 g of salt is dissolved in 175 mL of water is

$$\frac{35 \text{ g salt}}{175 \text{ mL water}} = 0.2 \text{ g/mL}$$

Now It's Your Turn

Calculate the concentrations of each solution below. Solution A has 55 g of sugar dissolved in 500 mL of water. Solution B has 36 g of sugar dissolved in 144 mL of water. Which solution is the more dilute one? Which is the more concentrated?

Smelly solutions? Follow your nose and learn more on page 76.

Answers to MATHBREAK

Solution A: 55 g/500 mL = 0.11 g/mL

Solution B: 36 g/144 mL = 0.25 g/mL

Solution A is more dilute.

Solution B is more concentrated.

Concentration: How Much Solute Is Dissolved? A measure of the amount of solute dissolved in a solvent is **concentration.** Concentration can be expressed in grams of solute per milliliter of solvent. Knowing the exact concentration of a solution is very important in chemistry and medicine because using the wrong concentration can be dangerous.

Solutions can be described as being *concentrated* or *dilute*. Look at **Figure 13.** Both solutions have the same amount of solvent, but the solution on the left contains less solute than the solution on the right. The solution on the left is dilute while the solution on the right is concentrated. Keep in mind that the terms *concentrated* and *dilute* do not specify the amount of solute that is actually dissolved. Try your hand at calculating concentration and describing solutions as concentrated or dilute in the MathBreak at left.

Figure 13 *The dilute solution on the left contains less solute than the concentrated solution on the right.*

A solution that contains all the solute it can hold at a given temperature is said to be *saturated*. An *unsaturated* solution contains less solute than it can hold at a given temperature. More solute can dissolve in an unsaturated solution.

Solubility: How Much Solute Can Dissolve? If you add too much sugar to a glass of lemonade, not all of the sugar can dissolve. Some of the sugar collects on the bottom of the glass. To determine the maximum amount of sugar that can dissolve, you would need to know the solubility of sugar. The **solubility** of a solute is the amount of solute needed to make a saturated solution using a given amount of solvent at a certain temperature. Solubility is usually expressed in grams of solute per 100 mL of solvent. **Figure 14** on the next page shows the solubility of several different substances in water at different temperatures.

CONNECT TO EARTH SCIENCE

Many caves, such as Carlsbad Caverns in New Mexico and Mammoth Cave in Kentucky, were formed by calcium carbonate alternately dissolving in water and being deposited when water evaporated. Have students find out how these caves formed and make a model or a poster or write a report. Sheltered English

---

Figure 14 Solubility of Different Substances

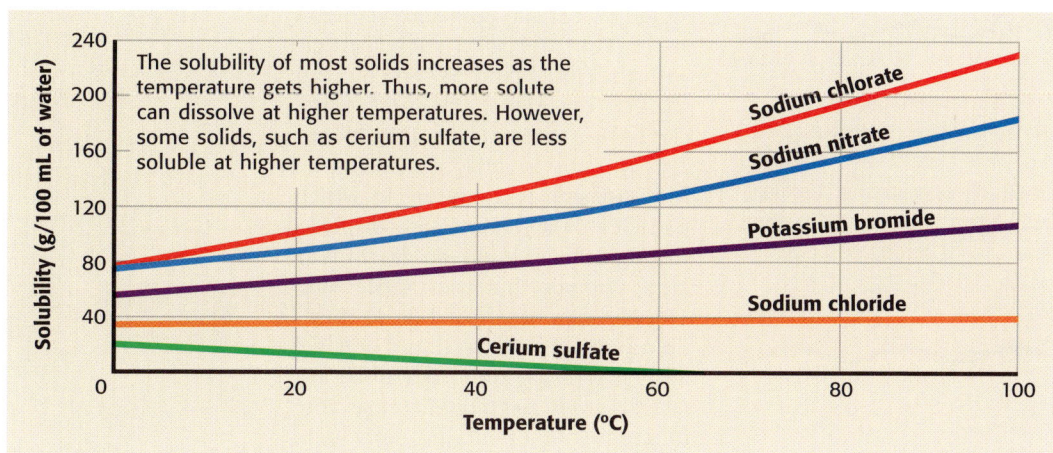

The solubility of most solids increases as the temperature gets higher. Thus, more solute can dissolve at higher temperatures. However, some solids, such as cerium sulfate, are less soluble at higher temperatures.

(Graph: Solubility (g/100 mL of water) vs. Temperature (°C), showing curves for Sodium chlorate, Sodium nitrate, Potassium bromide, Sodium chloride, and Cerium sulfate.)

Unlike the solubility of most solids in liquids, the solubility of gases in liquids decreases as the temperature is raised. Bubbles of gas appear in hot water long before the water begins to boil. The gases that are dissolved in the water cannot remain dissolved as the temperature increases because the solubility of the gases is lower at higher temperatures.

What Affects How Quickly Solids Dissolve in Liquids?
Many familiar solutions are formed when a solid solute is dissolved in water. Several factors affect how fast the solid will dissolve. Look at **Figure 15** to see three methods used to make a solute dissolve faster. You can see why you will enjoy a glass of lemonade sooner if you stir granulated sugar into the lemonade before adding ice!

Figure 15 *Mixing, heating, and crushing iron(III) chloride increase the speed at which it will dissolve.*

Mixing by stirring or shaking causes the solute particles to separate from one another and spread out more quickly among the solvent particles.

Heating causes particles to move more quickly. The solvent particles can separate the solute particles and spread them out more quickly.

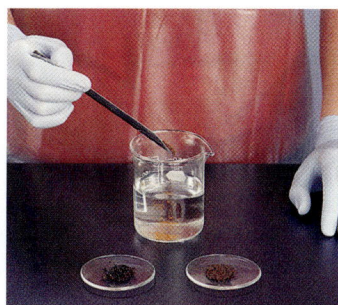

Crushing the solute increases the amount of contact between the solute and the solvent. The particles of solute mix with the solvent more quickly.

67

LabBook PG 140
A Sugar Cube Race!

USING THE GRAPH
Using the graph in **Figure 14**, ask students to compare the solubility of the different solids at different temperatures. Which solid's solubility decreases as temperature increases? (cerium sulfate)

Point out to students that some elements and compounds, such as iodine and calcium carbonate, are not soluble (or are only slightly soluble) in water. Iodine is, however, soluble in alcohol, forming a solution called tincture of iodine.

DISCUSSION
Ask students how the solubility of gases varies with temperature. (Solubility decreases as temperature increases.)

Ask students how that explains why a glass of soda goes "flat" when it sits in a warm room. (The CO_2 gas that gives the soda its "fizz" comes out of the solution as the soda warms up.)

Homework

PORTFOLIO
Concept Mapping
Have students describe three ways to increase the speed at which a solid will dissolve in a liquid. (mixing or stirring, heating the solvent, or crushing the solute)

Then ask them to name some home appliances that are used to speed the solution process. (mixers, food processors, heating elements or burners on a stove, microwaves)

Have students create a concept map to display their answers.
Sheltered English

Teaching Transparency 212
"Solubility Graph"

GOING FURTHER

Darken the room, and turn on the light from a projector. Clap two chalkboard erasers together in the beam of light. Ask students if they can explain what they are observing. (chalk particles suspended in the air scatter the light)

The mixture of chalk dust in air is a suspension. Eventually, the particles of chalk settle out of the air.

Answer to APPLY

The bottle must be shaken to remix the suspension. If the medicine is not mixed, the amount of the medicine taken may not contain the correct dosage of each component and therefore may not be effective.

LabBook   **PG 142**
Unpolluting Water

CONNECT TO EARTH SCIENCE

Research Have students research what happens when a major volcanic eruption sends particulate matter into the atmosphere. They could start with the 1991 eruption of Mount Pinatubo, in the Philippines. Use Teaching Transparency 134, "Three Types of Volcanoes," to help students begin their project.

Teaching Transparency 134 "Three Types of Volcanoes"   LINK TO EARTH SCIENCE

Biology CONNECTION

Blood is a suspension. The suspended particles, mainly red blood cells, white blood cells, and platelets, are actually suspended in a solution called plasma. Plasma is 90 percent water and 10 percent dissolved solutes, including sugar, vitamins, and proteins.

Figure 16 Dirty air is a suspension that could damage a car's engine. The air filter in a car separates dust from air to keep the dust from getting into the engine.

Suspensions

When you shake up a snow globe, you are mixing the solid snow particles with the clear liquid. When you stop shaking the globe, the snow particles settle to the bottom of the globe. This mixture is called a suspension. A **suspension** is a mixture in which particles of a material are dispersed throughout a liquid or gas but are large enough that they settle out. The particles are insoluble, so they do not dissolve in the liquid or gas. Suspensions are often described as *heterogeneous mixtures* because the different components are easily seen. Other examples of suspensions include muddy water and Italian salad dressing.

The particles in a suspension are fairly large, and they scatter or block light. This often makes a suspension difficult to see through. But the particles are too heavy to remain mixed without being stirred or shaken. If a suspension is allowed to sit undisturbed, the particles will settle out, as in a snow globe.

A suspension can be separated by passing it through a filter. The liquid or gas passes through, but the solid particles are large enough to be trapped by the filter, as shown in **Figure 16.**

APPLY

Shake Well Before Use

Many medicines, such as remedies for upset stomach, are suspensions. The directions on the label instruct you to shake the bottle well before use. Why must you shake the bottle? What problem could arise if you don't?

SCIENCE HUMOR

Q: What did the chemist say to the suspension?

A: "Settle down!"

IS THAT A FACT!

Blood is a suspension, but cells contain colloids of solid particles suspended in water.

Colloids

Some mixtures have properties of both solutions and suspensions. These mixtures are known as colloids (KAWL oyDz). A **colloid** is a mixture in which the particles are dispersed throughout but are not heavy enough to settle out. The particles in a colloid are relatively small and are fairly well mixed. Solids, liquids, and gases can be used to make colloids. You might be surprised at the number of colloids you encounter each day. Milk, mayonnaise, stick deodorant—even the gelatin and whipped cream in **Figure 17**—are colloids. The materials that compose these products do not separate between uses because their particles do not settle out.

Although the particles in a colloid are much smaller than the particles in a suspension, they are still large enough to scatter a beam of light shined through the colloid, as shown in **Figure 18.** Finally, unlike a suspension, a colloid cannot be separated by filtration. The particles are small enough to pass through a filter.

Figure 17
This dessert includes two delicious examples of colloids—fruity gelatin and whipped cream.

Figure 18 *The particles in the colloid fog scatter light, making it difficult for drivers to see the road ahead.*

LabBook
Make a colloid found in your kitchen on page 141 of the LabBook.

SECTION REVIEW

1. List two methods of making a solute dissolve faster.

2. Identify the solute and solvent in a solution made from 15 mL of oxygen and 5 mL of helium.

3. **Comparing Concepts** What are three differences between solutions and suspensions?

internetconnect

SCiLINKS
NSTA

TOPIC: Mixtures
GO TO: www.scilinks.org
sciLINKS NUMBER: HSTP095

▼ **Answers to Section Review**

1. Answers may include mixing, heating, or crushing the solute.

2. Helium is the solute, and oxygen is the solvent.

3. Sample answer: Unlike particles in a solution, particles in a suspension are large enough to settle out, block light, and be trapped by a filter. Particles in a solution do none of those.

---

4) Close

Quiz

1. Which of the following is not a solution: air in a scuba tank, muddy water, a soft drink, or salt water? (muddy water)

2. When solid iodine is dissolved in alcohol, which is the solute? the solvent? (iodine—solute, alcohol—solvent)

3. Why might a lake in a tropical area contain more dissolved minerals than a lake in Maine? (The temperature of the tropical lake is probably higher, and many minerals are more soluble in warmer water than in colder water.)

LabBook PG 141
Making Butter

ALTERNATIVE ASSESSMENT

MATERIALS
FOR EACH SMALL GROUP:
• paper cups and straws
• container of warm water
• small amounts of vinegar, vegetable oil, ground coffee, instant coffee, and sand

Safety Caution: Caution students to wear safety goggles, gloves, and an apron.

Ask students to determine how many solutions and suspensions they can make by mixing these materials. Have them make two lists: solutions and suspensions. (Possible answers: Solutions—water and sugar, water and instant coffee; Suspensions—water and oil; sand and water)

Reinforcement Worksheet
"It's All Mixed Up"

Flame Tests
Teacher's Notes

Time Required

One or two 45-minute class periods

Lab Ratings

EASY			HARD

TEACHER PREP 🍾🍾🍾
STUDENT SET-UP 🍾🍾
CONCEPT LEVEL 🍾🍾🍾🍾
CLEAN UP 🍾🍾🍾

MATERIALS

The materials listed are for each group of 2–3 students. The unknown solution should be clear. Use only dilute hydrochloric acid—concentrations lower than 1.0 M. When diluting an acid, always add the acid to the water.

Preparation Notes

Prepare three chloride solutions, such as KCl, $CaCl_2$, and NaCl. Make enough of one of the three solutions to serve as the "unknown" solution. Prepare a mild concentration (approximately 10 g per 500 mL) of each test solution. You will need 5 to 10 mL of each solution per group. Make the wire holder with Nichrome® wire or paper clips and ice-cream sticks or corks. Bend one end of the wire into a small loop like a bubble wand. Tape the other end of the wire to the stick, or insert it into the cork.

Datasheets for LabBook

Discovery Lab

USING SCIENTIFIC METHODS

Flame Tests

Fireworks make fantastic combinations of color. The colors are the results of burning different compounds. Imagine that you are the lead chemist for a fireworks company. You must identify an unknown compound so that it may be used in the correct fireworks show. You will need to use your knowledge that every compound has a unique set of properties.

MATERIALS

- 4 small test tubes
- test-tube rack
- masking tape
- 4 chloride test solutions
- spark igniter
- Bunsen burner
- wire and holder
- dilute hydrochloric acid in a small beaker
- distilled water in a small beaker

Make a Prediction

1 Can you identify the unknown compound by heating it in a flame? Explain.

Conduct an Experiment

Caution: Be very careful in handling all chemicals. Tell your teacher immediately if you spill a chemical.

2 Arrange the test tubes in the test-tube rack. Use masking tape to label the tubes with the following names: "Calcium chloride," "Potassium chloride," "Sodium chloride," and "Unknown."

3 Copy the table below into your ScienceLog. Then ask your teacher for your portions of the solutions.

Test Results	
Compound	**Color of flame**
Calcium chloride	
Potassium chloride	*DO NOT WRITE IN BOOK*
Sodium chloride	
Unknown	

4 Light the burner. Clean the wire by first dipping it into the dilute hydrochloric acid. Then dip it into the distilled water. Holding the wooden handle, heat the wire in the blue flame of the burner. Stop heating when the wire is glowing and it no longer colors the flame.
Caution: Use extreme care around an open flame.

Safety Caution

Remind students to review all safety cautions and icons before beginning this lab activity. Students should touch only the wooden handle of the wire holder device because the wire will become hot and could cause burns. Students should be careful with the dilute hydrochloric acid. If contact occurs, they should flush their skin immediately with water. Long hair and loose clothing should be restricted around an open flame. Remind students to keep wooden sticks away from the open flame. In case of an acid spill, first dilute the spill with water. Then mop up the spill with wet cloths or a wet mop while wearing disposable plastic gloves.

Collect Data

5 Dip the clean wire into the first test solution. Hold the wire at the tip of the inner cone of the burner flame. Observe the changes that happen. In the table, record the color given to the flame.

6 Clean the wire by repeating step 4.

7 Repeat steps 5 and 6 for the other solutions.

8 Follow your teacher's instructions for cleanup and disposal.

Analyze the Results

9 Is the flame color a test for the metal or for the chloride in each compound? Explain your answer.

10 What is the identity of your unknown solution? How do you know?

Draw Conclusions

11 Why is it necessary to clean the wire carefully before testing each solution?

12 Infer whether the compound sodium fluoride would make the same color as sodium chloride in a flame test. Explain your answer.

13 Each of the compounds you tested is made from chlorine, which is a poisonous gas at room temperature. Why is it safe to use these compounds without a gas mask?

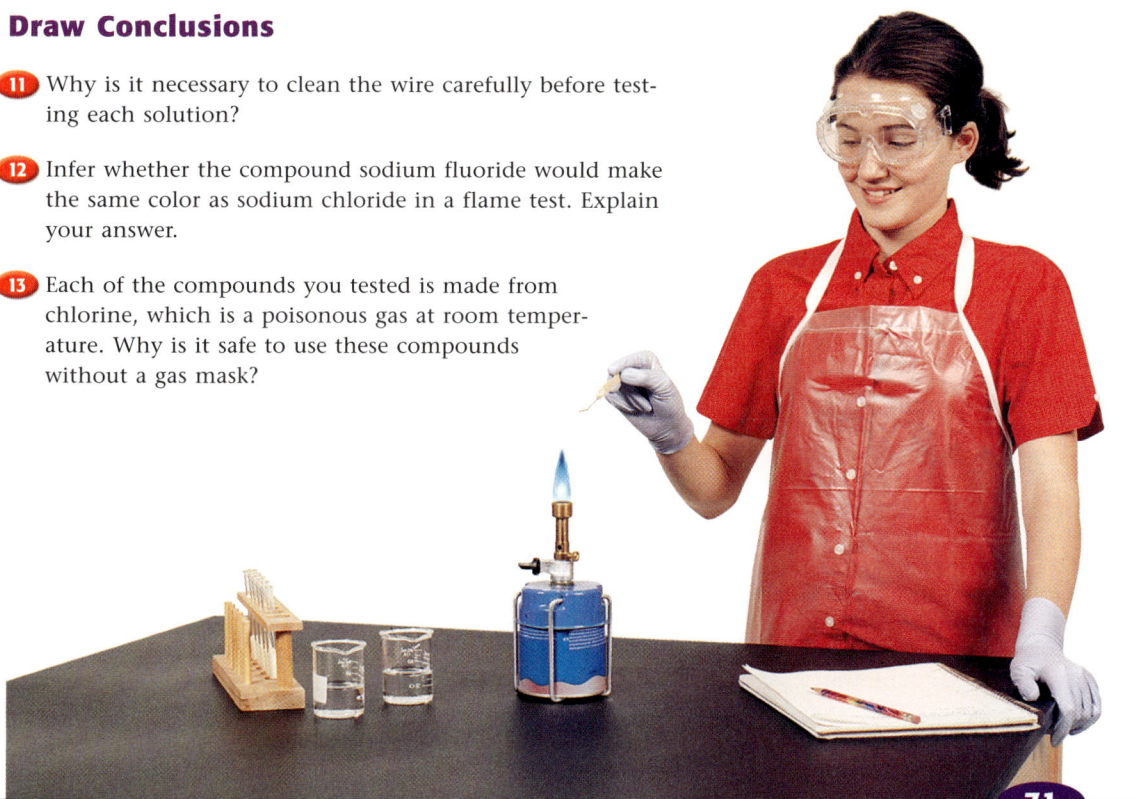

Answers

9. The flame test is a test for the metal in each compound. Because each compound contains chloride, the color difference must be due to the different metals. Any color contribution from the chloride would be the same in each trial.

10. Answers will depend on the teacher's choice for the unknown compound. Students will know its identity because it will produce the same color flame as one of the other three solutions.

11. The wire must be cleaned so the color observed is from the solution being tested, not from a mixture of two solutions.

12. Yes; the sodium fluoride compound would likely burn the same color as the sodium chloride compound because the flame test is a test for the metal in a compound and both compounds contain sodium.

13. Compounds have chemical and physical properties that are different from those of the elements the compounds are formed from.

Science Skills Worksheet "Using Logic"

71

Disposal Information

Hydrochloric acid: Titrate with 0.1 M NaOH as required until the pH is between 6 and 8, and pour the liquid down the drain.

Calcium chloride solution: Adjust the pH of the waste liquid with 1.0 M acid or base until the pH is between 5 and 9. Pour the neutralized liquid down the drain.

Potassium chloride and sodium chloride solutions: These can be washed down the sink with plenty of water, provided your school drains are connected to a sanitary sewer system with a treatment plant.

CLASSROOM TESTED & APPROVED

Kenneth J. Horn Fallston Middle School Fallston, Maryland

Chapter Highlights

Chapter Highlights

VOCABULARY DEFINITIONS

SECTION 1

element a pure substance that cannot be separated or broken down into simpler substances by physical or chemical means

pure substance a substance in which there is only one type of particle; includes elements and compounds

metals elements that are shiny and are good conductors of thermal and electrical energy; most metals are malleable and ductile

nonmetals elements that are dull (not shiny) and that are poor conductors of thermal and electrical energy

metalloids elements that have properties of both metals and nonmetals; sometimes referred to as semiconductors

SECTION 2

compound a pure substance composed of two or more elements that are chemically combined

---

SECTION 1

Vocabulary
- **element** (p. 54)
- **pure substance** (p. 54)
- **metals** (p. 57)
- **nonmetals** (p. 57)
- **metalloids** (p. 57)

Section Notes
- A substance in which all the particles are alike is a pure substance.
- An element is a pure substance that cannot be broken down into anything simpler by physical or chemical means.
- Each element has a unique set of physical and chemical properties.
- Elements are classified as metals, nonmetals, or metalloids, based on their properties.

SECTION 2

Vocabulary
- **compound** (p. 58)

Section Notes
- A compound is a pure substance composed of two or more elements chemically combined.
- Each compound has a unique set of physical and chemical properties that are different from the properties of the elements that compose it.
- The elements that form a compound always combine in a specific ratio according to their masses.
- Compounds can be broken down into simpler substances by chemical changes.

✓ Skills Check

Math Concepts

CONCENTRATION The concentration of a solution is a measure of the amount of solute dissolved in a solvent. For example, a solution is formed by dissolving 85 g of sodium nitrate in 170 mL of water. The concentration of the solution is calculated as follows:

$$\frac{85 \text{ g sodium nitrate}}{170 \text{ mL water}} = 0.5 \text{ g/mL}$$

Visual Understanding

THREE CATEGORIES OF ELEMENTS Elements are classified as metals, nonmetals, or metalloids, based on their properties. The chart on page 57 provides a summary of the properties that distinguish each category.

SEPARATING MIXTURES Mixtures can be separated through physical changes based on differences in the physical properties of their components. Review the illustrations on page 63 for some techniques for separating mixtures.

Lab and Activity Highlights

Flame Tests PG 70

A Sugar Cube Race! PG 140

Making Butter PG 141

Unpolluting Water PG 142

Datasheets for LabBook (blackline masters for these labs)

SECTION 3

Vocabulary

mixture *(p. 62)*

solution *(p. 64)*

solute *(p. 64)*

solvent *(p. 64)*

concentration *(p. 66)*

solubility *(p. 66)*

suspension *(p. 68)*

colloid *(p. 69)*

Section Notes

• A mixture is a combination of two or more substances, each of which keeps its own characteristics.

• Mixtures can be separated by physical means, such as filtration and evaporation.

• The components of a mixture can be mixed in any proportion.

• A solution is a mixture that appears to be a single substance but is composed of a solute dissolved in a solvent. Solutions do not settle, cannot be filtered, and do not scatter light.

• Concentration is a measure of the amount of solute dissolved in a solvent.

• The solubility of a solute is the amount of solute needed to make a saturated solution using a given amount of solvent at a certain temperature.

• Suspensions are heterogeneous mixtures that contain particles large enough to settle out, be filtered, and block or scatter light.

• Colloids are mixtures that contain particles too small to settle out or be filtered but large enough to scatter light.

Labs

A Sugar Cube Race! *(p. 140)*

Making Butter *(p. 141)*

Unpolluting Water *(p. 142)*

SECTION 3

mixture a combination of two or more substances that are not chemically combined

solution a mixture that appears to be a single substance but is composed of particles of two or more substances that are distributed evenly amongst each other

solute the substance that is dissolved to form a solution

solvent the substance in which a solute is dissolved to form a solution

concentration a measure of the amount of solute dissolved in a solvent

solubility the ability to dissolve in another substance; more specifically, the amount of solute needed to make a saturated solution using a given amount of solvent at a certain temperature

suspension a mixture in which particles of a material are dispersed throughout a liquid or gas but are large enough that they settle out

colloid a mixture in which the particles are dispersed throughout but are not heavy enough to settle out

Vocabulary Review Worksheet

Blackline masters of these Chapter Highlights can be found in the **Study Guide**.

internet connect

GO TO: go.hrw.com

Visit the **HRW** Web site for a variety of learning tools related to this chapter. Just type in the keyword:

KEYWORD: HSTMIX

SCILINKS™

NSTA

GO TO: www.scilinks.org

Visit the **National Science Teachers Association** on-line Web site for Internet resources related to this chapter. Just type in the *sci*LINKS number for more information about the topic:

TOPIC: The *Titanic* — *sci*LINKS NUMBER: HSTP080

TOPIC: Elements — *sci*LINKS NUMBER: HSTP085

TOPIC: Compounds — *sci*LINKS NUMBER: HSTP090

TOPIC: Mixtures — *sci*LINKS NUMBER: HSTP095

73

Lab and Activity Highlights

LabBank

Labs You Can Eat, An Iron-ic Cereal Experience

EcoLabs & Field Activities, Ozone News Zone

Whiz-Bang Demonstrations, Dense Suspense

Inquiry Labs, Separation Anxiety

Long-Term Projects & Research Ideas, A Coin-cidence

Interactive Explorations CD-ROM

CD 1, Exploration 4, "What's the Matter?"

Chapter Review
Answers

USING VOCABULARY

1. compound
2. solubility
3. suspension
4. element; compound
5. nonmetals
6. solute

UNDERSTANDING CONCEPTS

Multiple Choice

7. c	11. c
8. b	12. a
9. b	13. a
10. b	14. c

Short Answer

15. Elements cannot be separated into simpler substances, but compounds can be separated by chemical means.
16. Nail polish is the solute. Acetone is the solvent.

Concept Mapping

17. An answer to this exercise can be found at the front of this book.

CRITICAL THINKING AND PROBLEM SOLVING

18. Sample answer: Pass the mixture through a screen that allows the salt and pepper to pass through but traps the pebbles. Mix the salt and pepper with water to dissolve the salt. Filter the mixture to trap the pepper. Evaporate the water to recover the salt.
19. The powder is a compound. The change in color and the formation of a gas imply that a chemical change took place. Compounds can be broken down by chemical changes.

Concept Mapping Transparency 4

Chapter Review

USING VOCABULARY

Complete the following sentences by choosing the appropriate term from the vocabulary list to fill in each blank:

1. A __?__ has a definite ratio of components.

2. The amount of solute needed to form a saturated solution is the __?__ of the solute.

3. A __?__ can be separated by filtration.

4. A pure substance must be either a(n) __?__ or a(n) __?__.

5. Elements that are brittle and dull are __?__.

6. The substance that dissolves to form a solution is the __?__.

UNDERSTANDING CONCEPTS

Multiple Choice

7. Which of the following increases the solubility of a gas in a liquid?
   a. increasing the temperature
   b. stirring
   c. decreasing the temperature
   d. decreasing the amount of liquid

8. Which of the following best describes chicken noodle soup?
   a. element   c. compound
   b. mixture   d. solution

9. Which of the following does not describe elements?
   a. all the particles are alike
   b. can be broken down into simpler substances
   c. have unique sets of properties
   d. can join together to form compounds

10. A solution that contains a large amount of solute is best described as
    a. unsaturated.   c. dilute.
    b. concentrated.   d. weak.

11. Which of the following substances can be separated into simpler substances only by chemical means?
    a. sodium   c. water
    b. salt water   d. gold

12. Which of the following would not increase the rate at which a solid dissolves?
    a. decreasing the temperature
    b. crushing the solid
    c. stirring
    d. increasing the temperature

13. An element that conducts thermal energy well and is easily shaped is a
    a. metal.
    b. metalloid.
    c. nonmetal.
    d. None of the above

14. In which classification of matter are the components chemically combined?
    a. alloy   c. compound
    b. colloid   d. suspension

Short Answer

15. What is the difference between an element and a compound?

16. When nail polish is dissolved in acetone, which substance is the solute and which is the solvent?

20. The exact concentration tells you exactly how much solute is dissolved in the solvent. *Concentrated* and *dilute* are descriptive terms that do not tell you the amount of solute.
21. Each piece of fruit in a fruit salad keeps the appearance and flavor of that fruit, demonstrating that components of a mixture keep their original properties. You can separate each type of fruit using a fork, which is a physical process, demonstrating that a mixture can be separated by physical means. The amount of each fruit used in a salad can be changed, demonstrating that a mixture can be formed from any ratio of components.

Concept Mapping

17. Use the following terms to create a concept map: matter, element, compound, mixture, solution, suspension, colloid.

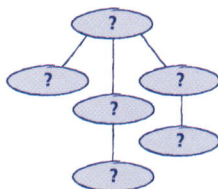

CRITICAL THINKING AND PROBLEM SOLVING

18. Describe a procedure to separate a mixture of salt, finely ground pepper, and pebbles.

19. A light green powder is heated in a test tube. A gas is given off, while the solid becomes black. In which classification of matter does the green powder belong? Explain your reasoning.

20. Why is it desirable to know the exact concentration of solutions rather than whether they are concentrated or dilute?

21. Explain the three properties of mixtures using a fruit salad as an example.

22. To keep the "fizz" in carbonated beverages after they have been opened, should you store them in a refrigerator or in a cabinet? Explain.

MATH IN SCIENCE

23. What is the concentration of a solution prepared by mixing 50 g of salt with 200 mL of water?

24. How many grams of sugar must be dissolved in 150 mL of water to make a solution with a concentration of 0.6 g/mL?

INTERPRETING GRAPHICS

25. Use Figure 14 on page 67 to answer the following questions:
    a. Can 50 g of sodium chloride dissolve in 100 mL of water at 60°C?
    b. How much cerium sulfate is needed to make a saturated solution in 100 mL of water at 30°C?
    c. Is sodium chloride or sodium nitrate more soluble in water at 20°C?

26. Dr. Sol Vent tested the solubility of a compound. The data below was collected using 100 mL of water. Graph Dr. Vent's results. To increase the solubility, would you increase or decrease the temperature? Explain.

Temperature (°C)	10	25	40	60	95
Dissolved solute (g)	150	70	34	25	15

27. What type of mixture is shown in the photo below? Explain.

Reading Check-up Take a minute to review your answers to the Pre-Reading Questions found at the bottom of page 52. Have your answers changed? If necessary, revise your answers based on what you have learned since you began this chapter.

---

22. Carbonated beverages should be stored in a refrigerator. Gases are more soluble at lower temperatures, so more gas will stay dissolved in the beverage if it is kept cold.

MATH IN SCIENCE

23. 50 g/200 mL = 0.25 g/mL
24. 150 mL × 0.6 g/mL = 90 g

INTERPRETING GRAPHICS

25. a. no
    b. approximately 8 g
    c. sodium nitrate
26. (Teacher Note: The graph should have dissolved solute on the *y*-axis and temperature on the *x*-axis. The curve will decrease from left to right.) You should decrease the temperature to increase the solubility. As the temperature decreases, more solute can dissolve.
27. The mixture shown is a suspension. The mixture appears cloudy, which means the particles are large. A layer of the material is forming at the bottom of the jar. This shows that the mixture is settling out, which is a property of suspensions.

Blackline masters of this Chapter Review can be found in the **Study Guide.**

Background

A perfume is any substance used as a pleasant fragrance. Many cosmetics contain perfumes. Low-priced perfumes called *oderants* are added to many products, including paper, plastics, and rubber products, to hide unpleasant odors or to make the products more attractive to consumers.

Discussion

Perfume scents are often grouped according to their dominant odor. The major groups are floral, spicy, woody, and herbal. Using some of the scents listed below, have a class "smelling test" in which the students smell each scent and guess which category the scent belongs in. Discuss the results. Common scents you can use include the following:

- floral: jasmine, lily of the valley, rose, gardenia
- spicy: clove, cinnamon, nutmeg
- woody: sandalwood, cedar
- herbal: clover, rosemary
- other: vanilla, balsam, patchouli

Note: After students smell two or three scents, their olfactory functions may become desensitized. To restore their ability, allow students to sniff coffee beans after each scent.

Science, Technology, and Society

Perfume: Fragrant Solutions

Making perfume is an ancient art. It was practiced, for example, by the ancient Egyptians, who rubbed their bodies with a substance made by soaking fragrant woods and resins in water and oil. From certain references and formulas in the Bible, we know that the ancient Israelites also practiced the art of perfume making. Other sources indicate that this art was also known to the early Chinese, Arabs, Greeks, and Romans.

▲ *Perfumes have been found in the tombs of Egyptians who lived more than 3,000 years ago.*

Only the E-scent-ials

Over time, perfume making has developed into a complicated art. A fine perfume may contain more than 100 different ingredients. The most familiar ingredients come from fragrant plants or flowers, such as sandalwood or roses. These plants get their pleasant odor from their essential oils, which are stored in tiny, baglike parts called sacs. The parts of plants that are used for perfumes include the flowers, roots, and leaves. Other perfume ingredients come from animals and from man-made chemicals.

Making Scents

Perfume makers first remove essential oils from the plants using distillation or reactions with solvents. Then the essential oils are blended with other ingredients to create perfumes. Fixatives, which usually come from animals, make the other odors in the perfume last longer. Oddly enough, most natural fixatives smell awful! For example, civet musk is a foul-smelling liquid that the civet cat sprays on its enemies.

Taking Notes

When you take a whiff from a bottle of perfume, the first odor you detect is called the top note. It is a very fragrant odor that evaporates rather quickly. The middle note, or modifier, adds a different character to the odor of the top note. The base note, or end note, is the odor that lasts the longest.

▲ *Not all perfume ingredients smell good. The foul-smelling oil from the African civet cat is used as a fixative in some perfumes.*

Smell for Yourself
► Test a number of different perfumes and colognes to see if you can identify three different notes in each.

76

Answer to Smell for Yourself

You may wish to use the same scents mentioned above for this activity. Alternatively, you might bring in some popular colognes for students to test. These scents will most likely be more complex because they may include as many as 100 different ingredients.

Did You Know . . .

- Other fixatives include castor from the beaver, musk from the male musk deer, and ambergris from the sperm whale.
- The word *perfume* comes from the Latin phrase *per fumum*, which means "through smoke."

Science Fiction

"The Strange Case of Dr. Jekyll and Mr. Hyde"

by Robert Louis Stevenson

A vicious, detestable man murders an old gentleman. A wealthy and respectable scientist commits suicide. Are these two tragedies connected in some way?

Dr. Henry Jekyll is an admirable member of society. He is a doctor and a scientist. Although wild as a young man, Jekyll has become cold and analytical as he has aged and has pursued his scientific theories. Now he wants to understand the nature of human identity. He wants to explore the different parts of the human personality that usually fit together smoothly to make a complete person. His theory is that if he can separate his personality into "good" and "evil" parts, he can get rid of his evil side and lead a happy, useful life. So Jekyll develops a chemical mixture that will allow him to test his theory. The results are startling!

Who is the mysterious Mr. Hyde? He is not a scientist. He is a man of action and anger, who sparks fear in the hearts of those he comes in contact with. Where did he come from? What does he do? How can local residents be protected from his wrath?

Robert Louis Stevenson's story of the decent doctor Henry Jekyll and the violent Edward Hyde is a classic science-fiction story. When Jekyll mixes his "salts" and drinks his chemical mixture, he changes his life—and Edward Hyde's—completely. To find out more, read Stevenson's "The Strange Case of Dr. Jekyll and Mr. Hyde" in the *Holt Anthology of Science Fiction.*

77

SCIENCE FICTION

"The Strange Case of Dr. Jekyll and Mr. Hyde"
by Robert Louis Stevenson

When Dr. Jekyll mixes his "salts" and drinks his chemical mixture, he changes his life—and Edward Hyde's—completely.

Teaching Strategy

Reading Level This novelette is challenging for middle school readers. It may be helpful for you to review the story in advance and review some of the unfamiliar words and phrases with students ahead of time.

Background

The science connection in this story is that Dr. Jekyll is a "man of science" who conducts experiments with chemical compounds to explore his own personality. He wants to separate out the purely good part of himself from the brutish and evil. Jekyll experiments until he finds a mixture of "salts" that creates the split. But once he has released Mr. Hyde, it becomes increasingly difficult for Jekyll to control him!

About the Author Stevenson was a sickly child and was plagued by health problems all his life. Some critics argue that Hyde represents the anger and frustration that Stevenson felt as a result of his ill health. Other critics compare the story to Mary Shelley's *Frankenstein* because both deal with the idea that humans can be perfected. Still others see a reflection of the conflict between good and evil, which people have talked about and written about for more than 4,000 years. All three of these themes are in the story, and any one of them would be a good starting point for a class discussion.

Chapter Organizer

CHAPTER ORGANIZATION	TIME MINUTES	OBJECTIVES	LABS, INVESTIGATIONS, AND DEMONSTRATIONS
Chapter Opener pp. 78–79	45	National Standards: UCP 2, SAI 1, HNS 2	**Start-Up Activity,** Where Is It? p. 79
Section 1 Development of the Atomic Theory	135	▶ Describe some of the experiments that led to the current atomic theory. ▶ Compare the different models of the atom. ▶ Explain how the atomic theory has changed as scientists have discovered new information about the atom. UCP 1, 2, SAI 2, ST 2, SPSP 5, HNS 1–3	**Whiz-Bang Demonstrations,** As a Matter of Space
Section 2 The Atom	135	▶ Compare the charge, location, and relative mass of protons, neutrons, and electrons. ▶ Calculate the number of particles in an atom using the atomic number, mass number, and overall charge. ▶ Calculate the atomic mass of elements. UCP 1–3, HNS 2; Labs UCP 1, 2	**Making Models,** Made to Order, p. 94 **Datasheets for LabBook,** Made to Order **Whiz-Bang Demonstrations,** Candy Lights **Long-Term Projects & Research Ideas,** How Low Can They Go?

See page **T23** *for a complete correlation of this book with the*

NATIONAL SCIENCE EDUCATION STANDARDS.

TECHNOLOGY RESOURCES

Guided Reading Audio CD
English or Spanish, Chapter 4

One-Stop Planner CD-ROM with Test Generator

CNN Scientists in Action, Nobel Prize Physicists, Segment 14

CLASSROOM WORKSHEETS, TRANSPARENCIES, AND RESOURCES	SCIENCE INTEGRATION AND CONNECTIONS	REVIEW AND ASSESSMENT
Directed Reading Worksheet **Science Puzzlers, Twisters & Teasers**		
Directed Reading Worksheet, Section 1 **Transparency 244,** Thomson's Cathode-Ray Tube Experiment **Transparency 107,** Gold Crystal Structure **Transparency 245,** Rutherford's Gold Foil Experiment **Math Skills for Science Worksheet,** Using Proportions and Cross-Multiplication	**Cross-Disciplinary Focus,** p. 83 in ATE **Math and More,** p. 84 in ATE **Connect to Earth Science,** p. 84 in ATE **Math and More,** p. 85 in ATE **Real-World Connection,** p. 85 in ATE **Careers:** Experimental Physicist—Melissa Franklin, p. 101	**Section Review,** p. 83 **Self-Check,** p. 85 **Homework,** p. 85 in ATE **Section Review,** p. 86 **Quiz,** p. 86 in ATE **Alternative Assessment,** p. 86 in ATE
Directed Reading Worksheet, Section 2 **Transparency 246,** Parts of an Atom **Math Skills for Science Worksheet,** Arithmetic with Decimals **Transparency 247,** Forces in the Atom **Reinforcement Worksheet,** Atomic Timeline **Critical Thinking Worksheet,** Incredible Shrinking Scientist!	**Real-World Connection,** p. 88 in ATE **Astronomy Connection,** p. 90 **Connect to Paleontology,** p. 90 in ATE **Apply,** p. 91 **Real-World Connection,** p. 91 in ATE **MathBreak,** Atomic Mass, p. 92 **Math and More,** p. 92 in ATE **Across the Sciences:** Water on the Moon? p. 100	**Section Review,** p. 89 **Section Review,** p. 93 **Quiz,** p. 93 in ATE **Alternative Assessment,** p. 93 in ATE

internet connect

go.hrw.com

Holt, Rinehart and Winston On-line Resources

go.hrw.com

For worksheets and other teaching aids related to this chapter, visit the HRW Web site and type in the keyword: **HSTATS**

SCI LINKS NSTA

National Science Teachers Association

www.scilinks.org

Encourage students to use the *sci*LINKS numbers listed in the internet connect boxes to access information and resources on the **NSTA** Web site.

END-OF-CHAPTER REVIEW AND ASSESSMENT

Chapter Review in Study Guide
Vocabulary and Notes in Study Guide
Chapter Tests with Performance-Based Assessment, Chapter 4 Test
Chapter Tests with Performance-Based Assessment, Performance-Based Assessment 4
Concept Mapping Transparency 11

Chapter Resources & Worksheets

Visual Resources

TEACHING TRANSPARENCIES

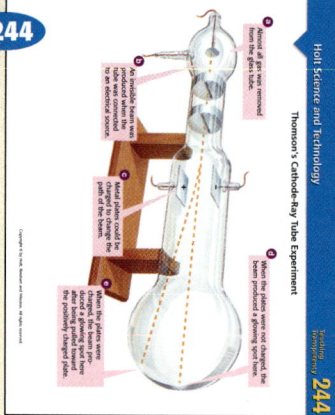

#244 Holt Science and Technology — Teaching Transparency 244 — Thomson's Cathode-Ray Tube Experiment

#245 Holt Science and Technology — Teaching Transparency 245 — Rutherford's Gold Foil Experiment

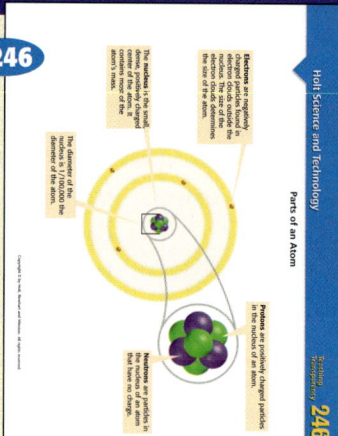

#246 Holt Science and Technology — Teaching Transparency 246 — Parts of an Atom

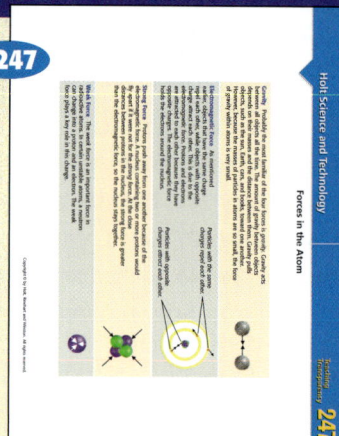

#247 Holt Science and Technology — Teaching Transparency 247 — Forces in the Atom

TEACHING TRANSPARENCIES

#107 Holt Science and Technology — Gold Crystal Structure 107

The atomic structure of gold

The crystal structure of gold

Crystals of the mineral gold

LINK TO EARTH SCIENCE

CONCEPT MAPPING TRANSPARENCY

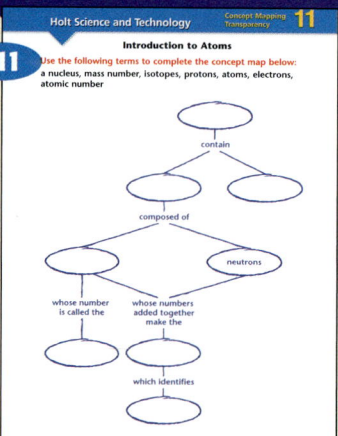

Holt Science and Technology — Concept Mapping Transparency 11

Introduction to Atoms

#11 Use the following terms to complete the concept map below: a nucleus, mass number, isotopes, protons, atoms, electrons, atomic number

contain

composed of

neutrons

whose number is called the

whose numbers added together make the

which identifies

Meeting Individual Needs

DIRECTED READING

#4 DIRECTED READING WORKSHEET

Introduction to Atoms

Chapter Introduction

As you begin this chapter, answer the following.
1. Read the title of the chapter. List three things that you already know about this subject.

2. Write two questions about this subject that you would like answered by the time you finish this chapter.

3. How does the title of the Start-Up Activity relate to the subject of the chapter?

Start-Up Activity (p. 000)

4. How do you think rolling marbles in this activity will help you identify the mystery object?

REINFORCEMENT & VOCABULARY REVIEW

#4 REINFORCEMENT WORKSHEET

Atomic Timeline

Complete this worksheet after you have finished reading Chapter X, Section 2. The table below contains a number of statements and scientists connected to major discoveries in the development of atomic theory. On a separate sheet of paper, construct a timeline, and label the following dates: 440 B.C., 1803, 1897, 1911, 1913, and twentieth century. Cut the table below along the dotted lines, and tape or glue each box of information at the correct point along the timeline.

There are small, negatively charged particles inside an atom. 1897 (Thomson)	Electron paths are unpredictable. twentieth century (Schrödinger and Heisenberg)
There is a small, dense, positively charged nucleus. 1911 (Rutherford)	Electrons travel in definite paths. 1913 (Bohr)
Electrons orbit the nucleus. 1911 (Rutherford)	Electrons move in empty space in the atom. 1911 (Rutherford)
Electrons jump between levels from path to path. 1913 (Bohr)	His theory of atomic structure became the "plum-pudding" model. 1897 (Thomson)
He conducted the cathode-ray tube experiment. 1897 (Thomson)	Electrons are found in electron clouds, not paths. twentieth century (Schrödinger and Heisenberg)
Atoms of different elements are different. 1803 (Dalton)	Atoms of the same element are exactly alike. 1803 (Dalton)
Atoms contain mostly empty space. 1911 (Rutherford)	Atoms constantly move. 1803 (Dalton)
Atoms are small, hard particles. 440 B.C. (Democritus)	All substances are made of atoms. 1803 (Dalton)
He conducted experiments in combining elements. 1803 (Dalton)	He conducted the gold foil experiment. 1911 (Rutherford)
Atoms are "uncuttable." 440 B.C. (Democritus)	Atoms form different materials by joining together. 1803 (Dalton)
Democritus 440 B.C.	Rutherford 1911
Thomson 1897	Dalton 1803
Bohr 1913	Schrödinger and Heisenberg twentieth century

#4 VOCABULARY REVIEW WORKSHEET

Atomic Anagrams

Try this anagram after you finished Chapter X. Use the definitions below to unscramble the vocabulary words.

1. weighted average of isotopes — SAMS MICTOA
2. smallest particle into which an element can be divided and still be the same substance — MOAT
3. scientific explanation — RYTHOE
4. type of particle — TORPNO
5. made up of protons and neutrons — UCSELUN
6. type of particle — TRONUNE
7. atoms with the same number of protons but different number of neutrons — SOOTPIES
8. type of particle — CLEENBOT
9. number of protons in a nucleus — BRUMEN MICOTA
10. representation of an object or system — OLDEM
11. regions where electrons are likely to be found — SCUDLO
12. SI unit used to measure the mass of particles — MUA
13. sum of protons and neutrons — SAMS BRUNEM

SCIENCE PUZZLERS, TWISTERS & TEASERS

#4 SCIENCE PUZZLERS, TWISTERS & TEASERS

Introduction to Atoms

Mystery Guests

1. The three particles of an atom appeared recently on a talk show, and they stood behind a screen to hide their identities. (Not that you could see them anyway.) Identify their statements below, based on what you know about their characteristics.

   a. I don't mean to be negative all the time, but, well, I'm always on the go.

   b. Me? I stay positive. It's the only way I know how to be.

   c. I have almost no mass—no weight to throw around. And just once I'd like to be at the center of things.

   d. I stay neutral on most nuclear issues.

   e. When we (other particles just like me) outnumber the electrons, the whole atom has a more positive energy.

Sound Alikes

2. Each clue below will lead you to one or two short words. Combine the syllables to find the hidden terms, which are used in the study of atoms.

   a. Frozen water / A breakfast grain / A famous comedian named Bob; A TV hospital show; Chicago

   b. A long skinny fish; one is electric / When the teacher speaks in class / The opposite of "offn"

Chapter 4 • Introduction to Atoms

Review & Assessment

STUDY GUIDE

#4 VOCABULARY & NOTES WORKSHEET
Introduction to Atoms

#4 CHAPTER REVIEW WORKSHEET
Introduction to Atoms

CHAPTER TESTS WITH PERFORMANCE-BASED ASSESSMENT

#4 INTRODUCTION TO ATOMS
Chapter 4 Test

#4 INTRODUCTION TO ATOMS — SKILL BUILDER
Chapter 4 Performance-Based Assessment

Lab Worksheets

WHIZ-BANG DEMONSTRATIONS

#4 TEACHER-LED DEMONSTRATION — DISCOVERY LAB
Candy Lights

LONG-TERM PROJECTS & RESEARCH IDEAS

#4 STUDENT WORKSHEET — DESIGN YOUR OWN
How Low Can They Go?

DATASHEETS FOR LABBOOK

#4 Made to Order

Applications & Extensions

CRITICAL THINKING & PROBLEM SOLVING

#4 CRITICAL THINKING WORKSHEET
Incredible Shrinking Scientist!

SCIENTISTS IN ACTION

#14 Science in the News: Critical Thinking Worksheets
Segment 14
Nobel Prize Physicists

Chapter Background

SECTION 1

Development of the Atomic Theory

▶ Democritus

Democritus (c. 460–c. 370 B.C.) was a Greek philosopher and leading advocate of the theory that all phenomena in nature could be understood in terms of the movements of particles called atoms (from the Greek word *atomos*, meaning "indivisible").

- The views of Democritus sharply contrasted those of Aristotle and others, who held to the theory that all matter could be reduced to a combination of four elements: earth, water, air, and fire.

IS THAT A FACT!

- ◀ Democritus's ideas were not widely accepted because Aristotle, who was better known and respected, did not accept the idea of atoms. Only fragments of Democritus's writings survive, and most of our knowledge of his ideas comes from negative remarks about his theories in other people's writings.

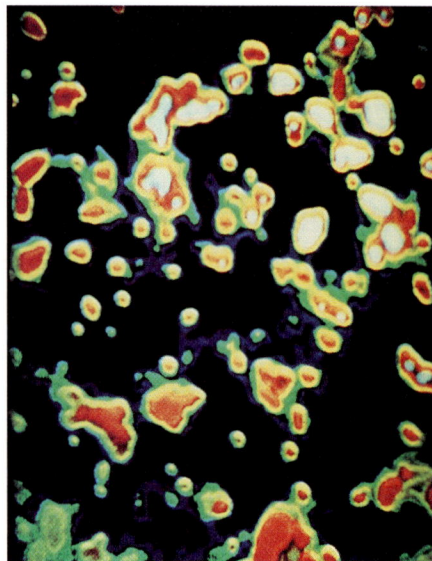

▶ From Greek to Modern Atomic Theory

Democritus and other Greek philosophers laid the groundwork for the modern atomic theory, but it was not until the sixteenth and seventeenth centuries that interest in atoms and atomic structure was renewed. During that time, the work of Sir Isaac Newton, Robert Boyle, and Pierre Gassendi helped to further the development of the atomic theory.

- In the nineteenth century, experiments by John Dalton, Amedeo Avogadro, James Clerk Maxwell, and Rudolf Clausius began to reveal the nature and structure of atoms.

- Sir Joseph John Thomson's discovery of electrons in 1897 and his later research on protons and gases indicated that atoms were not the smallest indivisible units of matter, as previously thought. He showed that subatomic particles with either a negative or positive charge form at least part of the structure of an atom. Thomson won the Nobel Prize in Physics in 1906.

- While Thomson was director of the Cavendish Laboratory, at Cambridge University, in Cambridge, England, one of his graduate students was Ernest Rutherford. Rutherford went on to win the Nobel Prize in Chemistry in 1908 for his work on radioactivity.

SECTION 2

The Atom

▶ **How Small Are They?**

Determining the diameter of an atom is difficult because atoms are not small, hard spheres. Measurement often varies, depending on the method used. On average, the diameter of an atom ranges from about 7×10^{-9} cm to 5×10^{-8} cm.

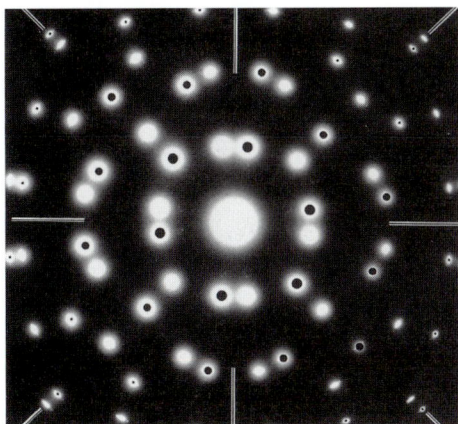

▶ **Quarks and Gluons: the Smallest . . . So Far**

Protons and neutrons are composed of smaller particles called *quarks*. The existence of quarks was suggested in 1963 by two physicists, Murray Gell-Mann and George Zweig.

- Hadrons are all particles that feel the strong force. They include baryons, quark triplets; antibaryons; and mesons, quark-antiquark pairs. Baryons include protons and neutrons.

- A gluon is believed to be a subatomic particle that "glues" quarks together with the strong nuclear force.

IS THAT A FACT!

- ◆ When gluons bind to each other, they are referred to as "glueballs."

- ◆ The term *quark* originated from a line in James Joyce's novel *Finnegans Wake*: "Three quarks for Muster Mark."

▶ **Isotopes**

Two major types of isotopes exist: stable and unstable. Stable isotopes persist in nature, whereas unstable isotopes undergo radioactive decay toward a more stable state.

- There are approximately 280 stable isotopes of the natural elements. A natural element is usually predominantly one stable isotope with smaller amounts of other stable and unstable isotopes.

- Radioactive isotopes can be natural or artificial. The naturally occurring radioisotopes have existed since Earth's formation.

- The first artificial radioisotopes were produced in 1934 by Frederic and Irene Joliot-Curie. Since then, more than 1,800 artificial radioisotopes have been produced using a variety of nuclear bombardment techniques.

For background information about teaching strategies and issues, refer to the *Professional Reference for Teachers.*

CHAPTER
4

Introduction to Atoms

Sections

Pre-Reading Questions

Students may not know the answers to these questions before reading the chapter, so accept any reasonable response.

Suggested Answers

1. Scientists have described the atom in several ways: as a solid sphere; as a plum pudding; and as a dense, central nucleus with electrons moving around it. The later models have electrons moving in paths at certain distances around the nucleus or have electron clouds around the nucleus, where electrons are most likely found.

2. Atoms are composed of protons and neutrons, which make up the nucleus in the center of the atom, and electrons, which are most likely found in regions around the nucleus called electron clouds.

3. All atoms have a small, dense, positively charged nucleus with electrons in electron clouds around the nucleus. Atoms are neutral because the number of protons equals the number of electrons.

Pre-Reading Questions

1. What are some ways that scientists have described the atom?

2. What are the parts of the atom, and how are they arranged?

3. How are atoms of all elements alike?

78

ATOMIC BUBBLES

You probably have made bubbles with a plastic wand and a soapy liquid. To trace the paths of atoms, some scientists also made bubbles, but they did not use a wand. They used a bubble chamber. A bubble chamber is filled with a hot, pressurized liquid that forms bubbles when a charged particle moves through it. Why are scientists interested in bubbles? The bubbles give them information about particles called atoms that make up all objects. In this chapter, you will learn about atoms and experiments that led to the modern atomic theory. You will also learn about the parts and structure of an atom.

---

internet connect

go.hrw.com

HRW On-line Resources

go.hrw.com

For worksheets and other teaching aids, visit the HRW Web site and type in the keyword: **HSTATS**

SC/LINKS NSTA

www.scilinks.com

Use the *sci*LINKS numbers at the end of each chapter for additional resources on the **NSTA** Web site.

Smithsonian Institution

www.si.edu/hrw

Visit the Smithsonian Institution Web site for related on-line resources.

CNNfyi.com

www.cnnfyi.com

Visit the CNN Web site for current events coverage and classroom resources.

WHERE IS IT?

Some theories about the internal structure of atoms were formed by observing the effects of aiming very small moving particles at atoms. In this activity, you will form an idea about the location and size of a hidden object by rolling marbles at it.

Procedure

1. Place a **rectangular piece of cardboard** on **four books or blocks** so that each corner of the cardboard rests on a book or block.

2. Ask your teacher to place the **unknown object** under the cardboard. Be sure that you do not see the object.

3. Place a **large piece of paper** on top of the cardboard.

4. Carefully roll a **marble** under the cardboard. Record on the paper the position where the marble enters and exits. Also record the direction it travels.

5. Keep rolling the marble from different directions to collect data about the shape and location of the object.

6. Write down all your observations in your ScienceLog.

Analysis

7. Form a conclusion about the object's shape, size, and location. Record your conclusion in your ScienceLog.

79

START-UP
Activity

WHERE IS IT?

MATERIALS
For Each Group:
• rectangular piece of cardboard
• unknown object
• piece of plain paper
• books or blocks
• marble
• pencil

Teacher's Notes

The size of the cardboard should be large enough to prevent students from seeing the hidden object after you place it under the cardboard. Students will need to cover the cardboard completely with paper in order to mark the information they need to gather. Small pieces of wood cut into simple geometric shapes would work well as the objects used by each group.

Remind students to roll the marble gently as they try to establish where and what the object is.

Answer to START-UP Activity

7. Accept all reasonable answers.

Focus

Development of the Atomic Theory

Students learn the early history of the atomic theory. Students also trace changes in atomic theory as scientists have discovered more about atomic structure.

🔔 Bellringer

Display the following quote by Democritus (c. 460–c. 370 B.C.). Have students write in their ScienceLog what they think the statement means. Do not divulge the source.

"Color exists by convention, sweet by convention, bitter by convention; in reality nothing exists but atoms and the void."

Discuss Democritus and his statement with students.

1 Motivate

ACTIVITY

Give students a newspaper and magnifying lens. Have them use the magnifier to examine photographs in the newspaper. Have them describe what they see. Students should notice that the pictures are made up of thousands of tiny dots of ink. Explain that many objects that appear to be whole are actually made up of many smaller parts. It was this idea that led the first philosophers and scientists to theorize that all matter is made up of tiny, indivisible parts. These tiny bits of matter became known as atoms. <mark>Sheltered English</mark>

Terms to Learn

atom	model
theory	nucleus
electrons	electron clouds

What You'll Do

- ◆ Describe some of the experiments that led to the current atomic theory.
- ◆ Compare the different models of the atom.
- ◆ Explain how the atomic theory has changed as scientists have discovered new information about the atom.

Development of the Atomic Theory

The photo at right shows uranium atoms magnified 3.5 million times by a scanning tunneling microscope. An **atom** is the smallest particle into which an element can be divided and still be the same substance. Atoms make up elements; elements combine to form compounds. Because all matter is made of elements or compounds, atoms are often called the building blocks of matter.

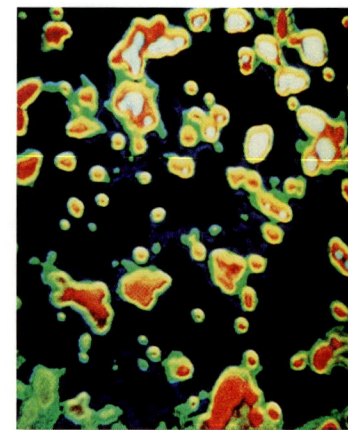

Before the scanning tunneling microscope was invented, in 1981, no one had ever seen an atom. But the existence of atoms is not a new idea. In fact, atomic theory has been around for more than 2,000 years. A **theory** is a unifying explanation for a broad range of hypotheses and observations that have been supported by testing. In this section, you will travel through history to see how our understanding of atoms has developed. Your first stop—ancient Greece.

Democritus Proposes the Atom

Imagine that you cut the silver coin shown in **Figure 1** in half, then cut those halves in half, and so on. Could you keep cutting the pieces in half forever? Around 440 B.C., a Greek philosopher named Democritus (di MAHK ruh tuhs) proposed that you would eventually end up with an "uncuttable" particle. He called this particle an *atom* (from the Greek word *atomos*, meaning "indivisible"). Democritus proposed that all atoms are small, hard particles made of a single material formed into different shapes and sizes. He also claimed that atoms are always moving and that they form different materials by joining together.

Figure 1 *This coin was in use during Democritus's time. Democritus thought the smallest particle in an object like this silver coin was an atom.*

80

**MISCONCEPTION /// ALERT **

The photo at the top of this page does not show individual uranium atoms but rather clumps of uranium atoms. The colors were added to the photograph to make the clumps easier to see.

IS THAT A FACT!

It would take 1.05×10^{17} gold atoms to cover the entire surface of a dollar bill. That's 105 quadrillion gold atoms!

Aristotle Disagrees Aristotle (ER is TAHT uhl), a Greek philosopher who lived from 384 to 322 B.C., disagreed with Democritus's ideas. He believed that you would never end up with an indivisible particle. Although Aristotle's ideas were eventually proved incorrect, he had such a strong influence on popular belief that Democritus's ideas were largely ignored for centuries.

Dalton Creates an Atomic Theory Based on Experiments

By the late 1700s, scientists had learned that elements combine in specific proportions based on mass to form compounds. For example, hydrogen and oxygen always combine in the same proportion to form water. John Dalton, a British chemist and school teacher, wanted to know why. He performed experiments with different substances. His results demonstrated that elements combine in specific proportions because they are made of individual atoms. Dalton, shown in **Figure 2,** published his own atomic theory in 1803. His theory stated the following:

- **All substances are made of atoms. Atoms are small particles that cannot be created, divided, or destroyed.**
- **Atoms of the same element are exactly alike, and atoms of different elements are different.**
- **Atoms join with other atoms to make new substances.**

Figure 2 John Dalton developed his atomic theory from observations gathered from many experiments.

Not Quite Correct Toward the end of the nineteenth century scientists agreed that Dalton's theory explained many of their observations. However, as new information was discovered that could not be explained by Dalton's ideas, the atomic theory was revised to more correctly describe the atom. As you read on, you will learn how Dalton's theory has changed, step by step, into the current atomic theory.

BRAIN FOOD

In 342 or 343 B.C., King Phillip II of Macedon appointed Aristotle to be a tutor for his son, Alexander. Alexander later conquered Greece and the Persian Empire (in what is now Iran) and became known as Alexander the Great.

IS THAT A FACT!

Along with contributing to the atomic theory, John Dalton was also the first to describe colorblindness. Dalton himself was colorblind. The paper that contains his article describing the condition was published in 1794.

READING STRATEGY

Making a Prediction Before students read this section, ask them to predict whether Democritus's theory about the atom is correct or even partly correct. Then ask them which parts of his theory might have been changed in the 2,500 years since it was first proposed. Have students write their predictions in their ScienceLog.

ACTIVITY

Dalton Discussion Discuss how Dalton's atomic theory explains his observations. Use diagrams of different compounds (water, carbon dioxide, table salt) to show how elements combine in proportions and how the same elements always combine in the same proportions to make the same compounds. Talk about the meaning of the terms *indestructible* and *indivisible* and how such properties affect chemical changes.

MEETING INDIVIDUAL NEEDS

Learners Having Difficulty Some students may benefit from making and using flashcards that connect the scientists profiled in this section with their accomplishments. Have students make their own sets of flashcards. One side of the card should feature the name of the scientist, and the other side should include a small illustration representing the model or experiment associated with the scientist.
Sheltered English

Directed Reading Worksheet Section 1

The terms *positive* and *negative* are arbitrary. Particles do not have little plus and minus signs on them. The terms were first used by Benjamin Franklin to describe phenomena he observed. They were quickly adopted by other eighteenth-century scientists. Today, the terms are the standard terms applied to a variety of particles and interactions.

ACTIVITY

Students should study **Figure 3**, then read the section on electric charge. Have them work in pairs to do this activity on static electricity.

MATERIALS

FOR EACH PAIR:
- round balloon 23 cm in diameter
- 3 small saucerlike dishes
- 15 g (1 tbsp) each of salt, sugar, and confetti paper

1. Inflate the balloon, and tie a knot at the end.
2. Stroke the balloon on clean, oil-free hair.
3. Hold the balloon over the dish of salt. Observe the result.
4. Clean off the balloon if necessary, repeat step 2, and hold the balloon over the sugar.
5. Repeat the same procedure for the confetti.

The balloon that was rubbed on the hair gained electrons. Ask students to explain their results.

Thomson Finds Electrons in the Atom

In 1897, a British scientist named J. J. Thomson made a discovery that identified an error in Dalton's theory. Using simple equipment (compared with modern equipment), Thomson discovered that there are small particles *inside* the atom. Therefore, atoms *can* be divided into even smaller parts.

Thomson experimented with a cathode-ray tube, as shown in **Figure 3.** He discovered that a positively charged plate (marked with a positive sign in the illustration) attracts the beam. Thomson concluded that the beam was made of particles with a negative electric charge.

Figure 3 Thomson's Cathode-Ray Tube Experiment

a Almost all gas was removed from the glass tube.

b An invisible beam was produced when the tube was connected to a source of electrical energy.

c Metal plates could be charged to change the path of the beam.

d When the plates were not charged, the beam produced a glowing spot here.

e When the plates were charged, the beam produced a glowing spot here after being pulled toward the positively charged plate.

Just What Is Electric Charge?

Have you ever rubbed a balloon on your hair? The properties of your hair and the balloon seem to change, making them attract one another. To describe these observations, scientists say that the balloon and your hair become "charged." There are two types of electric charges—positive and negative. Objects with opposite charges attract each other, while objects with the same charge push each other away.

82

Teaching Transparency 244 "Thomson's Cathode-Ray Tube Experiment"

IS THAT A FACT!

Thales of Miletus, the earliest known Greek philosopher and scientist, is said to have been the first to observe static electricity. He rubbed a piece of amber with a wool cloth and observed that lightweight objects were attracted to it.

Negative Corpuscles Thomson repeated his experiment several times and found that the particle beam behaved in exactly the same way each time. He called the particles in the beam corpuscles (KOR puhs uhls). His results led him to conclude that corpuscles are present in every type of atom and that all corpuscles are identical. The negatively charged particles found in all atoms are now called **electrons.**

Like Plums in a Pudding Thomson revised the atomic theory to account for the presence of electrons. Because Thomson knew that atoms have no overall charge, he realized that positive charges must be present to balance the negative charges of the electrons. But Thomson didn't know the location of the electrons or of the positive charges. So he proposed a model to describe a possible structure of the atom. A **model** is a representation of an object or system. A model is different from a theory in that a model presents a picture of what the theory explains.

Thomson's model, illustrated in **Figure 4,** came to be known as the plum-pudding model, named for an English dessert that was popular at the time. Today you might call Thomson's model the chocolate-chip-ice-cream model; electrons in the atom could be compared to the chocolate chips found throughout the ice cream!

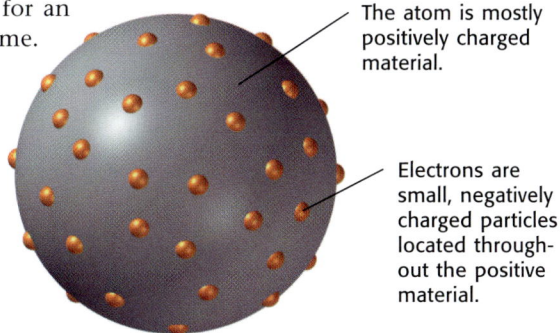

The atom is mostly positively charged material.

Electrons are small, negatively charged particles located throughout the positive material.

Figure 4 *Thomson's plum-pudding model of the atom is shown above. A modern version of Thomson's model might be chocolate-chip ice cream.*

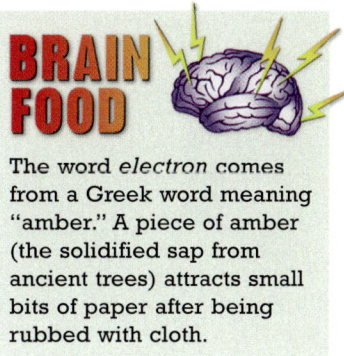

SECTION REVIEW

1. What discovery demonstrated that atoms are not the smallest particles?

2. What did Dalton do in developing his theory that Democritus did not do?

3. **Analyzing Methods** Why was it important for Thomson to repeat his experiment?

internet connect

SCI LINKS
NSTA

TOPIC: Development of the Atomic Theory
GO TO: www.scilinks.org
*sci*LINKS NUMBER: HSTP255

83

MEETING INDIVIDUAL NEEDS

Writing **Advanced Learners**
Rutherford was influenced by the ideas of Japanese physicist Hantaro Nagaoka, who in 1904 suggested that electrons circled in orbits within the atom. Have advanced learners research Japanese scientists such as Nagaoka, Hideki Yukawa, and Kenjiro Takayanagi and share with the class the contributions these scientists made to modern science.

MATH and MORE

The length of a dollar bill is 15.70 cm. The width is 6.65 cm. If it takes 500 million gold atoms laid end to end to measure the length of a dollar bill, how many gold atoms would it take to measure the width? (almost 212 million gold atoms)

CONNECT TO EARTH SCIENCE

Use Teaching Transparency 107, "Gold Crystal Structure," to show students the arrangement of gold atoms in a sample of gold and to help them understand how Rutherford's experiment worked. The positively charged alpha particles in Rutherford's experiment (Figures 5 and 6) did not actually collide with the nuclei of the gold atoms in the foil. Because alpha particles and gold nuclei both have positive charges, their like charges repelled each other before a collision between the particles could take place. (The Investigate! activity at the beginning of this chapter is modeled after Rutherford's gold foil experiment.)

Rutherford Opens an Atomic "Shooting Gallery"

Find out about Melissa Franklin, a modern atom explorer, on page 101.

In 1909, a former student of Thomson's named Ernest Rutherford decided to test Thomson's theory. He designed an experiment to investigate the structure of the atom. He aimed a beam of small, positively charged particles at a thin sheet of gold foil. These particles were larger than *protons,* even smaller positive particles identified in 1902. **Figure 5** shows a diagram of Rutherford's experiment. To find out where the particles went after being "shot" at the gold foil, Rutherford surrounded the foil with a screen coated with zinc sulfide, a substance that glowed when struck by the particles.

Figure 5 Rutherford's Gold Foil Experiment

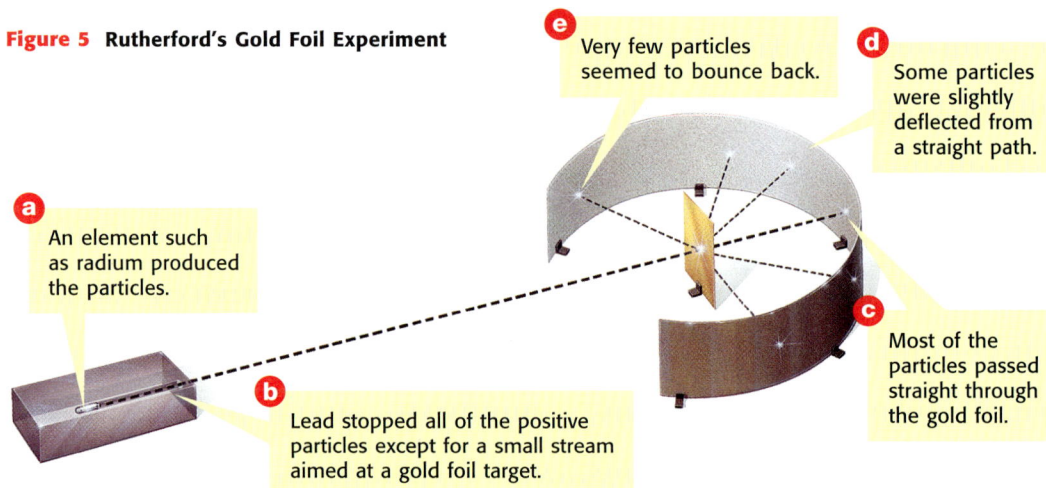

e Very few particles seemed to bounce back.

d Some particles were slightly deflected from a straight path.

a An element such as radium produced the particles.

c Most of the particles passed straight through the gold foil.

b Lead stopped all of the positive particles except for a small stream aimed at a gold foil target.

Rutherford Gets Surprising Results Rutherford thought that if atoms were soft "blobs" of material, as suggested by Thomson, then the particles would pass through the gold and continue in a straight line. Most of the particles did just that. But to Rutherford's great surprise, some of the particles were deflected (turned to one side) a little, some were deflected a great deal, and some particles seemed to bounce back. Rutherford reportedly said,

"It was quite the most incredible event that has ever happened to me in my life. It was almost as if you fired a fifteen-inch shell into a piece of tissue paper and it came back and hit you."

Teaching Transparency 107 "Gold Crystal Structure"

LINK TO EARTH SCIENCE

Teaching Transparency 245 "Rutherford's Gold Foil Experiment"

IS THAT A FACT!

The gold foil experiments were actually performed by Ernest Marsden and Hans Geiger (who later invented the Geiger counter), two research assistants working in Rutherford's lab.

Rutherford Presents a New Atomic Model Rutherford realized that the plum-pudding model of the atom did not explain his results. In 1911, he revised the atomic theory and developed a new model of the atom, as shown in **Figure 6.** To explain the deflection of the particles, Rutherford proposed that in the center of the atom is a tiny, extremely dense, positively charged region called the **nucleus** (NOO klee uhs). Most of the atom's mass is concentrated here. Rutherford reasoned that positively charged particles that passed close by the nucleus were pushed away by the positive charges in the nucleus. A particle that headed straight for a nucleus would be pushed almost straight back in the direction from which it came. From his results, Rutherford calculated that the diameter of the nucleus was 100,000 times smaller than the diameter of the gold atom. To imagine how small this is, look at **Figure 7.**

Figure 6
Rutherford's Model of the Atom

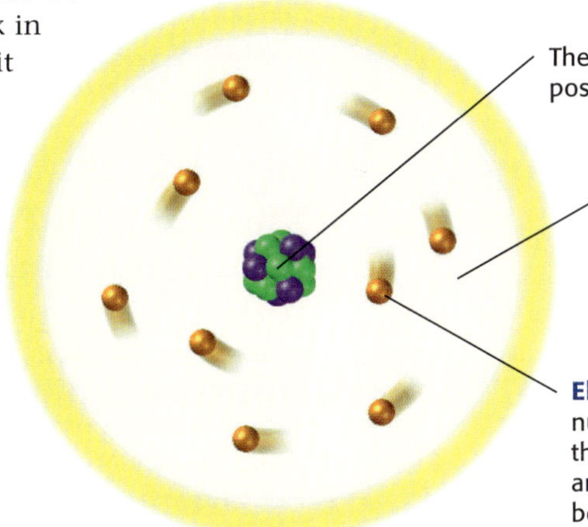
The atom has a small, dense, positively charged nucleus.
The atom is mostly empty space through which electrons travel.
Electrons travel around the nucleus like planets around the sun, but their exact arrangement could not be described.

Figure 7 *The diameter of this pinhead is 100,000 times smaller than the diameter of the stadium.*

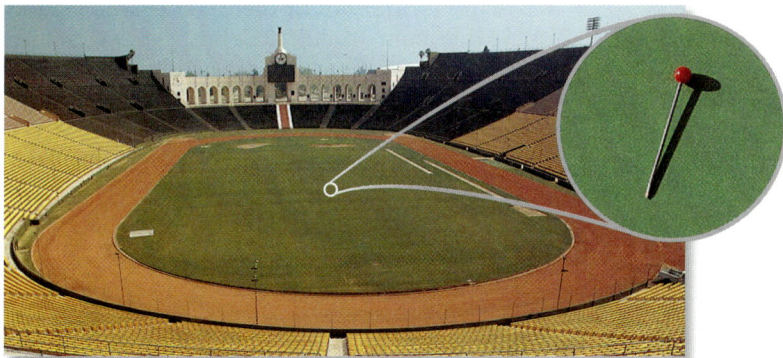

Self-Check

Why did Thomson think the atom contains positive charges? *(See page 168 to check your answer.)*

Answer to Self-Check

The particles Thomson discovered had negative charges. Because an atom has no charge, it must contain positively charged particles to cancel the negative charges.

internetconnect

SCILINKS NSTA
TOPIC: Development of the Atomic Theory
GO TO: www.scilinks.org
***sci*LINKS NUMBER:** HSTP255

MATH and MORE

Have students measure the diameter of a penny in millimeters and imagine that this penny represents the nucleus of an atom. Ask them to calculate the diameter of the whole atom to which this nucleus belongs. Have students report their answers in millimeters and meters. (A penny has a diameter of 19 mm. Thus, the diameter of the atom with a nucleus the size of a penny would be 19 mm × 100,000 = 1,900,000 mm, or 1,900 m.)

Math Skills Worksheet "Using Proportions and Cross-Multiplication"

REAL-WORLD CONNECTION

By the mid 1900s, many nations shared a deep concern about the misuse of nuclear energy. In response to this concern, the Atomic Energy Commission was formed in 1946 to oversee civilian uses of nuclear power. The first international conference on nuclear energy was held in Geneva, Switzerland, in 1955.

Homework

Writing Thomson's model of the atom was the first of many models created to explain the atom's structure. Each model is revised or replaced as scientists learn more. Current models may be changed when scientists discover something new. Have students write a paragraph explaining why making and using models of scientific discoveries is important. They should give at least two reasons.

3) Extend

GOING FURTHER

Electrons can move from level to level only by absorbing or losing a fixed amount of energy called a *quantum*. If the electron absorbs a quantum of energy, it moves to a higher level (farther away from the nucleus). If it releases energy, it moves to a lower level. Have students research the quantum phenomenon and present their results to the class on a poster or in a skit.

Teacher's Note: The Bohr model **(Figure 8)** is used in this chapter because it is easier to use and allows students to understand the number of electrons in each energy level.

4) Close

Quiz

Prepare large index cards with the names of the scientists discussed in this section—Democritus, Dalton, Thomson, Rutherford, and Bohr. On separate cards, write the major discovery or accomplishment associated with each scientist. Students should be able to match correctly the scientists with their discovery or accomplishment.

ALTERNATIVE ASSESSMENT

Concept Mapping Have students make a concept map comparing the different models of the atom.

internetconnect

SC*i*LINKS
NSTA

TOPIC: Modern Atomic Theory
GO TO: www.scilinks.org
*sci*LINKS NUMBER: HSTP260

Figure 8 Bohr's Model of the Atom

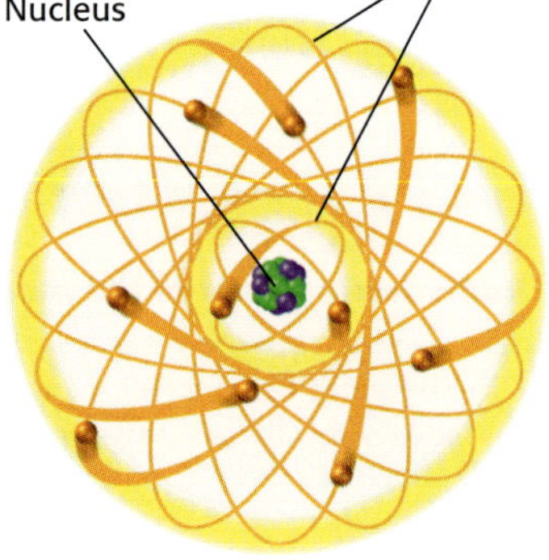

Electron paths

Nucleus

Figure 9 The Current Model of the Atom

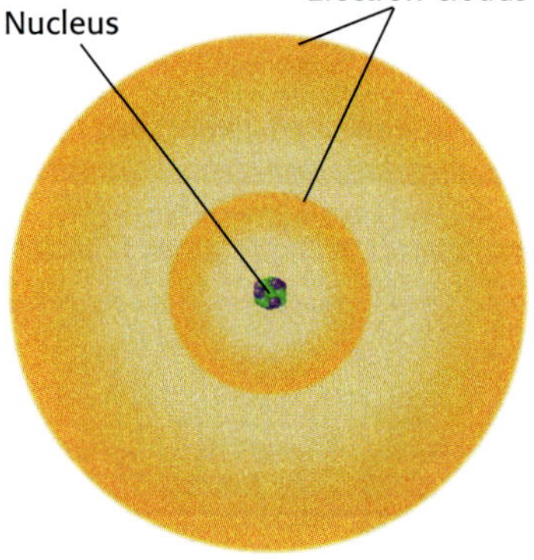

Electron clouds

Nucleus

internetconnect

SC*i*LINKS
NSTA

TOPIC: Modern Atomic Theory
GO TO: www.scilinks.org
*sci*LINKS NUMBER: HSTP260

Bohr States That Electrons Can Jump Between Levels

In 1913, Niels Bohr, a Danish scientist who worked with Rutherford, suggested that electrons travel around the nucleus in definite paths. These paths are located in levels at certain distances from the nucleus, as illustrated in **Figure 8.** Bohr proposed that no paths are located between the levels, but electrons can jump from a path in one level to a path in another level. Think of the levels as rungs on a ladder. You can stand *on* the rungs of a ladder but not *between* the rungs. Bohr's model was a valuable tool in predicting some atomic behavior, but the atomic theory still had room for improvement.

The Modern Theory: Electron Clouds Surround the Nucleus

Many twentieth-century scientists have contributed to our current understanding of the atom. An Austrian physicist named Erwin Schrödinger and a German physicist named Werner Heisenberg made particularly important contributions. Their work further explained the nature of electrons in the atom. For example, electrons do not travel in definite paths as Bohr suggested. In fact, the exact path of a moving electron cannot be predicted. According to the current theory, there are regions inside the atom where electrons are *likely* to be found—these regions are called **electron clouds.** Electron clouds are related to the paths described in Bohr's model. The electron-cloud model of the atom is illustrated in **Figure 9.**

SECTION REVIEW

1. In what part of an atom is most of its mass located?

2. What are two differences between the atomic theory described by Thomson and that described by Rutherford?

3. **Comparing Concepts** Identify the difference in how Bohr's theory and the modern theory describe the location of electrons.

▼ *Answers to Section Review*

1. the nucleus

2. Two differences in the theories are that Thomson's model had the negatively charged particles in the positive material but Rutherford's model had them moving around the positive material. Thomson's model does not have a nucleus in the atom, but Rutherford's model does.

3. Sample answer: Bohr's theory was that electrons move in definite paths around the nucleus. The modern theory states that the path of an electron cannot be known. Only the areas of the atom where electrons are likely to be found can be described.

The Atom

In the last section, you learned how the atomic theory developed through centuries of observation and experimentation. Now it's time to learn about the atom itself. In this section, you'll learn about the particles inside the atom, and you'll learn about the forces that act on those particles. But first you'll find out just how small an atom really is.

How Small Is an Atom?

The photograph below shows the pattern that forms when a beam of electrons is directed at a sample of aluminum. By analyzing this pattern, scientists can determine the size of an atom. Analysis of similar patterns for many elements has shown that aluminum atoms, which are average-sized atoms, have a diameter of about 0.00000003 cm. That's three hundred-millionths of a centimeter. That is so small that it would take a stack of 50,000 aluminum atoms to equal the thickness of a sheet of aluminum foil from your kitchen!

As another example, consider an ordinary penny. Believe it or not, a penny contains about 2×10^{22} atoms, which can be written as 20,000,000,000,000,000,000,000 atoms, of copper and zinc. That's twenty thousand billion billion atoms—over 3,000,000,000,000 times more atoms than there are people on Earth! So if there are that many atoms in a penny, each atom must be very small. You can get a better idea of just how small an atom is in **Figure 10.**

Figure 10 *If you could enlarge a penny until it was as wide as the continental United States, each of its atoms would be only about 3 cm in diameter—about the size of this table-tennis ball.*

BRAIN FOOD

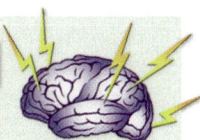

The size of atoms varies widely. Helium atoms have the smallest diameter, and francium atoms have the largest diameter. In fact, about 600 atoms of helium would fit in the space occupied by a single francium atom!

IS THAT A FACT!

One molecule of water is composed of three atoms—two of hydrogen and one of oxygen. One molecule of natural rubber is composed of approximately 295,000 atoms—175,000 carbon atoms and about 120,000 hydrogen atoms.

WEIRD SCIENCE

What do you get if you take particles of silica and oxygen and link them together in long strands? The lightest solid material known! Some of these silica aerogels have a density of only 0.003 g/cm³ and are used, among other things, for collecting cosmic dust in space.

SECTION 2

Focus

The Atom

This section describes what is known about the particles inside an atom. Students learn about the atomic number and mass number of an atom and about charge and isotopes. Finally, students will calculate the atomic mass of an element and figure the number of particles within an atom.

Bellringer

Have students answer the following question in their ScienceLog:

An *atom* is the smallest particle into which an element can be divided and still be that element. Now that scientists have learned that an atom is made up of even smaller particles, is this definition still accurate? Explain your answer.

1) Motivate

DISCUSSION

Students may have difficulty visualizing the very large and very small numbers that are used when discussing atoms. Give each student a penny to hold and look at while the class discusses the large number of atoms in the penny and the extremely small size of each atom.

Directed Reading Worksheet Section 2

Activity As students learn about the particles inside an atom, have them create and label diagrams of several different atoms in their ScienceLog. Students can use **Figures 11–15** as guidelines. The diagrams should show the different particles, where they are located, and other information, such as mass and charge.
Sheltered English

MISCONCEPTION ALERT

When students think of the smallest particle possible, they may picture a dust particle. As they read through this section, help them understand how small atomic size really is. Even one dust particle is made of millions of atoms!

REAL-WORLD CONNECTION

Attempting to comprehend the size of atoms and their components interests people other than chemists and physicists. Charles and Ray Eames were architects and designers who were also fascinated with size and numbers. This interest led them to make the award-winning film *Powers of Ten* (1977). The film is available on video and is a fascinating exploration into the "small" of atoms and the "large" of the universe.

Teaching Transparency 246
"Parts of an Atom"

What's Inside an Atom?

As tiny as an atom is, it consists of even smaller particles—protons, neutrons, and electrons—as shown in the model in **Figure 11.** (The particles represented in the figures are not shown in their correct proportions because the electrons would be too small to see.)

Figure 11 Parts of an Atom

Electrons are negatively charged particles found in electron clouds outside the nucleus. The size of the electron clouds determines the size of the atom.

Protons are positively charged particles in the nucleus of an atom.

The **nucleus** is the small, dense, positively charged center of the atom. It contains most of the atom's mass.

Neutrons are particles in the nucleus of an atom that have no charge.

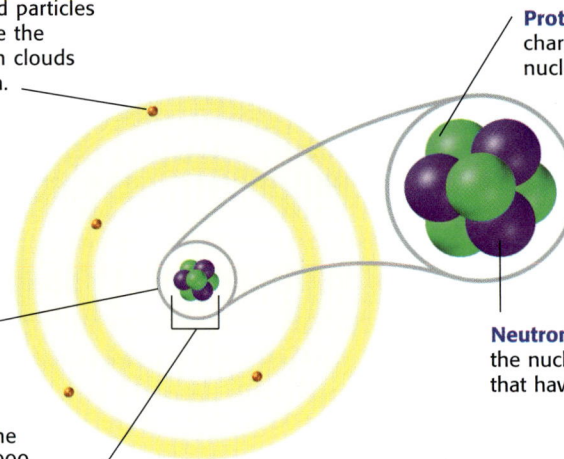

The diameter of the nucleus is 1/100,000 the diameter of the atom.

Proton Profile
Charge: positive
Mass: 1 amu
Location: nucleus

Neutron Profile
Charge: none
Mass: 1 amu
Location: nucleus

The Nucleus **Protons** are the positively charged particles of the nucleus. It was these particles that repelled Rutherford's "bullets." All protons are identical. The mass of a proton is approximately 1.7×10^{-24} g, which can also be written as 0.0000000000000000000000017 g. Because the masses of particles in atoms are so small, scientists developed a new unit for them. The SI unit used to express the masses of particles in atoms is the **atomic mass unit (amu).** Scientists assign each proton a mass of 1 amu.

Neutrons are the particles of the nucleus that have no charge. All neutrons are identical. Neutrons are slightly more massive than protons, but the difference in mass is so small that neutrons are also given a mass of 1 amu.

Protons and neutrons are the most massive particles in an atom, yet the nucleus has a very small volume. So the nucleus is very dense. If it were possible to have a nucleus the volume of an average grape, that nucleus would have a mass greater than 9 million metric tons!

88

IS THAT A FACT!

Carbon-12 is used by scientists to determine an atomic mass unit (amu). The amu is exactly one-twelfth the mass of a carbon-12 atom. Because carbon-12 has 6 protons and 6 neutrons in its nucleus, the mass of a proton and a neutron are each considered to be 1 amu.

SCIENCE HUMOR

A neutron walks into a diner and orders a glass of orange juice at the lunch counter. When the waiter brings the juice, the neutron asks, "How much do I owe you?"

The waiter replies, "For you, no charge."

Outside of the Nucleus *Electrons* are the negatively charged particles in atoms. Electrons are likely to be found around the nucleus within electron clouds. The charges of protons and electrons are opposite but equal in size. An atom is neutral (has no overall charge) because there are equal numbers of protons and electrons, so their charges cancel out. If the numbers of electrons and protons are not equal, the atom becomes a charged particle called an *ion* (IE ahn). Ions are positively charged if the protons outnumber the electrons, and they are negatively charged if the electrons outnumber the protons.

Electrons are very small in mass compared with protons and neutrons. It takes more than 1,800 electrons to equal the mass of 1 proton. In fact, the mass of an electron is so small that it is usually considered to be zero.

Electron Profile

Charge:	negative
Mass:	almost zero
Location:	electron clouds

SECTION REVIEW

1. What particles form the nucleus?

2. Explain why atoms are neutral.

3. **Summarizing Data** Why do scientists say that most of the mass of an atom is located in the nucleus?

How Do Atoms of Different Elements Differ?

There are over 110 different elements, each of which is made of different atoms. What makes atoms different from each other? To find out, imagine that it's possible to "build" an atom by putting together protons, neutrons, and electrons.

Starting Simply It's easiest to start with the simplest atom. Protons and electrons are found in all atoms, and the simplest atom consists of just one of each. It's so simple it doesn't even have a neutron. Put just one proton in the center of the atom for the nucleus. Then put one electron in the electron cloud, as shown in the model in **Figure 12.** Congratulations! You have just made the simplest atom—a hydrogen atom.

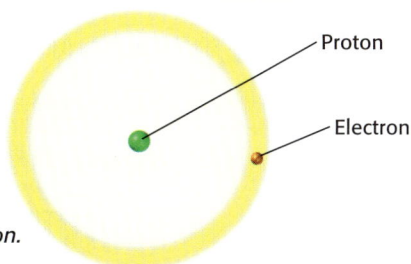

Proton

Electron

Figure 12 *The simplest atom has one proton and one electron.*

89

▼ Answers to Section Review

1. protons and neutrons

2. Atoms are neutral because the number of protons and the number of electrons are the same. Thus, the positive charges and negative charges cancel out.

3. Sample answer: Protons and neutrons are located in the nucleus, and electrons are outside the nucleus. Because an electron has almost zero mass and a proton and a neutron have a mass of 1 amu each, most of the mass of an atom is in the nucleus.

GUIDED PRACTICE

Display a large version of the periodic table, and distribute a copy of the periodic table to each student. Explain to students that this is a table of all the known elements. Help students find the atomic number of different elements using the periodic table (the atomic number is the number above the chemical symbol). Remind students that the atomic number is the number of protons in the nucleus of each atom of a particular element. Sheltered English

---

CONNECT TO
PALEONTOLOGY

The isotope carbon-14 is used in radiocarbon-dating of animal and plant fossils. Uranium-238, uranium-235, and thorium-232 are isotopes that scientists use to tell the age of rocks and meteorites.

Figure 13 *A helium nucleus must have neutrons in it to keep the protons from moving apart.*

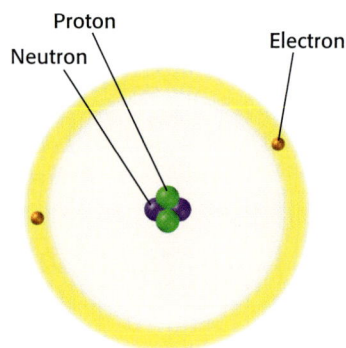

Proton
Neutron
Electron

Astronomy

CONNECTION

Hydrogen is the most abundant element in the universe. It is the fuel for the sun and other stars. It is currently believed that there are roughly 2,000 times more hydrogen atoms than oxygen atoms and 10,000 times more hydrogen atoms than carbon atoms.

Now for Some Neutrons Now build an atom containing two protons. Both of the protons are positively charged, so they repel one another. You cannot form a nucleus with them unless you add some neutrons. For this atom, two neutrons will do. Your new atom will also need two electrons outside the nucleus, as shown in the model in **Figure 13.** This is an atom of the element helium.

Building Bigger Atoms You could build a carbon atom using 6 protons, 6 neutrons, and 6 electrons; or you could build an oxygen atom using 8 protons, 9 neutrons, and 8 electrons. You could even build a gold atom with 79 protons, 118 neutrons, and 79 electrons! As you can see, an atom does not have to have equal numbers of protons and neutrons.

The Number of Protons Determines the Element How can you tell which elements these atoms represent? The key is the number of protons. The number of protons in the nucleus of an atom is the **atomic number** of that atom. All atoms of an element have the same atomic number. Every hydrogen atom has only one proton in its nucleus, so hydrogen has an atomic number of 1. Every carbon atom has six protons in its nucleus, so carbon has an atomic number of 6.

Are All Atoms of an Element the Same?

Back in the atom-building workshop, you make an atom that has one proton, one electron, and one neutron, as shown in **Figure 14.** The atomic number of this new atom is 1, so the atom is hydrogen. However, this hydrogen atom's nucleus has two particles; therefore, this atom has a greater mass than the first hydrogen atom you made. What you have is another isotope (IE suh TOHP) of hydrogen. **Isotopes** are atoms that have the same number of protons but have different numbers of neutrons. Atoms that are isotopes of each other are always the same element because the number of protons in each atom is the same.

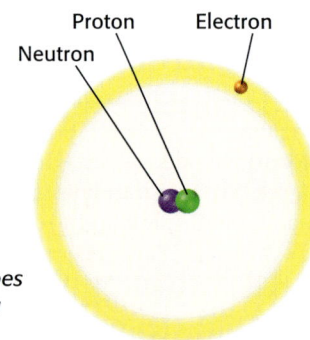

Proton
Electron
Neutron

Figure 14 *The atom in this model and the one in Figure 12 are isotopes because each has one proton but a different number of neutrons.*

SCIENCE HUMOR

Have you heard the one about the chemist who was reading a book about helium? He couldn't put it down.

Science Bloopers

In 1811, Bernard Courtois was washing seaweed ashes in sulfuric acid. A black precipitate appeared unexpectedly. The precipitate was an unknown element and was later named iodine (from the Greek word for "violet-like") because of the purple vapor it produces when heated.

Properties of Isotopes Each element has a limited number of isotopes that occur naturally. Some isotopes of an element have unique properties because they are unstable. An unstable atom is an atom whose nucleus can change its composition. This type of isotope is *radioactive*. However, isotopes of an element share most of the same chemical and physical properties. For example, the most common oxygen isotope has 8 neutrons in the nucleus, but other isotopes have 9 or 10 neutrons. All three isotopes are colorless, odorless gases at room temperature. Each isotope has the chemical property of combining with a substance as it burns and even behaves the same in chemical changes in your body.

How Can You Tell One Isotope from Another?

You can identify each isotope of an element by its mass number. The **mass number** is the sum of the protons and neutrons in an atom. Electrons are not included in an atom's mass number because their mass is so small that they have very little effect on the atom's total mass. Look at the boron isotope models shown in **Figure 15** to see how to calculate an atom's mass number.

APPLY

Isotopes and Light Bulbs

Oxygen reacts, or undergoes a chemical change, with the hot filament in a light bulb, quickly burning out the bulb. Argon does not react with the filament, so a light bulb filled with argon burns out more slowly than one filled with oxygen. Do all three naturally-occurring isotopes of argon have the same effect in light bulbs? Explain your reasoning.

Figure 15 *Each of these boron isotopes has five protons. But because each has a different number of neutrons, each has a different mass number.*

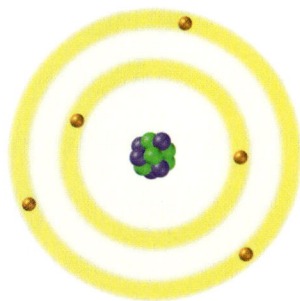

Protons: 5
Neutrons: 5
Electrons: 5
Mass number = protons + neutrons = 10

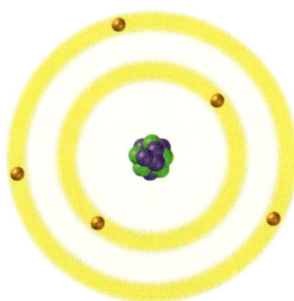

Protons: 5
Neutrons: 6
Electrons: 5
Mass number = protons + neutrons = 11

91

IS THAT A FACT!

There are at least 2,670 known isotopes. Tin has 38 isotopes, the most of any of the elements. Hydrogen has the fewest isotopes, with only three. The least-stable isotope is lithium-5, which decays in 4.4×10^{-22} seconds.

Answer to APPLY

All three isotopes of argon will have the same effect (longer life) in light bulbs. Isotopes share the same chemical properties, including nonreactivity.

REAL-WORLD CONNECTION

Isotopes have many applications in the field of nuclear medicine. Cobalt-60 is used to treat cancerous tumors, and iodine-131 is used in the treatment of hyperthyroidism. Have students research the use of radioactive isotopes in the detection and treatment of disease. Students can present their results in posters, concept maps, or reports.

ACTIVITY

Making Models

MATERIALS
FOR EACH STUDENT:
• colored dots (available at office-supply stores)
• construction paper
• markers

1. Distribute dots, paper, and markers to students.
2. Have students study the atoms shown in **Figures 12, 13, 14,** and **15.**
3. Instruct students to reconstruct each of the isotopes.
4. Make sure that students use a different-colored dot for the protons, the neutrons, and the electrons.
5. Remind students to notice that only the number of neutrons changes from isotope to isotope.
6. Have students make dot models of some other simple isotopes.

internet connect

SCiLINKS
NSTA

TOPIC: Isotopes
GO TO: www.scilinks.org
*sci*LINKS NUMBER: HSTP270

Answer to Activity

hydrogen-2: 1 proton and 1 neutron in the nucleus; 1 electron in the electron cloud

helium-3: 2 protons and 1 neutron in the nucleus; 2 electrons in the electron cloud

carbon-14: 6 protons and 8 neutrons in the nucleus; 6 electrons in the electron cloud

MATH and MORE

Do an example of calculating atomic mass with students. Use the information given for copper-63 and copper-65 on this page. Let students fill in some missing numbers. Label each of the values.

$(63 \times 0.69) = 43.47$ amu

$(65 \times 0.31) = 20.15$ amu

$(43.47) + (20.15) = 63.62$ amu

Now have students calculate the atomic mass of titanium. Its five isotopes are:

titanium-46 (8.0 percent)

titanium-47 (7.3 percent)

titanium-48 (73.8 percent)

titanium-49 (5.5 percent)

titanium-50 (5.4 percent)

47.9 amu

Math Skills Worksheet "Arithmetic with Decimals"

Answer to MATHBREAK

$(11 \times 0.80) = \quad 8.8$ amu

$(10 \times 0.20) = \underline{+\ 2.0}$ amu

$\quad\quad\quad\quad\quad 10.8$ amu

Reinforcement Worksheet "Atomic Timeline"

Activity

Draw diagrams of hydrogen-2, helium-3, and carbon-14. Show the correct number and location of each type of particle. For the electrons, simply write the total number of electrons in the electron cloud. Use colored pencils or markers to represent the protons, neutrons, and electrons.

TRY at HOME

$\div$ 5 $\div$ Ω ∞ $+ \Omega^{\vee}$ 9 $\infty^{\leq}$ Σ 2

MATH BREAK

Atomic Mass

To calculate the atomic mass of an element, multiply the mass number of each isotope by its percentage abundance in decimal form. Then add these amounts together to find the atomic mass. For example, chlorine-35 makes up 76 percent (its percentage abundance) of all the chlorine in nature, and chlorine-37 makes up the other 24 percent. The atomic mass of chlorine is calculated as follows:

$(35 \times 0.76) = \quad 26.6$

$(37 \times 0.24) = \underline{+8.9}$

$\quad\quad\quad\quad\quad 35.5$ amu

Now It's Your Turn

Calculate the atomic mass of boron, which occurs naturally as 20 percent boron-10 and 80 percent boron-11.

MISCONCEPTION ALERT

The atomic mass of copper shown on this page differs from that shown on the periodic table. The atomic mass of an element is calculated from the relative atomic masses of each isotope, not from the mass numbers. Relative atomic masses account for the fact that protons and neutrons are not exactly 1 amu and that electrons do have some mass.

Naming Isotopes To identify a specific isotope of an element, write the name of the element followed by a hyphen and the mass number of the isotope. A hydrogen atom with one proton and no neutrons has a mass number of 1. Its name is hydrogen-1. Hydrogen-2 has one proton and one neutron. The carbon isotope with a mass number of 12 is called carbon-12. If you know that the atomic number for carbon is 6, you can calculate the number of neutrons in carbon-12 by subtracting the atomic number from the mass number. For carbon-12, the number of neutrons is $12 - 6$, or 6.

12	Mass number
-6	Number of protons (atomic number)
6	Number of neutrons

Calculating the Mass of an Element

Most elements found in nature contain a mixture of two or more stable (nonradioactive) isotopes. For example, all copper is composed of copper-63 atoms and copper-65 atoms. The term *atomic mass* describes the mass of a mixture of isotopes. **Atomic mass** is the weighted average of the masses of all the naturally occurring isotopes of an element. A weighted average accounts for the percentages of each isotope that are present. Copper, including the copper in the Statue of Liberty (shown in **Figure 16**), is 69 percent copper-63 and 31 percent copper-65. The atomic mass of copper is 63.6 amu. You can try your hand at calculating atomic mass by doing the MathBreak at left.

Figure 16 *The copper used to make the Statue of Liberty includes both copper-63 and copper-65. Copper's atomic mass is 63.6 amu.*

What Forces Are at Work in Atoms?

You have seen how atoms are composed of protons, neutrons, and electrons. But what are the *forces* (the pushes or pulls between two objects) acting between these particles? Four basic forces are at work everywhere, including within the atom—gravity, the electromagnetic force, the strong force, and the weak force. These forces are discussed below.

Forces in the Atom

Gravity Probably the most familiar of the four forces is *gravity.* Gravity acts between all objects all the time. The amount of gravity between objects depends on their masses and the distance between them. Gravity pulls objects, such as the sun, Earth, cars, and books, toward one another. However, because the masses of particles in atoms are so small, the force of gravity within atoms is very small.

Electromagnetic Force As mentioned earlier, objects that have the same charge repel each other, while objects with opposite charge attract each other. This is due to the *electromagnetic force.* Protons and electrons are attracted to each other because they have opposite charges. The electromagnetic force holds the electrons around the nucleus.

Particles with the same charges repel each other.

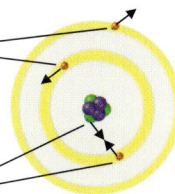

Particles with opposite charges attract each other.

Strong Force Protons push away from one another because of the electromagnetic force. A nucleus containing two or more protons would fly apart if it were not for the *strong force.* At the close distances between protons in the nucleus, the strong force is greater than the electromagnetic force, so the nucleus stays together.

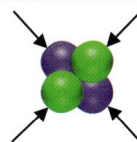

Weak Force The *weak force* is an important force in radioactive atoms. In certain unstable atoms, a neutron can change into a proton and an electron. The weak force plays a key role in this change.

SECTION REVIEW

1. List the charge, location, and mass of a proton, a neutron, and an electron.

2. Determine the number of protons, neutrons, and electrons in an atom of aluminum-27.

3. **Doing Calculations** The metal thallium occurs naturally as 30 percent thallium-203 and 70 percent thallium-205. Calculate the atomic mass of thallium.

internetconnect

SCiLINKS.
NSTA

TOPIC: Inside the Atom, Isotopes
GO TO: www.scilinks.org
*sci*LINKS NUMBER: HSTP265, HSTP270

93

Answers to Section Review

1. A proton has a 1+ charge, is found in the nucleus, and has a mass of 1 amu. A neutron has no charge, is found in the nucleus, and has a mass of 1 amu. An electron has a 1– charge, is found in the electron cloud outside the nucleus, and has almost no mass.

2. 13 protons, 14 neutrons, and 13 electrons

3. 204.4 amu

3 Extend

GOING FURTHER

Writing Help students understand the electromagnetic force, gravity, and the weak and strong forces. Ask them what would happen if these forces did not exist. How would matter be changed? What would happen to the solar system? the galaxy? the universe? Ask students to write a short story and describe a universe where these forces work only sometimes—or not at all.

4 Close

Quiz

Have students write a simple description of an atom for somebody who knows nothing about atoms. For example, students might explain atoms and atomic structure to a young child. Encourage students to use diagrams and examples in their descriptions.

ALTERNATIVE ASSESSMENT

Concept Mapping Have students create a concept map using the following terms:

proton, atomic mass unit, neutron, atomic number, isotopes, mass number, atomic mass

Teaching Transparency 247
"Forces in the Atom"

Critical Thinking Worksheet
"Incredible Shrinking Scientist!"

Made to Order
Teacher's Notes

Time Required

One 45-minute class period

Lab Ratings

EASY ——————→ HARD

TEACHER PREP 🧪🧪
STUDENT SET-UP 🧪
CONCEPT LEVEL 🧪🧪🧪
CLEAN UP 🧪

MATERIALS

The materials listed are for a pair of students. Foam balls of any color are acceptable as long as there are two colors. Flexible pipe cleaners may be used instead of toothpicks.

Safety Caution

Remind students to review all safety cautions and icons before beginning this lab activity.

Preparation Notes

Before you begin this lab, review the concepts of isotopes, atomic number, and mass number.

To create colored balls, use colored markers or spray paint. Alternatively, you can label white balls with "N" or "P." If you prefer to make two-dimensional models, use colored dots (from an office-supply store) to represent the different particles. Reinforce the idea that the particles should be compact—the strong force binds the particles together as tightly as possible.

Making Models Lab

Made to Order

Imagine that you are a new worker at the Elements-4-U Company, which makes elements. Your job is to build the atomic nucleus for each element ordered by your customers. You were hired because you know about the makeup of a nucleus and also because you understand how isotopes of an element are different from each other. Now it's time to get to work!

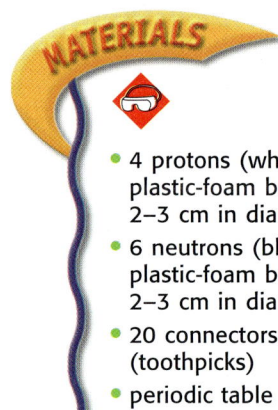

MATERIALS

- 4 protons (white plastic-foam balls, 2–3 cm in diameter)
- 6 neutrons (blue plastic-foam balls, 2–3 cm in diameter)
- 20 connectors (toothpicks)
- periodic table

Procedure

1. Copy the table, shown on the next page, into your ScienceLog. Leave room to add more elements.

2. Before you start the lab, put on your cover goggles. Your first task is to build the nucleus of hydrogen-1. Pick up one proton (a white plastic-foam ball). Congratulations! You have just built a model of a hydrogen-1 nucleus, the simplest nucleus possible.

3. Count the number of protons and neutrons in the nucleus. Fill in rows 1 and 2 for this element in the table.

4. Use the information in rows 1 and 2 to determine the atomic number and mass number of the element. Record this information in the table.

5. Draw a picture of your model in your ScienceLog.

94

Datasheets for LabBook

CLASSROOM TESTED & APPROVED

Sharon L. Woolf
Langston Hughes Middle School
Reston, Virginia

Isotope Data							
	Hydrogen-1	Hydrogen-2	Helium-3	Helium-4	Lithium-7	Beryllium-9	Beryllium-10
No. of protons							
No. of neutrons							
Atomic number							
Mass number							

DO NOT WRITE IN BOOK

6 Hydrogen-2 is an isotope of hydrogen that has one proton and one neutron. Using a toothpick, add a neutron to your hydrogen-1 nucleus. (In a nucleus, the protons and neutrons are held together by a force. The toothpicks in this activity stand for the force.) Repeat steps 3–5.

7 Helium-3 is an isotope of helium that has two protons and one neutron. Add one proton to your hydrogen-2 nucleus to create a helium-3 nucleus. Each particle should be connected to the other two particles so that they make a triangle, not a line. Protons and neutrons always form the smallest shape possible. Repeat steps 3–5.

8 For the next part, you will need to use information from the periodic table of the elements. Look at the illustration below. It shows the periodic table entry for carbon. For your job, the most important information in the periodic table is the atomic number. You can find the atomic number of any element at the top of its entry on the table. In the example, the atomic number of carbon is 6.

6
C
Carbon
12.0

Atomic number

9 Use the information in the periodic table to build models of the following isotopes of elements: helium-4, lithium-7, beryllium-9, and beryllium-10. Remember to put the protons and neutrons as close together as possible. Each particle should connect to at least two others. Repeat steps 3–5 for each isotope.

Analyze the Results

10 What is the relationship between the number of protons and the atomic number?

11 If you know the atomic number and the mass number of an isotope, how could you figure out the number of neutrons in the nucleus?

12 Look up uranium on the periodic table. What is the atomic number of uranium? How many neutrons does the isotope uranium-235 have?

Communicate Results

13 Compare your model with the models of other groups. How are they the same? How are they different?

Going Further
Working with another group, join your models together. What element (and isotope) have you made?

Answers

10. The number of protons is the same as the atomic number.

11. The number of neutrons equals the mass number minus the atomic number.

12. 92; 143 neutrons ($235 - 92 = 143$)

13. Accept all reasonable answers.

Going Further

If all of the protons and neutrons are used, the isotope created will be oxygen-20.

Science Skills Worksheet
"Using Models to Communicate"

95

Chapter Highlights

Chapter Highlights

VOCABULARY DEFINITIONS

SECTION 1

atom the smallest particle into which an element can be divided and still be the same substance

theory a unifying explanation for a broad range of hypotheses and observations that have been supported by testing

electrons the negatively charged particles found in all atoms

model a representation of an object or system

nucleus the tiny, extremely dense, positively charged region in the center of an atom

electron clouds the regions inside an atom where electrons are likely to be found

SECTION 1

Vocabulary

atom (p. 80)
theory (p. 80)
electrons (p. 83)
model (p. 83)
nucleus (p. 85)
electron clouds (p. 86)

Section Notes

- Atoms are the smallest particles of an element that retain the properties of the element.

- In ancient Greece, Democritus argued that atoms were the smallest particles in all matter.

- Dalton proposed an atomic theory that stated the following: Atoms are small particles that make up all matter; atoms cannot be created, divided, or destroyed; atoms of an element are exactly alike; atoms of different elements are different; and atoms join together to make new substances.

- Thomson discovered electrons. His plum-pudding model described the atom as a lump of positively charged material with negative electrons scattered throughout.

- Rutherford discovered that atoms contain a small, dense, positively charged center called the nucleus.

- Bohr suggested that electrons move around the nucleus at only certain distances.

- According to the current atomic theory, electron clouds are where electrons are most likely to be in the space around the nucleus.

✓ Skills Check

Math Concepts

ATOMIC MASS The atomic mass of an element takes into account the mass of each isotope and the percentage of the element that exists as that isotope. For example, magnesium occurs naturally as 79 percent magnesium-24, 10 percent magnesium-25, and 11 percent magnesium-26. The atomic mass is calculated as follows:

$$
\begin{aligned}
(24 \times 0.79) &= 19.0 \\
(25 \times 0.10) &= 2.5 \\
(26 \times 0.11) &= \underline{+2.8} \\
&24.3 \text{ amu}
\end{aligned}
$$

Visual Understanding

ATOMIC MODELS The atomic theory has changed over the past several hundred years. To understand the different models of the atom, look over Figures 2, 4, 6, 8, and 9.

PARTS OF THE ATOM Atoms are composed of protons, neutrons, and electrons. To review the particles and their placement in the atom, study Figure 11 on page 88.

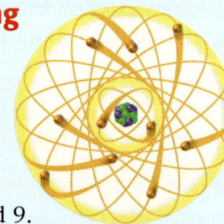

96

Lab and Activity Highlights

Made to Order **PG 94**

Datasheets for LabBook
(blackline masters for these labs)

SECTION 2

Vocabulary

protons (p. 88)

atomic mass unit (p. 88)

neutrons (p. 88)

atomic number (p. 90)

isotopes (p. 90)

mass number (p. 91)

atomic mass (p. 92)

Section Notes

- A proton is a positively charged particle with a mass of 1 amu.
- A neutron is a particle with no charge that has a mass of 1 amu.
- An electron is a negatively charged particle with an extremely small mass.

- Protons and neutrons make up the nucleus. Electrons are found in electron clouds outside the nucleus.
- The number of protons in the nucleus of an atom is the atomic number. The atomic number identifies the atoms of a particular element.
- Isotopes of an atom have the same number of protons but have different numbers of neutrons. Isotopes share most of the same chemical and physical properties.
- The mass number of an atom is the sum of the atom's neutrons and protons.

- The atomic mass is an average of the masses of all naturally occurring isotopes of an element.
- The four forces at work in an atom are gravity, the electromagnetic force, the strong force, and the weak force.

protons the positively charged particles of the nucleus

atomic mass unit the SI unit used to express the masses of particles in atoms

neutrons the particles of the nucleus that have no charge

atomic number the number of protons in the nucleus of an atom

isotopes atoms that have the same number of protons but have different numbers of neutrons

mass number the sum of the protons and neutrons in an atom

atomic mass the weighted average of the masses of all the naturally occurring isotopes of an element

Vocabulary Review Worksheet

Blackline masters of these Chapter Highlights can be found in the **Study Guide.**

---

internet connect

go.hrw.com

GO TO: go.hrw.com

Visit the HRW Web site for a variety of learning tools related to this chapter. Just type in the keyword:

KEYWORD: HSTATS

SCI LINKS SM

NSTA

GO TO: www.scilinks.org

Visit the **National Science Teachers Association** on-line Web site for Internet resources related to this chapter. Just type in the *sci*LINKS number for more information about the topic:

TOPIC: Development of the Atomic Theory	*sci*LINKS NUMBER: HSTP255
TOPIC: Modern Atomic Theory	*sci*LINKS NUMBER: HSTP260
TOPIC: Inside the Atom	*sci*LINKS NUMBER: HSTP265
TOPIC: Isotopes	*sci*LINKS NUMBER: HSTP270

97

Lab and Activity Highlights

LabBank

Whiz-Bang Demonstrations,
- As a Matter of Space
- Candy Lights

Long-Term Projects & Research Ideas,
How Low Can They Go?

Chapter Review
Answers

USING VOCABULARY

1. electron clouds
2. protons
3. Neutrons
4. mass number
5. atomic mass

UNDERSTANDING CONCEPTS

Multiple Choice

6. c
7. d
8. a
9. a
10. c
11. a
12. a
13. b

Short Answer

14. Sample answers: New information is discovered that proves a previous theory to be incorrect; a new theory explains existing information better.
15. electromagnetic force
16. Sample answer: The plum-pudding model describes the atom as a lump of positively charged material with negatively charged particles throughout. The positively charged material is like the dough, and the negative particles are like the plums in a plum pudding.

Chapter Review

USING VOCABULARY

The statements below are false. For each statement, replace the underlined word to make a true statement.

1. Electrons are found in the <u>nucleus</u> of an atom.

2. All atoms of the same element contain the same number of <u>neutrons</u>.

3. <u>Protons</u> have no electric charge.

4. The <u>atomic number</u> of an element is the number of protons and neutrons in the nucleus.

5. The <u>mass number</u> is an average of the masses of all naturally occurring isotopes of an element.

UNDERSTANDING CONCEPTS

Multiple Choice

6. The discovery of which particle proved that the atom is not indivisible?
   a. proton   c. electron
   b. neutron   d. nucleus

7. In his gold foil experiment, Rutherford concluded that the atom is mostly empty space with a small, massive, positively charged center because
   a. most of the particles passed straight through the foil.
   b. some particles were slightly deflected.
   c. a few particles bounced back.
   d. All of the above

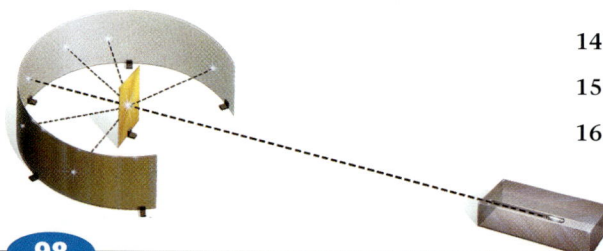

8. How many protons does an atom with an atomic number of 23 and a mass number of 51 have?
   a. 23   c. 51
   b. 28   d. 74

9. An atom has no overall charge if it contains equal numbers of
   a. electrons and protons.
   b. neutrons and protons.
   c. neutrons and electrons.
   d. None of the above

10. Which statement about protons is true?
    a. Protons have a mass of 1/1,840 amu.
    b. Protons have no charge.
    c. Protons are part of the nucleus of an atom.
    d. Protons circle the nucleus of an atom.

11. Which statement about neutrons is true?
    a. Neutrons have a mass of 1 amu.
    b. Neutrons circle the nucleus of an atom.
    c. Neutrons are the only particles that make up the nucleus.
    d. Neutrons have a negative charge.

12. Which of the following determines the identity of an element?
    a. atomic number   c. atomic mass
    b. mass number   d. overall charge

13. Isotopes exist because atoms of the same element can have different numbers of
    a. protons.   c. electrons.
    b. neutrons.   d. None of the above

Short Answer

14. Why do scientific theories change?

15. What force holds electrons in atoms?

16. In two or three sentences, describe the plum-pudding model of the atom.

98

Concept Mapping

17. Use the following terms to create a concept map: atom, nucleus, protons, neutrons, electrons, isotopes, atomic number, mass number.

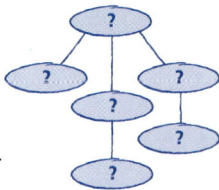

18. Particle accelerators, like the one shown below, are devices that speed up charged particles in order to smash them together. Sometimes the result of the collision is a new nucleus. How can scientists determine whether the nucleus formed is that of a new element or that of a new isotope of a known element?

19. John Dalton made a number of statements about atoms that are now known to be incorrect. Why do you think his atomic theory is still found in science textbooks?

MATH IN SCIENCE

20. Calculate the atomic mass of gallium consisting of 60 percent gallium-69 and 40 percent gallium-71.

21. Calculate the number of protons, neutrons, and electrons in an atom of zirconium-90, which has an atomic number of 40.

INTERPRETING GRAPHICS

22. Study the models below, and answer the questions that follow:

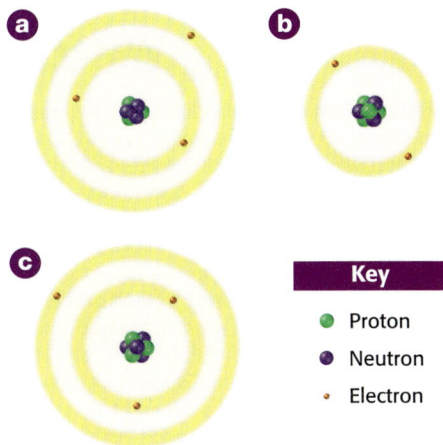

Key	
●	Proton
●	Neutron
·	Electron

a. Which models represent isotopes of the same element?

b. What is the atomic number for (a)?

c. What is the mass number for (b)?

23. Predict how the direction of the moving particle in the figure below will change, and explain what causes the change to occur.

Reading Check-up

Take a minute to review your answers to the Pre-Reading Questions found at the bottom of page 78. Have your answers changed? If necessary, revise your answers based on what you have learned since you began this chapter.

99

Concept Mapping

17. An answer to this exercise can be found at the front of this book.

CRITICAL THINKING AND PROBLEM SOLVING

18. Scientists must determine the atomic number, or the number of protons, in the newly formed nucleus. The nucleus is that of a new element only if the number of protons is different from all known elements.

19. Sample answer: Dalton's atomic theory was the first one based on experimental evidence. It helps show how a theory develops as new information is discovered.

MATH IN SCIENCE

20. $(69 \times 0.60) + (71 \times 0.40) =$ 69.8 amu

21. 40 protons, 50 neutrons, and 40 electrons

INTERPRETING GRAPHICS

22. a. a and c
b. 3
c. 7

23. The particle will move downward, away from the group of particles. The moving particle has a positive charge, and the group of particles is composed of positively charged and neutral particles. The electromagnetic force causes like charges to repel, so the moving particle will be pushed away from the group of particles.

Background

- Not only are the elements hydrogen and oxygen essential for human life, they are also necessary for space travel and exploration. Solar energy striking the moon could be used to break down the water into hydrogen and oxygen.

- These two elements are a valuable rocket fuel. Lack of a refueling station has been a limiting factor in space travel thus far. The amount of water on the moon may be vast and could make space travel far more feasible.

- The sun and other stars are always giving off protons, electrons, and neutrons. The solar wind from our sun bombards the moon with neutrons and other particles.

- Many of these particles bounce straight back into space when they collide with something, but some remain and bounce around on the surface. As a result, the moon is abuzz with neutrons whirring around. Fast neutrons usually fly right back into space.

- When an epithermal neutron bounces into something that is nearly the same size and weight as itself, it slows into a thermal neutron.

ACROSS THE SCIENCES

PHYSICAL SCIENCE • ASTRONOMY

Water on the Moon?

When the astronauts of the Apollo space mission explored the surface of the moon in 1969, all they found was rock powder. None of the many samples of moon rocks they carried back to Earth contained any hint of water. Because the astronauts didn't see water on the moon and scientists didn't detect any in the lab, scientists believed there was no water on the moon.

Then in 1994, radio waves suggested another possibility. On a 4-month lunar jaunt, an American spacecraft called *Clementine* beamed radio waves toward various areas of the moon, including a few craters that never receive sunlight. Mostly, the radio waves were reflected by what appeared to be ground-up rock. However, in part of one huge, dark crater, the radio waves were reflected as if by . . . *ice.*

Hunting for Hydrogen Atoms

Scientists were intrigued by *Clementine's* evidence. Two years later, another spacecraft, *Lunar Prospector,* traveled to the moon. Instead of trying to detect water with radio waves, *Prospector* scanned the moon's surface with a device called a *neutron spectrometer* (NS). A neutron spectrometer counts the number of slow neutrons bouncing off a surface. When a neutron hits something about the same mass as itself, it slows down. As it turns out, the only thing close to the mass of a neutron is an *atom* of the lightest of all elements, hydrogen. So when the NS located high concentrations of slow-moving neutrons on the moon, it indicated to scientists that the neutrons were crashing into hydrogen atoms.

As you know, water consists of two atoms of hydrogen and one atom of oxygen. The presence of hydrogen atoms on the moon is more evidence that water may exist there.

▲ *The Lunar Prospector spacecraft may have found water on the moon.*

How Did It Get There?

Some scientists speculate that the water molecules came from comets (which are 90 percent water) that hit the moon more than 4 billion years ago. Water from comets may have landed in the frigid, shadowed craters of the moon, where it mixed with the soil and froze. The Aitken Basin, at the south pole of the moon, where much of the ice was detected, is more than 12 km deep in places. Sunlight never touches most of the crater. And it is very cold—temperatures there may fall to −229°C. The conditions seem right to lock water into place for a very long time.

Think About Lunar Life

▶ Do some research on conditions on the moon. What conditions would humans have to overcome before we could establish a colony there?

100

Answers for Think About Lunar Life

Among other things, travelers to the moon must overcome the great expense of taking all of their life-support materials with them, including fuel, water, oxygen, food, and protective clothing. Moon dwellers would also face temperature extremes and intense radiation.

They would be in danger of being hit by meteorites. They would also have to find a safe way to get around on the moon.

Accept all reasonable answers for solving these problems.

CAREERS

EXPERIMENTAL PHYSICIST

In the course of a single day, you could find **Melissa Franklin** operating a huge drill, giving a tour of her lab to a 10-year-old, putting together a gigantic piece of electronic equipment, or even telling a joke. Then you'd see her really get down to business—studying the smallest particles of matter in the universe.

Melissa Franklin is an experimental physicist. "I am trying to understand the forces that describe how everything in the world moves—especially the smallest things," she explains. "I want to find the things that make up all matter in the universe and then try to understand the forces between them."

Other scientists rely on her to test some of the most important hypotheses in physics. For instance, Franklin and her team recently contributed to the discovery of a particle called the top quark. (Quarks are the tiny particles that make up protons and neutrons.)

Physicists had theorized that the top quark might exist but had no evidence. Franklin and more than 450 other scientists worked together to prove the existence of the top quark. Finding it required the use of a massive machine called a particle accelerator. Basically, a particle accelerator smashes particles together, and then scientists look for the remains of the collision. The physicists had to build some very complicated machines to detect the top quark, but the discovery was worth the effort. Franklin and the other researchers have earned the praise of scientists all over the world.

Getting Her Start

"I didn't always want to be a scientist, but what happens is that when you get hooked, you really get hooked. The next thing you know, you're driving forklifts and using overhead cranes while at the same time working on really tiny, incredibly complicated electronics. What I do is a combination of exciting things. It's better than watching TV."

It isn't just the best students who grow up to be scientists. "You can understand the ideas without having to be a math genius," Franklin says. Anyone can have good ideas, she says, absolutely anyone.

Don't Be Shy!

▶ Franklin also has some good advice for young people interested in physics. "Go and bug people at the local university. Just call up a physics person and say, 'Can I come visit you for a couple of hours?' Kids do that with me, and it's really fun." Why don't you give it a try? Prepare for the visit by making a list of questions you would like answered.

▲ *This particle accelerator was used in the discovery of the top quark.*

Chapter Organizer

CHAPTER ORGANIZATION	TIME MINUTES	OBJECTIVES	LABS, INVESTIGATIONS, AND DEMONSTRATIONS
Chapter Opener pp. 102–103	45	National Standards: UCP 1, 2, SAI 1, ST 2, SPSP 5, HNS 1, 3, PS 1b	**Start-Up Activity,** It's a Bird!, p. 103
Section 1 **Arranging the Elements**	90	▶ Describe how elements are arranged in the periodic table. ▶ Compare metals, nonmetals, and metalloids based on their properties and on their location in the periodic table. ▶ Describe the difference between a period and a group. UCP 1, ST 2, SPSP 5, HNS 1–3, PS 1b; Labs UCP 1, SAI 1	**Demonstration,** p. 104 in ATE **QuickLab,** Conduction Connection, p. 109 **Making Models,** Create a Periodic Table, p. 120 **Datasheets for LabBook,** Create a Periodic Table
Section 2 **Grouping the Elements**	90	▶ Explain why elements in a group often have similar properties. ▶ Describe the properties of the elements in the groups of the periodic table. UCP 2, 3, ST 2, SPSP 1, 5, HNS 3, PS 1b, 3e	**Interactive Explorations CD-ROM,** Element of Surprise *A **Worksheet** is also available in the **Interactive Explorations Teacher's Edition**.* **Inquiry Labs,** The Chemical Side of Light **Whiz-Bang Demonstrations,** Waiter, There's Carbon in My Sugar Bowl! **Long-Term Projects & Research Ideas,** It's Element-ary

*See page **T23** for a complete correlation of this book with the*

NATIONAL SCIENCE EDUCATION STANDARDS.

TECHNOLOGY RESOURCES

Guided Reading Audio CD
English or Spanish, Chapter 5

Interactive Explorations CD-ROM
CD 1, Exploration 5, Element of Surprise

CNN. Scientists in Action, Tracking Mercury in the Everglades, Segment 15

One-Stop Planner CD-ROM with Test Generator

CLASSROOM WORKSHEETS, TRANSPARENCIES, AND RESOURCES	SCIENCE INTEGRATION AND CONNECTIONS	REVIEW AND ASSESSMENT
Directed Reading Worksheet **Science Puzzlers, Twisters & Teasers**		
Directed Reading Worksheet, Section 1 **Transparency 248,** The Periodic Table of the Elements **Reinforcement Worksheet,** Placing All Your Elements on the Table	**Cross-Disciplinary Focus,** p. 107 in ATE **Real-World Connection,** p. 108 in ATE **Apply,** p. 110	**Section Review,** p. 111 **Quiz,** p. 111 in ATE **Alternative Assessment,** p. 111 in ATE
Directed Reading Worksheet, Section 2 **Transparency 79,** What's in a Bone? **Math Skills for Science Worksheet,** Checking Division with Multiplication **Science Skills Worksheet,** Finding Useful Sources **Critical Thinking Worksheet,** Believe It or Not	**Multicultural Connection,** p. 113 in ATE **Connect to Life Science,** p. 113 in ATE **Real-World Connection,** p. 114 in ATE **Math and More,** p. 115 in ATE **Environment Connection,** p. 116 **Cross-Disciplinary Focus,** p. 116 in ATE **Real-World Connection,** p. 116 in ATE **Multicultural Connection,** p. 117 in ATE **Connect to Astronomy,** p. 118 in ATE **Science, Technology, and Society:** The Science of Fireworks, p. 126 **Weird Science:** Buckyballs, p. 127	**Self-Check,** p. 114 **Section Review,** p. 115 **Section Review,** p. 119 **Quiz,** p. 119 in ATE **Alternative Assessment,** p. 119 in ATE

internet connect

go.hrw.com

Holt, Rinehart and Winston On-line Resources

go.hrw.com

For worksheets and other teaching aids related to this chapter, visit the HRW Web site and type in the keyword: **HSTPRT**

SCI LINKS NSTA

National Science Teachers Association

www.scilinks.org

Encourage students to use the *sci*LINKS numbers listed in the internet connect boxes to access information and resources on the **NSTA** Web site.

END-OF-CHAPTER REVIEW AND ASSESSMENT

Chapter Review in Study Guide

Vocabulary and Notes in Study Guide

Chapter Tests with Performance-Based Assessment, Chapter 5 Test

Chapter Tests with Performance-Based Assessment, Performance-Based Assessment 5

Concept Mapping Transparency 12

Chapter Resources & Worksheets

Visual Resources

TEACHING TRANSPARENCIES

#248

Holt Science and Technology

The Periodic Table of the Elements

248

TEACHING TRANSPARENCIES

#79

Holt Science and Technology

79

What's in a Bone?

Blood vessels

Spongy bone

Marrow

LINK TO LIFE SCIENCE

CONCEPT MAPPING TRANSPARENCY

#12

Holt Science and Technology

12

The Periodic Table

Use the following terms to complete the concept map below: elements, periods, metals, electrons, nonmetals, periodic table, families

is a chart of

arranged in

that are classified as

rows called

columns called

metalloids

in which the number of

increases from left to right

Meeting Individual Needs

DIRECTED READING

#5

DIRECTED READING WORKSHEET

The Periodic Table

Chapter Introduction

As you begin this chapter, answer the following.

1. Read the title of the chapter. List three things that you already know about this subject.

2. Write two questions about this subject that you would like answered by the time you finish this chapter.

3. How does the title of the Start-Up Activity relate to the subject of the chapter?

Section 1: Arranging the Elements (p. 104)

4. Why do you think scientists might have been frustrated by the organization of the elements before 1869?

REINFORCEMENT & VOCABULARY REVIEW

#5

REINFORCEMENT WORKSHEET

Placing All Your Elements on the Table

Complete this worksheet after you have finished reading Chapter 13, Section 2. You can tell a lot about the properties of an element just by looking at its location on the periodic table. This worksheet will help you better understand the connection between the periodic table and the properties of the elements. Follow the directions below and use crayons or colored pencils to color the periodic table at the bottom of the page.

1. Shade the square for hydrogen yellow.
2. Shade the groups with very inactive metals red.
3. Shade and label the noble gases black.
4. Shade the transition metals green.
5. Using black, mark the zigzag line that shows the position of the metalloids.
6. Shade the metalloids purple.
7. Use blue to shade all of the nonmetals that are not noble gases.
8. Shade the metals in Groups 13–16 brown.
9. Circle and label the actinides in yellow.
10. Circle and label the lanthanides in red.
11. Circle and label alkali metals in blue.
12. Circle and label the alkaline-earth metals in purple.
13. Circle and label the halogens in green.

The alkaline-earth metals react similarly because they all have the same number of electrons in their outer energy level.

14. Which group contains the alkaline-earth metals?
15. How many electrons are in the outer energy level of the atoms of the alkaline-earth metals?
16. Hydrogen is shaded in a different color than the rest of the elements in Group 1. Why do chemists set hydrogen apart from the other elements in the periodic table?

#5

VOCABULARY REVIEW WORKSHEET

Bringing It to the (Periodic) Table

Complete the following puzzle after you finish reading Chapter 13. On the next page is a partially filled in quotation by Dmitri Mendeleev. Fill in the term described by each clue below. Then put the numbered letters into the corresponding squares on the next page to find out what Mendeleev said. The answers to questions 9–11 are chemical symbols.

1. a force that opposes motion

2. the remains of plants and animals that lived millions of years ago

3. the kind of energy you have when you are swimming

4. the process that captures the sun's energy for foodmaking in plants

5. energy resources that cannot be replaced

6. units used to express energy

7. a well-defined group of objects that transfer energy among each other

8. The sum of kinetic and potential energies

9. potential energy related to an object's weight and distance to the ground

10. the ability to do work

11. resources that are continually replaced in nature

12. the energy of shape or position

SCIENCE PUZZLERS, TWISTERS & TEASERS

#5

SCIENCE PUZZLERS, TWISTERS & TEASERS

The Periodic Table

Periodic Crime

1. This is an eyewitness account of a crime recently committed. Which element committed the crime?

"He was definitely a metal, but really soft, like you could cut him with a knife. As he ran past us, we squirted him with a water gun. He burst into flame! It was unbelievable. We almost had him cornered, but he pulled out a vile of chlorine gas and in the blink of an eye, he disappeared. All that was left was a pile of table salt."

Elements in the Round

2. Moving from the outside of the circle to the center of the circle, choose one letter from each ring to find the names of eight common elements. Write the names of the elements on the lines provided. Each letter will be used only once.

Chapter 5 • The Periodic Table

Review & Assessment

STUDY GUIDE

#5 VOCABULARY & NOTES WORKSHEET
The Periodic Table

By studying the Vocabulary and Notes listed for each section below, you can gain a better understanding of this chapter.

SECTION 1
Vocabulary
In your own words, write a definition for each of the following terms in the space provided.

1. periodic
2. periodic law
3. period
4. group

Notes
Read the following section highlights. Then in your own words, write the highlights in your ScienceLog.

- Mendeleev developed the first periodic table. He arranged elements in order of increasing atomic mass. The properties of elements repeated in an orderly pattern, allowing Mendeleev to predict properties for elements that had not yet been discovered.
- Moseley rearranged the elements in order of increasing atomic number.
- The periodic law states that the chemical and physical properties of elements are periodic functions of their atomic numbers.
- Elements in the periodic table are divided into metals, metalloids, and nonmetals.
- Each element has a chemical symbol that is recognized around the world.
- A horizontal row of elements is called a period. The elements gradually change from metallic to nonmetallic from left to right across each period.
- A vertical column of elements is called a group or family. Elements in a group usually have similar properties.

#5 CHAPTER REVIEW WORKSHEET
The Periodic Table

USING VOCABULARY
To complete the following sentences, choose the correct term from each pair of terms listed below, and write the term in the space provided.

1. Elements in the same vertical column in the periodic table belong to the same _____ . (group or period)
2. Elements in the same horizontal row in the periodic table belong to the same _____ . (group or period)
3. The most reactive metals are _____ . (alkali metals or alkaline-earth metals)
4. Elements that are unreactive are called _____ . (noble gases or halogens)

UNDERSTANDING CONCEPTS
Multiple Choice

5. An element that is in a very reactive gas is most likely a member of the
   a. noble gases.
   b. alkali metals.
   c. halogens.
   d. actinides.

6. Which statement is true?
   a. Alkali metals are generally found in their uncombined form.
   b. Alkali metals are Group 1 elements.
   c. Alkali metals should be stored under water.
   d. Alkali metals are unreactive.

7. Which statement about the periodic table is false?
   a. There are more metals than nonmetals.
   b. The metalloids are located in Groups 13 through 16.
   c. The elements at the far left of the table are nonmetals.
   d. Elements are arranged by increasing atomic number.

8. One property of most nonmetals is that they are
   a. shiny.
   b. poor conductors of electric current.
   c. flattened when hit with a hammer.
   d. solids at room temperature.

9. Which is a true statement about elements?
   a. Every element occurs naturally.
   b. All elements are found in their uncombined form in nature.
   c. Each element has a unique atomic number.
   d. All of the elements exist in approximately equal quantities.

CHAPTER TESTS WITH PERFORMANCE-BASED ASSESSMENT

#5 THE PERIODIC TABLE
Chapter 5 Test

USING VOCABULARY
To complete the following sentences, choose the correct term from each pair of terms listed, and write the term in the blank.

1. A horizontal row of elements in the periodic table is called a _____ . (period or group)
2. The elements in Group 1 are the _____ metals, which react violently with water. (alkali or alkaline-earth)
3. A vertical column of elements in the periodic table is called a _____ . (period or group)
4. Neon and argon are known as _____ . (halogens or noble gases)
5. Elements in the same _____ often have similar chemical and physical properties. (period or group)

UNDERSTANDING CONCEPTS
Multiple Choice
Circle the correct answer.

6. Most of the elements in the periodic table are
   a. metals.
   b. metalloids.
   c. poor conductors of electric current.
   d. nonmetals.

7. Moseley rearranged the elements in Mendeleev's periodic table in terms of
   a. chemical symbols.
   b. atomic mass.
   c. density.
   d. atomic number.

8. Alkaline-earth metals _____ than alkali metals.
   a. are more reactive
   b. have greater density
   c. have lower atomic numbers
   d. are more explosive

9. The element _____ is a metalloid.
   a. silicon, Si
   b. carbon, C
   c. lead, Pb
   d. phosphorus, P

10. Most metals
    a. are easily shattered.
    b. are bad conductors of electric current.
    c. are made of atoms with many electrons in their outer energy level.
    d. are made of atoms with few electrons in their outer energy level.

#5 THE PERIODIC TABLE
SKILL BUILDER
Chapter 5 Performance-Based Assessment

Objective
You've read about the periodic table. Now you will have a chance to observe the chemical properties of elements in action! In this activity you will cause a reaction between the iron in steel wool and the oxygen in the air.

Know the Score!
As you work through the activity, keep in mind that you will be earning a grade for the following:
- how you work with materials and equipment (30%)
- the quality and clarity of your observations (40%)
- how you use the periodic table to explain those observations (30%)

MATERIALS
- graduated cylinder
- protective gloves
- vinegar
- small bowl
- fine steel wool, 0000 grade
- watch or clock
- thermometer
- rubber band

Procedure

1. Describe the steel wool before the reaction.
2. Use the graduated cylinder to measure 50 mL of vinegar. Pour the vinegar into the bowl.
3. Place the steel wool in the vinegar and leave it there for 2 minutes.
4. With a gloved hand, remove the steel wool from the vinegar. Describe the steel wool after the reaction.
5. Wrap the steel wool around the thermometer. Use a rubber band to hold the steel wool in place.
6. Record the thermometer's starting temperature.
7. Keep the steel wool around the thermometer for 10 minutes.
8. Record the thermometer's ending temperature.

Lab Worksheets

INQUIRY LABS

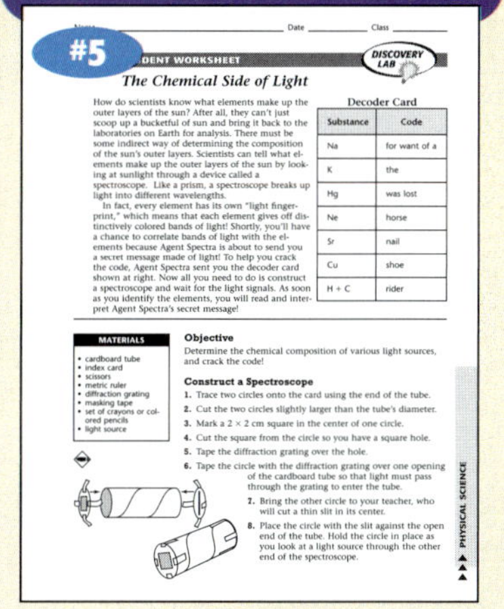

#5 STUDENT WORKSHEET
DISCOVERY LAB
The Chemical Side of Light

How do scientists know what elements make up the outer layers of the sun? After all, they can't just scoop up a bucketful of sun and bring it back to the laboratories on Earth for analysis. There must be some indirect way of determining the composition of the sun's outer layers. Scientists can tell what elements make up the outer layers of the sun by looking at sunlight through a device called a spectroscope. Like a prism, a spectroscope breaks up light into different wavelengths.

In fact, every element has its own "light fingerprint," which means that each element gives off distinctively colored bands of light! Shortly, you'll have a chance to complete a table of the elements and the colors of light they emit. This "decoder card" will reveal the elements because Agent Spectra is about to send you a secret message made of light! To help you crack the code, Agent Spectra sent you the decoder card shown at right. Now all you need to do is construct a spectroscope and wait for the light signals. As soon as you identify the elements, you will read and interpret Agent Spectra's secret message!

Decoder Card

Substance	Code
Na	for want of a
K	the
Mg	was lost
Ne	horse
Sr	nail
Co	shoe
H + C	rider

Objective
Determine the chemical composition of various light sources, and crack the code!

MATERIALS
- cardboard tube
- index card
- scissors
- metric ruler
- diffraction grating
- masking tape
- set of crayons or colored pencils
- light source

Construct a Spectroscope

1. Trace two circles onto the card using the end of the tube.
2. Cut the two circles slightly larger than the tube's diameter.
3. Mark a 2 × 2 cm square in the center of one circle.
4. Cut the square from the circle so you have a square hole.
5. Tape the diffraction grating over the hole.
6. Tape the circle with the diffraction grating over one opening of the cardboard tube so that light must pass through the grating to enter the tube.
7. Bring the other circle to your teacher, who will cut a thin slit in its center.
8. Place the circle with the slit against the open end of the tube. Hold the circle in place as you look at a light source through the other end of the spectroscope.

WHIZ-BANG DEMONSTRATIONS

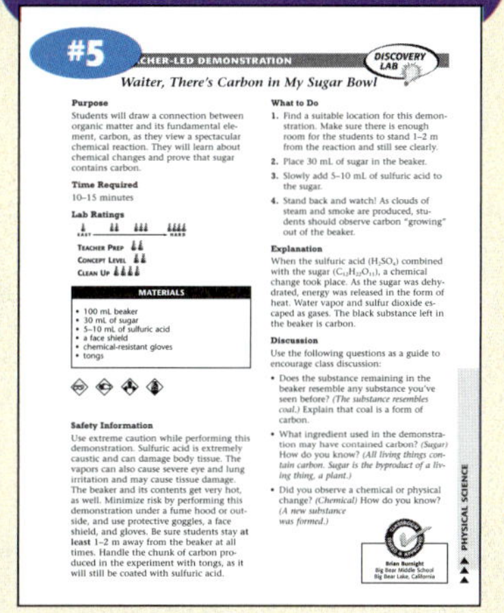

#5 TEACHER-LED DEMONSTRATION
DISCOVERY LAB
Waiter, There's Carbon in My Sugar Bowl

Purpose
Students will draw a connection between organic matter and its fundamental element, carbon, as they view a spectacular chemical reaction. They will learn about chemical changes and prove that sugar contains carbon.

Time Required
10–15 minutes

Lab Ratings
Teacher Prep
Concept Level
Clean Up

MATERIALS
- 100 mL beaker
- 30 mL of sugar
- 5–10 mL of sulfuric acid
- face shield
- chemical-resistant gloves
- tongs

What to Do

1. Find a suitable location for this demonstration. Make sure there is enough room for the students to stand 1–2 m from the reaction and still see clearly.
2. Place 30 mL of sugar in the beaker.
3. Slowly add 5–10 mL of sulfuric acid to the sugar.
4. Stand back and watch! As clouds of steam and smoke are produced, students should observe carbon "growing" out of the beaker.

Explanation
When the sulfuric acid (H_2SO_4) combined with the sugar ($C_{12}H_{22}O_{11}$), a chemical change took place. As the sugar was dehydrated, energy was released in the form of heat. Water vapor and sulfur dioxide escaped as gases. The black substance left in the beaker is carbon.

Discussion
Use the following questions as a guide to encourage class discussion:
- Does the substance remaining in the beaker resemble any substance you've seen before? (The substance resembles coal.) Explain that coal is a form of carbon.
- What ingredient used in the demonstration may have contained carbon? (Sugar: How do you know? (All living things contain carbon. Sugar is the byproduct of a living thing, a plant.)
- Did you observe a chemical or physical change? (Chemical) How do you know? (A new substance was formed.)

Safety Information
Use extreme caution while performing this demonstration. Sulfuric acid is extremely caustic and can damage body tissue. The vapors can also cause severe eye and lung irritation and may cause tissue damage. The beaker and its contents get very hot, as well. Minimize risk by performing this demonstration under a fume hood or outside. You and your students stay at least 1–2 m away from the beaker at all times. Handle the chunk of carbon produced in the experiment with tongs, as it will still be coated with sulfuric acid.

HELPFUL HINT
Try searching for the name of the metal, plus the word toxicity. For example, you might search for cadmium toxicity or lead poisoning.

LONG-TERM PROJECTS & RESEARCH IDEAS

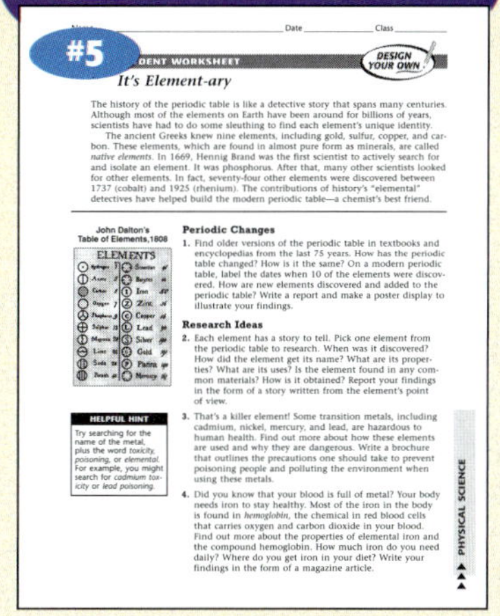

#5 STUDENT WORKSHEET
DESIGN YOUR OWN

The history of the periodic table is like a detective story that spans many centuries. Although most of the elements on Earth have been around for billions of years, scientists have had to do some sleuthing to find each element's unique identity.

The ancient Greeks knew nine elements, including gold, sulfur, copper, and carbon. These elements, which are found in almost pure form in minerals, are called native elements. In 1669, Hennig Brand was the first scientist to actively search for and isolate an element. It was phosphorus. After that, many other scientists looked for other elements. In fact, seventy-four other elements were discovered between 1737 (cobalt) and 1925 (rhenium). The contributions of history's "elemental" detectives have helped build the modern periodic table—a chemist's best friend.

John Dalton's Table of Elements, 1808
ELEMENTS

Periodic Changes

1. Find older versions of the periodic table in textbooks and encyclopedias from the last 75 years. How has the periodic table changed? How is it the same? On a modern periodic table, label the dates when 10 of the elements were discovered. How are new elements discovered and added to the periodic table? Write a report and make a poster display to illustrate your findings.

Research Ideas

2. Each element has a story to tell. Pick one element from the periodic table to research. When was it discovered? How did the element get its name? What are its properties? What are its uses? Is the element found in any common materials? How is it obtained? Report your findings in the form of a story written from the element's point of view.

3. That's a killer element! Some transition metals, including cadmium, nickel, mercury, and lead, are hazardous to human health. Find out more about how these elements are used and why they are dangerous. Write a brochure that outlines the precautions one should take to prevent poisoning people and polluting the environment when using these metals.

4. Did you know that your blood is full of metal? Your body needs iron to stay healthy. Most of the iron in the body is found in hemoglobin, the chemical in red blood cells that carries oxygen and carbon dioxide in your blood. Find out more about the properties of elemental iron and the compound hemoglobin. How much iron do you need daily? Where do you get iron in your diet? Write your findings in the form of a magazine article.

DATASHEETS FOR LABBOOK

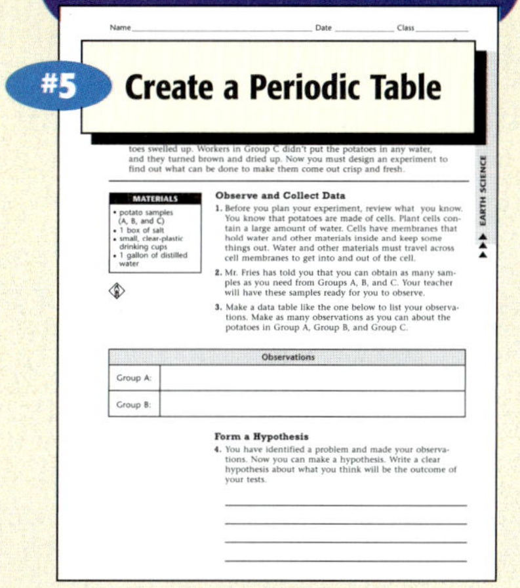

#5 **Create a Periodic Table**

toes swelled up. Workers in Group C didn't put the potatoes in any water, and they turned brown and dried up. Now you must design an experiment to find out what can be done to make them come out crisp and fresh.

MATERIALS
- potato samples (A, B, and C)
- 1 box of salt
- 1 box of sugar
- small, clear-plastic drinking cups
- 1 gallon of distilled water

Observe and Collect Data

1. Before you plan your experiment, review what you know. You know that potatoes are made of cells. Plant cells contain a large amount of water. Cells have membranes that hold water and other materials inside and keep some things out. Water and other materials must travel across cell membranes to get into and out of the cell.

2. Mr. Stamp has told you that you can obtain as many samples as you need from Groups A, B, and C. Your teacher will have these samples ready for you to observe.

3. Make a data table like the one below to list your observations. Make as many observations as you can about the potatoes in Group A, Group B, and Group C.

Observations

Group A:
Group B:

Form a Hypothesis

4. You have identified a problem and made your observations. Now you can make a hypothesis. Write a clear hypothesis about what you think will be the outcome of your tests.

Applications & Extensions

CRITICAL THINKING & PROBLEM SOLVING

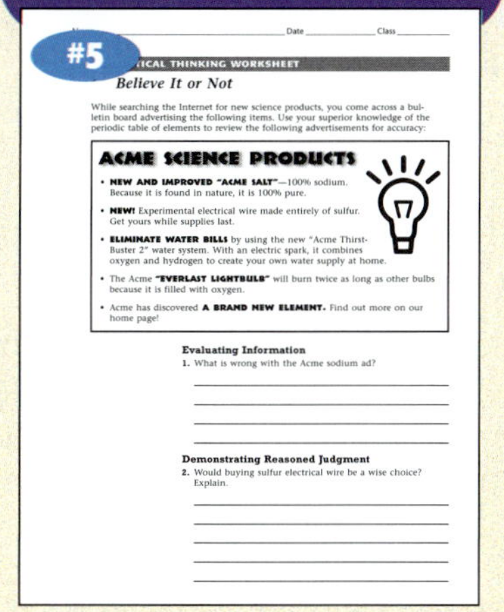

#5 CRITICAL THINKING WORKSHEET
Believe It or Not

While searching the Internet for new science products, you come across a bulletin board advertising the following items. Use your superior knowledge of the periodic table of elements to review the following advertisements for accuracy:

ACME SCIENCE PRODUCTS

- NEW AND IMPROVED "ACME SALT"—100% sodium. Because it is found in nature, it is 100% pure.
- NEW! Experimental electrical wire made entirely of sulfur. Get yours while supplies last.
- ELIMINATE WATER BILLS by using the new "Acme Thirst-Buster 2" water system. With an electric spark, it combines oxygen and hydrogen to create your own water supply at home.
- The Acme "EVERLAST LIGHTBULB" will burn twice as long as other bulbs because it is filled with oxygen.
- Acme has discovered A BRAND NEW ELEMENT. Find out more on our home page!

Evaluating Information

1. What is wrong with the Acme sodium ad?

Demonstrating Reasoned Judgment

2. Would buying sulfur electrical wire be a wise choice? Explain.

SCIENTISTS IN ACTION

#15 Science in the News: Critical Thinking Worksheets
Segment 15
Tracking Mercury in the Everglades

1. Explain why there is so much mercury in the atmosphere.

2. How does mercury travel from its source thousands of miles across the ocean to the Florida Everglades?

3. What are some of the possible effects of the mercury buildup in the Everglades?

4. What could Florida officials do to keep people from becoming sick with mercury poisoning?

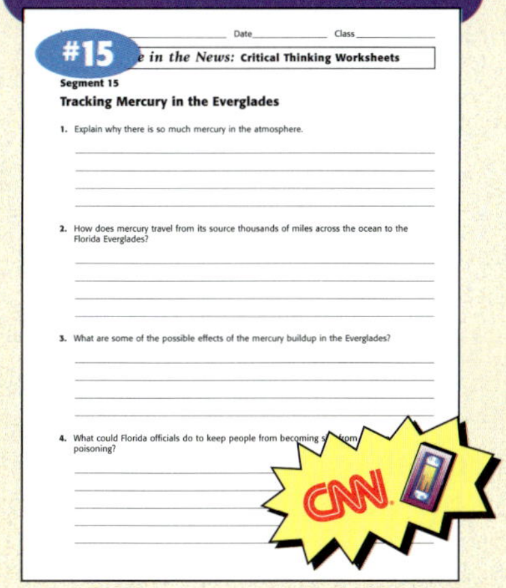

INTERACTIVE EXPLORATIONS

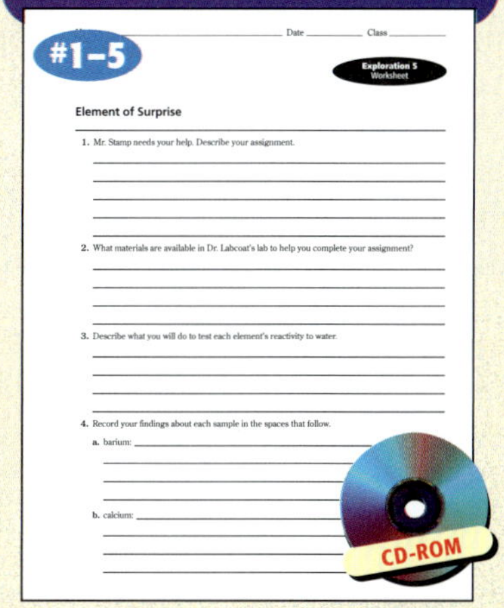

#1–5 Exploration 5 Worksheet

Element of Surprise

1. Mr. Stamp needs your help. Describe your assignment.

2. What materials are available in Dr. Labcoat's lab to help you complete your assignment?

3. Describe what you will do to test each element's reactivity to water.

4. Record your findings about each sample in the spaces that follow.
   a. barium:
   b. calcium:

Chapter Background

SECTION 1

Arranging the Elements

▶ Before the Periodic Table

Elements such as gold, silver, tin, copper, lead, and mercury have been known for thousands of years.

- The first modern discovery of an element was in 1669 when German alchemist Hennig Brand discovered phosphorus by precipitating it out of urine.

- Sixty-three elements had been discovered by 1869. As more and more elements were discovered, scientists recognized similarities and patterns in the properties of elements, and some scientists proposed classification schemes.

- In 1817, Johann Döbereiner (1780–1849) realized that the elements calcium, strontium, and barium had similar properties and that the atomic weight of strontium was about halfway between that of the other two elements.

▶ The Law of Triads

In 1829, Döbereiner made the first significant observation of chemical periodicity. After discovering two more triads—chlorine, bromine, and iodine (the halogen group) and lithium, sodium, and potassium (alkali metals)—Döbereiner proposed his law of triads: in nature there are triads of elements in which the middle element has properties that are an average of the other two when the elements are grouped by atomic weight.

- Between 1829 and 1858, several scientists worked on the idea of triads. They discovered that the chemical relationship extended beyond groups of three. Fluorine was added to the halogen group; oxygen, sulfur, selenium, and tellurium were grouped into another family; and nitrogen, phosphorus, arsenic, antimony, and bismuth were grouped into another family.

IS THAT A FACT!

- ◄ The first comprehensive arrangement of the elements showing the periodicity of chemical and physical properties was published in 1862 by French geologist A. E. Beguyer de Chancourtois. De Chancourtois positioned the elements on a cylinder in order of increasing atomic weight.

- ◄ When de Chancourtois arranged the elements so there were 16 on the cylinder per turn, he noted that closely related elements lined up vertically.

▶ The Law of Octaves

English chemist John Newlands (1837–1898) noticed that several pairs of similar elements were separated in atomic weight by some multiple of eight. In 1864, Newlands proposed his law of octaves: any element will exhibit properties similar to the eighth element following it in the table.

▶ The Father of the Periodic Table?

Two chemists, a German named Lothar Meyer (1830–1895) and a Russian named Dmitri Mendeleev (1834–1907), produced almost identical tables of the elements at almost the same time, completely independent of each other.

- Meyer was probably first; in 1868, he gave his extended table to a colleague to evaluate. Meyer published his table in 1870.

- Unfortunately for Meyer, Mendeleev published his table in 1869 and received credit for the first modern periodic table of the elements.

IS THAT A FACT!

- ◄ Mendeleev's (and Meyer's) table was a pioneering development because Mendeleev and other scientists used it to predict the existence of elements that had not yet been discovered.

- ◄ When Mendeleev first demonstrated his table, most scientists were skeptical. But then the element gallium was discovered in 1875 and was found to closely match Mendeleev's predicted properties!

SECTION 2

Grouping the Elements

▶ What Goes Where?

Mendeleev's table showed that the elements could be grouped in periods, but it didn't explain why. As the modern periodic table took shape, scientists realized that the underlying order was based on atomic structure, namely the number of protons in each atom.

- The known elements fall into three main categories, or classes, called metals, metalloids (semiconductors), and nonmetals.

▶ The Noble Gases

One of the most important additions to the periodic table was the discovery of the noble gases. English physicists John William Strutt, Lord Rayleigh (1842–1919), and William Ramsay (1852–1916) discovered argon in 1894.

- In 1895, Ramsay also discovered that helium exists on Earth. Then in 1898, Ramsay (and his assistant, Morris W. Travers) discovered three more noble gases—neon, krypton, and xenon.

- The final noble gas, radon, was discovered by German scientist Friedrich Ernst Dorn (1848–1916) in 1900.

IS THAT A FACT!

- ◤ Argon makes up about 1 percent of Earth's atmosphere, but it remained completely undetected until 1894 because it lacks chemical reactivity under normal conditions.

▶ The Modern Periodic Table

In the early 1940s, chemist Glenn Seaborg and his team worked on the Manhattan Project, America's secret effort to make the atomic bomb. Seaborg and his colleagues discovered the element plutonium in 1940.

- In the 1940s and 1950s, Seaborg's team synthesized and identified all the transuranic elements with atomic numbers 94 to 102.

- Seaborg also rearranged the periodic table by placing the actinide series below the lanthanide series. This was the last major change to the modern periodic table.

IS THAT A FACT!

- ◤ Some elements are highly reactive and combine explosively. When these reactive elements combine, the compounds they form are usually very stable. It takes a lot of energy to decompose the compound back into elements.

▶ Transuranic Elements

The chemical elements with atomic numbers greater than 92, known as transuranic elements, have been created in laboratories by bombarding heavy elements with neutrons or other subatomic particles. Plutonium (atomic number 94) occurs in small amounts in nature.

- All transuranic elements are radioactive, and some exist for only short amounts of time before they decay into other, lighter elements.

- Scientists continue to synthesize heavier transuranic elements, such as numbers 114, 116, and 118.

IS THAT A FACT!

- ◤ When elements after uranium were discovered, scientists first named them for the planets beyond Uranus: Neptune (neptunium) and Pluto (plutonium). This worked fine until an element was discovered after plutonium and there were no other planets!

- ◤ Early chemists, called alchemists, tried to change common metals, such as lead, into precious metals, such as gold. They weren't aware that this can be done only by removing protons from the lead atom.

> **For background information about teaching strategies and issues, refer to the *Professional Reference for Teachers.***

Pre-Reading Questions

Students may not know the answers to these questions before reading the chapter, so accept any reasonable response.

Suggested Answers

1. Elements are organized by their atomic number.

2. The properties of elements repeat in a pattern as you move across the table.

3. The elements in a group have the same number of electrons in their outer level.

CHAPTER
5

The Periodic Table

Sections

Pre-Reading Questions

1. How are elements organized in the periodic table?

2. Why is the table of the elements called "periodic"?

3. What one property is shared by elements in a group?

A BUILDING AS A PIECE OF ART!

Would you believe that this strange-looking building is an art museum? It is! It's the Guggenheim Museum in Bilbao, Spain. The building is made of limestone blocks, glass, and the element titanium. Titanium was chosen because it is strong, lightweight, and very resistant to corrosion and rust. In fact, the half-millimeter-thick fish-scale titanium panels covering most of the building are guaranteed to last 100 years! In this chapter, you will learn about some other elements on the periodic table and their properties.

102

---

internet connect

HRW On-line Resources	**SCiLINKS** NSTA	**CNNfyi.com**
go.hrw.com	**www.scilinks.com**	**www.si.edu/hrw**	**www.cnnfyi.com**
For worksheets and other teaching aids, visit the HRW Web site and type in the keyword: **HSTPRT**	Use the *sci*LINKS numbers at the end of each chapter for additional resources on the **NSTA** Web site.	Visit the Smithsonian Institution Web site for related on-line resources.	Visit the CNN Web site for current events coverage and classroom resources.

START-UP Activity

PLACEMENT PATTERN

In this activity, you will determine the pattern behind a new seating chart your teacher has created.

Procedure

1. In your ScienceLog, draw a seating chart for the classroom arrangement given to you by your teacher. Write the name of each of your classmates in the correct place on the chart.

2. Write information about yourself, such as your name, date of birth, hair color, and height, in the space that represents you on the chart.

3. Starting with the people around you, gather the same information about them. Write each person's information in the proper space on the seating chart.

Analysis

4. In your ScienceLog, identify a pattern to the information you gathered that might explain the order of the people in the seating chart. If you cannot find a pattern, collect more information and look again.

5. Test your pattern by gathering information from a person you did not talk to before.

6. If the new information does not support your pattern, reanalyze your data and collect more information to determine another pattern.

103

START-UP Activity

PLACEMENT PATTERN

Teacher's Notes

To do this activity, you will need to make a seating chart before the class period. Possible organizational ideas for the arrangement include placing students by birth date, by height, or alphabetically by their first names. This activity can be repeated using different patterns, including patterns that are periodic.

Answers to START-UP Activity

4. Students should be able to describe the new seating pattern and, using the information they collected, explain how they arrived at their result.

6. Some students may have difficulty determining the new seating pattern. Encourage other students to assist them in analyzing the data and finding the pattern.

Focus

Arranging the Elements

This section gives a short history of the periodic table. Students learn about the modern periodic table and are shown how to interpret it, and they learn how characteristics of elements led to their being grouped in a logical way.

🔔 Bellringer

Ask students to think of all the ways a deck of cards could be laid out so that the cards form some sort of identifiable pattern. Have them write as many different patterns as they can in their ScienceLog.

1) Motivate

DEMONSTRATION

Ask three volunteers to stand at the front of the class. Put two of them together, and ask the third to step off to the side for a moment. Ask the class what similar characteristics the two students share—in other words, why would these two be grouped together? List student responses on the board. Encourage students to look for as many similarities as possible.

Now separate the two; ask the third student to stand next to one of them. Repeat the exercise. Compare the two lists of characteristics. Discuss with the class the similarities and differences in the lists.

Directed Reading Worksheet Section 1

Terms to Learn

periodic   period
periodic law   group

What You'll Do

- Describe how elements are arranged in the periodic table.
- Compare metals, nonmetals, and metalloids based on their properties and on their location in the periodic table.
- Describe the difference between a period and a group.

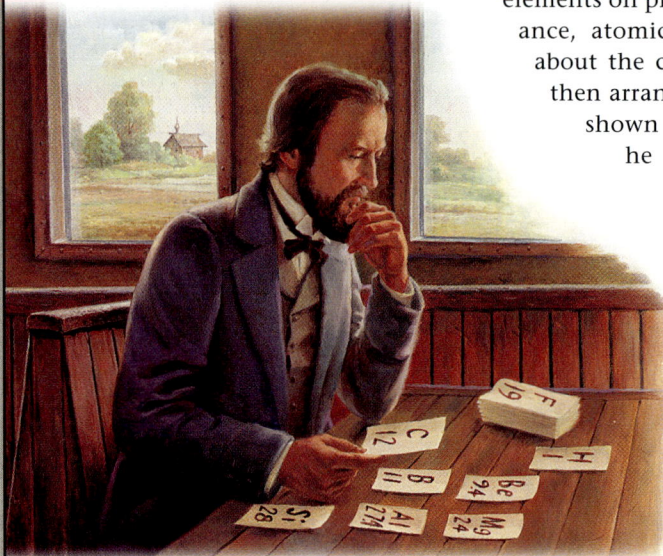

Arranging the Elements

Imagine you go to a new grocery store to buy a box of cereal. You are surprised by what you find. None of the aisles are labeled, and there is no pattern to the products on the shelves! You think it might take you days to find your cereal.

Some scientists probably felt a similar frustration before 1869. By that time, more than 60 elements had been discovered and described. However, it was not until 1869 that the elements were organized in any special way.

Discovering a Pattern

In the 1860s, a Russian chemist named Dmitri Mendeleev began looking for patterns among the properties of the elements. He wrote the names and properties of the elements on pieces of paper. He included density, appearance, atomic mass, melting point, and information about the compounds formed from the element. He then arranged and rearranged the pieces of paper, as shown in **Figure 1.** After much thought and work, he determined that there was a repeating pattern to the properties of the elements when the elements were arranged in order of increasing atomic mass.

Figure 1 *By playing "chemical solitaire" on long train rides, Mendeleev organized the elements according to their properties.*

The Properties of Elements Are Periodic Mendeleev saw that the properties of the elements were **periodic,** meaning they had a regular, repeating pattern. Many things that are familiar to you are periodic. For example, the days of the week are periodic because they repeat in the same order every 7 days.

When the elements were arranged in order of increasing atomic mass, similar chemical and physical properties were observed in every eighth element. Mendeleev's arrangement of the elements came to be known as a periodic table because the properties of the elements change in a periodic way.

Predicting Properties of Missing Elements Look at the section of Mendeleev's periodic table shown in **Figure 2.** Notice the question marks. Mendeleev recognized that there were elements missing and boldly predicted that elements yet to be discovered would fill the gaps. He also predicted the properties of the missing elements by using the pattern of properties in the periodic table. When one of the missing elements, gallium, was discovered a few years later, its properties matched Mendeleev's predictions very well. Since that time, all of the missing elements on Mendeleev's periodic table have been discovered. In the chart below, you can see Mendeleev's predictions for another missing element—germanium—and the actual properties of that element.

```
                                   Ni—Co—59
 H=1                                    Cu=63,4
         Be=9,4      Mg—24          Zn—65,2
         B—11        Al—27,4        ?—68
         C—12        Si—28          ?—70
         N—14        P—31           As—75
         O—16        S—32           Se—79,4
         F—19        Cl—35,5        Br—80
 Li=7  Na=23         K=39           Rb=85,4
                     Ca=40          Sr=87,6
                     ?—45           Ce=92
                    ?Er=56          La=94
                    ?Yt=60          Di=95
                    ?In=75,6        Th=118?
```

Figure 2 *Mendeleev used question marks to indicate some elements that he believed would later be identified.*

Properties of Germanium		
	Mendeleev's predictions	**Actual properties**
Atomic mass	72	72.6
Density	5.5 g/cm^3	5.3 g/cm^3
Appearance	dark gray metal	gray metal
Melting point	high melting point	937°C

Changing the Arrangement

Mendeleev noticed that a few elements in the table were not in the correct place according to their properties. He thought that the calculated atomic masses were incorrect and that more accurate atomic masses would eventually be determined. However, new measurements of the atomic masses showed that the masses were in fact correct.

The mystery was solved in 1914 by a British scientist named Henry Moseley (MOHZ lee). From the results of his experiments, Moseley was able to determine the number of protons—the atomic number—in an atom. When he rearranged the elements by atomic number, every element fell into its proper place in an improved periodic table.

Since 1914, more elements have been discovered. Each discovery has supported the periodic law, considered to be the basis of the periodic table. The **periodic law** states that the chemical and physical properties of elements are periodic functions of their atomic numbers. The modern version of the periodic table is shown on the following pages.

BRAIN FOOD

Moseley was 26 when he made his discovery. His work allowed him to predict that only three elements were yet to be found between aluminum and gold. The following year, as he fought for the British in World War I, he was killed in action at Gallipoli, Turkey. The British government no longer assigns scientists to combat duty.

105

2 Teach

ACTIVITY

Obtain samples of several elements that have visibly different properties and that can be displayed safely. Suggestions include:

- sulfur (in powder form)
- helium (in a clear latex balloon)
- iron (nails)
- aluminum (foil or gutter nails)
- nitrogen (a clear latex balloon filled with air serves the purpose)
- carbon (charcoal briquettes—some crushed, some whole)
- copper (electrical wire)
- silver (necklace or silver-plated tableware).

Ask students if they can identify any of the samples, and discuss why they have made a particular identification. Help students identify all the elements and label them.

Now ask students to identify *observed* characteristics for each element, such as color, physical state (solid/liquid/gas), and luster (shininess or glossiness). List on the board all the characteristics for each element. Give hints about characteristics they might miss.

BRAIN FOOD

Students should review Chapter 4, "Introduction to Atoms," especially the material on protons, atomic number, and atomic mass.

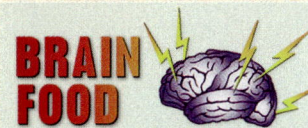

IS THAT A FACT!

In the late 1800s, scientists began studying the color spectra produced by elements when they are heated. Even the spectra have patterns within them, and the patterns are all different—no two elements have the same spectrum.

READING 📖 STRATEGY

Discussion Assist students in recognizing the layout pattern for the periodic table of the elements. Have them count across, group by group, to see that there are a total of 18 groups. Do the same for the seven periods. Discuss the triads that Döbereiner found (see Background material on page 101E), the expanded triads, and the noble gases. Emphasize that the lanthanides and actinides are parts of periods 6 and 7 and are not periods by themselves.

USING THE FIGURE

Look at the **Periodic Table of the Elements** on pages 106 and 107. The use of the colors on this periodic table will be continued throughout the book. Any time a square from the periodic table is shown, the color pattern for the type of element it is and its state will match what is shown here. This will help remind students which elements are in a group as they learn about each individual group.

The groups on this periodic table are numbered in the format most currently accepted by the International Union of Pure and Applied Chemistry (IUPAC). Be aware that older copies of the periodic table may have Roman numerals and letters to designate the various groups.

Periodic Table of the Elements

Each square on the table includes an element's name, chemical symbol, atomic number, and atomic mass.

Atomic number ———— 6
Chemical symbol ———— **C**
Element name ———— Carbon
Atomic mass ———— 12.0

The background color indicates the type of element. Carbon is a nonmetal.

The color of the chemical symbol indicates the physical state at room temperature. Carbon is a solid.

Background
Metals
Metalloids
Nonmetals

Chemical Symbol
Solid
Liquid
Gas

Period 1
| 1 |
| H |
| Hydrogen |
| 1.0 |

	Group 1	Group 2
Period 2	3 Li Lithium 6.9	4 Be Beryllium 9.0
Period 3	11 Na Sodium 23.0	12 Mg Magnesium 24.3

	Group 1	Group 2	Group 3	Group 4	Group 5	Group 6	Group 7	Group 8	Group 9
Period 4	19 K Potassium 39.1	20 Ca Calcium 40.1	21 Sc Scandium 45.0	22 Ti Titanium 47.9	23 V Vanadium 50.9	24 Cr Chromium 52.0	25 Mn Manganese 54.9	26 Fe Iron 55.8	27 Co Cobalt 58.9
Period 5	37 Rb Rubidium 85.5	38 Sr Strontium 87.6	39 Y Yttrium 88.9	40 Zr Zirconium 91.2	41 Nb Niobium 92.9	42 Mo Molybdenum 95.9	43 Tc Technetium (97.9)	44 Ru Ruthenium 101.1	45 Rh Rhodium 102.9
Period 6	55 Cs Cesium 132.9	56 Ba Barium 137.3	57 La Lanthanum 138.9	72 Hf Hafnium 178.5	73 Ta Tantalum 180.9	74 W Tungsten 183.8	75 Re Rhenium 186.2	76 Os Osmium 190.2	77 Ir Iridium 192.2
Period 7	87 Fr Francium (223.0)	88 Ra Radium (226.0)	89 Ac Actinium (227.0)	104 Rf Rutherfordium (261.1)	105 Db Dubnium (262.1)	106 Sg Seaborgium (263.1)	107 Bh Bohrium (262.1)	108 Hs Hassium (265)	109 Mt Meitnerium (266)

A row of elements is called a period.

A column of elements is called a group or family.

Lanthanides
| 58 Ce Cerium 140.1 | 59 Pr Praseodymium 140.9 | 60 Nd Neodymium 144.2 | 61 Pm Promethium (144.9) | 62 Sm Samarium 150.4 |

Actinides
| 90 Th Thorium 232.0 | 91 Pa Protactinium 231.0 | 92 U Uranium 238.0 | 93 Np Neptunium (237.0) | 94 Pu Plutonium 244.1 |

These elements are placed below the table to allow the table to be narrower.

internet**connect**
SCiLINKS
NSTA
TOPIC: The Periodic Table
GO TO: www.scilinks.org
*sci*LINKS **NUMBER:** HSTP280

This zigzag line reminds you where the metals, nonmetals, and metalloids are.

	Group 13	Group 14	Group 15	Group 16	Group 17	Group 18
						2 **He** Helium 4.0
	5 **B** Boron 10.8	6 **C** Carbon 12.0	7 **N** Nitrogen 14.0	8 **O** Oxygen 16.0	9 **F** Fluorine 19.0	10 **Ne** Neon 20.2
	13 **Al** Aluminum 27.0	14 **Si** Silicon 28.1	15 **P** Phosphorus 31.0	16 **S** Sulfur 32.1	17 **Cl** Chlorine 35.5	18 **Ar** Argon 39.9

Group 10	Group 11	Group 12	Group 13	Group 14	Group 15	Group 16	Group 17	Group 18
28 **Ni** Nickel 58.7	29 **Cu** Copper 63.5	30 **Zn** Zinc 65.4	31 **Ga** Gallium 69.7	32 **Ge** Germanium 72.6	33 **As** Arsenic 74.9	34 **Se** Selenium 79.0	35 **Br** Bromine 79.9	36 **Kr** Krypton 83.8
46 **Pd** Palladium 106.4	47 **Ag** Silver 107.9	48 **Cd** Cadmium 112.4	49 **In** Indium 114.8	50 **Sn** Tin 118.7	51 **Sb** Antimony 121.8	52 **Te** Tellurium 127.6	53 **I** Iodine 126.9	54 **Xe** Xenon 131.3
78 **Pt** Platinum 195.1	79 **Au** Gold 197.0	80 **Hg** Mercury 200.6	81 **Tl** Thallium 204.4	82 **Pb** Lead 207.2	83 **Bi** Bismuth 209.0	84 **Po** Polonium (209.0)	85 **At** Astatine (210.0)	86 **Rn** Radon (222.0)
110 **Uun*** Ununnilium (271)	111 **Uuu*** Unununium (272)	112 **Uub*** Ununbium (277)		114 **Uuq*** Ununquadium (285)		116 **Uuh*** Ununhexium (289)		118 **Uuo*** Ununoctium (293)

A number in parenthesis is the mass number of the most stable form of that element.

63 **Eu** Europium 152.0	64 **Gd** Gadolinium 157.3	65 **Tb** Terbium 158.9	66 **Dy** Dysprosium 162.5	67 **Ho** Holmium 164.9	68 **Er** Erbium 167.3	69 **Tm** Thulium 168.9	70 **Yb** Ytterbium 173.0	71 **Lu** Lutetium 175.0
95 **Am** Americium (243.1)	96 **Cm** Curium (247.1)	97 **Bk** Berkelium (247.1)	98 **Cf** Californium (251.1)	99 **Es** Einsteinium (252.1)	100 **Fm** Fermium (257.1)	101 **Md** Mendelevium (258.1)	102 **No** Nobelium (259.1)	103 **Lr** Lawrencium (262.1)

*The official names and symbols for the elements greater than 109 will eventually be approved by a committee of scientists.

107

MEETING INDIVIDUAL NEEDS

Learners Having Difficulty
Make a board game consisting of a dozen or so flip pages. Select 24–36 elements from the periodic table. Have students write short descriptions of these elements on the tops of one set of flip pages. Ask other students to find images or photographs of something made from those elements. Under the second flip page is the answer.

Students play the game by trying to guess the element from its description or its uses. The game works best if a wide variety of elements are available. Keeping score is up to you. Sheltered English

REAL-WORLD CONNECTION

Homes built between 1965 and 1973 may contain aluminum wiring, which can be very dangerous. This type of wiring has conductors made of aluminum that may corrode at any connection. Corrosion causes increased electrical resistance, and this may cause the wire to overheat and start a fire. By 1973, manufacturers had corrected this problem, making aluminum wiring used after 1973 much safer. To be safe, people who buy homes that were wired between 1965 and 1973 should check the wiring and replace it if necessary.

Finding Your Way Around the Periodic Table

At first glance, you might think studying the periodic table is like trying to explore a thick jungle without a guide—it would be easy to get lost! However, the table itself contains a lot of information that will help you along the way.

Classes of Elements Elements are classified as metals, nonmetals, and metalloids, according to their properties. The number of electrons in the outer energy level of an atom also helps determine which category an element belongs in. The zigzag line on the periodic table can help you recognize which elements are metals, which are nonmetals, and which are metalloids.

Metals

Most elements are metals. Metals are found to the left of the zigzag line on the periodic table. Atoms of most metals have few electrons in their outer energy level, as shown at right.

Most metals are solid at room temperature. Mercury, however, is a liquid. Some additional information on properties shared by most metals is shown below.

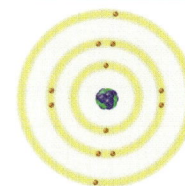

A model of a magnesium atom

Most metals are **good conductors** of thermal energy. This iron griddle conducts thermal energy from a stovetop to cook your favorite foods.

Most metals are **malleable,** meaning that they can be flattened with a hammer without shattering. Aluminum is flattened into sheets to make cans and foil.

Most metals are **ductile,** which means that they can be drawn into thin wires. All metals are good conductors of electric current. The wires in the electrical devices in your home are made from the metal copper.

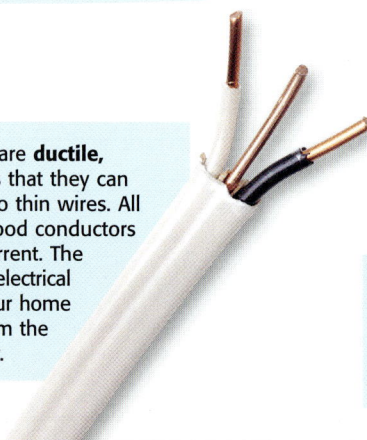

Metals tend to be **shiny.** You can see a reflection in a mirror because light reflects off the shiny surface of a thin layer of silver behind the glass.

108

IS THAT A FACT!

Mercury is the only metal that is liquid at room temperature. It was not thought to be a metal until it was frozen in 1759. The metal cesium is almost a liquid metal. It has a melting point of 28.4°C, so on a hot day in Phoenix, Arizona, (or anywhere else) cesium metal would melt into a puddle.

Nonmetals

Nonmetals are found to the right of the zigzag line on the periodic table. Atoms of most nonmetals have an almost complete set of electrons in their outer level, as shown at right. (Atoms of one group of nonmetals, the noble gases, have a complete set of electrons, with most having eight electrons in their outer energy level.)

More than half of the nonmetals are gases at room temperature. The properties of nonmetals are the opposite of the properties of metals, as shown below.

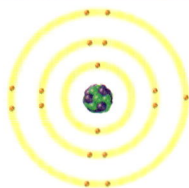

A model of a chlorine atom

Sulfur, like most nonmetals, is **not shiny.**

Nonmetals are **not malleable or ductile.** In fact, solid nonmetals, like carbon (shown here in the graphite of the pencil lead), are brittle and will break or shatter when hit with a hammer.

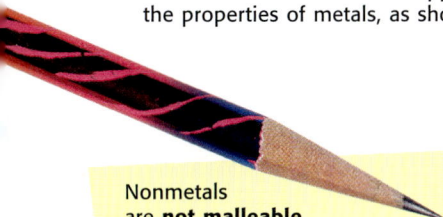

Nonmetals are **poor conductors** of thermal energy and electric current. If the gap in a spark plug is too wide, the nonmetals nitrogen and oxygen in the air will stop the spark, and a car's engine will not run.

QuickLab

Conduction Connection

1. Fill a **plastic-foam cup** with **hot water.**

2. Stand a piece of **copper wire** and a **graphite lead** from a mechanical pencil in the water.

3. After 1 minute, touch the top of each object. Record your observations.

4. Which material conducted thermal energy the best? Why?

Metalloids

Metalloids, also called semiconductors, are the elements that border the zigzag line on the periodic table. Atoms of metalloids have about a half-complete set of electrons in their outer energy level, as shown at right.

Metalloids have some properties of metals and some properties of nonmetals, as shown below.

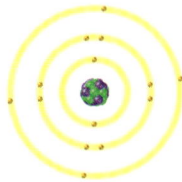

A model of a silicon atom

Boron is almost as **hard** as diamond, but it is also **very brittle.** At high temperatures, boron is a good conductor of electric current.

Tellurium is **shiny,** but it is also **brittle** and is easily smashed into a powder.

109

IS THAT A FACT!

Metalloids are also called semiconductors because they conduct electric current more easily than nonmetals but less easily than metals. The semiconductors silicon and germanium are extremely important in your everyday life. These elements are used to create microprocessor chips for computers.

QuickLab

MATERIALS

FOR EACH STUDENT:
- copper wire
- pencil lead
- plastic–foam cup
- hot water

Safety Caution: Remind students to review all safety cautions and icons before beginning this lab activity and the Reteaching activity below.

Teacher Notes: The wire should be approximately the same thickness as the pencil lead. Test the procedure; adjust the time if necessary.

Answer to QuickLab

4. The wire conducted thermal energy better than the pencil lead. The wire is made of the metal copper; pencil lead is made of graphite, a form of the nonmetal carbon. Metals conduct thermal energy better than nonmetals.

RETEACHING

Reinforce the meanings of the terms *malleable* and *brittle* as they apply to metals and nonmetals.

Safety Caution: Students should wear safety goggles when doing this activity.

Pass out to pairs or small groups of students small pieces of lead, such as fishing weights, and small hammers. Students can hammer and shape the lead to demonstrate malleability. Students may discover that malleability is not perfect, as the lead may break.

To demonstrate brittleness, wrap a charcoal briquette in an old towel and strike the wrapped briquette with a hammer. Select a piece of the briquette, and snap it to further demonstrate brittleness.

GOING FURTHER

Have students check the ingredients of foods and other products in their homes and write down the ingredients that have recognizable chemicals (such as sodium fluoride). In class, have students place self-adhesive notes containing product names on the corresponding elements on a wall-chart periodic table. Discuss with students the wide variety of elements in everyday use.

Answer to Activity

One column should include the following: H–hydrogen, B–boron, C–carbon, N–nitrogen, O–oxygen, F–fluorine. The chemical symbols for these elements are the first letter of the name of the element.

The second column should include the following: He–helium, Li–lithium, Be–beryllium, Ne–neon. The chemical symbols for these elements are the first two letters of the name of the element.

Answer to APPLY

The periodic table has the same shape, atomic numbers, and chemical symbols. The names of the elements are in a different language (Japanese). One reason it is important that the same chemical symbols are used around the world is so people can understand one another when discussing chemical substances.

Activity

Draw a line down a sheet of paper to divide it into two columns. Look at the elements with atomic numbers 1 through 10 on the periodic table. Write all the chemical symbols and names that follow one pattern in one column on your paper and all chemical symbols and names that follow a second pattern in the second column. Write a sentence describing each pattern you found. **TRY at HOME**

Each Element Is Identified by a Chemical Symbol Each square on the periodic table contains information about an element, including its atomic number, atomic mass, name, and chemical symbol. An international committee of scientists is responsible for approving the names and chemical symbols of the elements. The names of the elements come from many sources. For example, some elements are named after important scientists (mendelevium, einsteinium), and others are named for geographical regions (germanium, californium).

The chemical symbol for each element usually consists of one or two letters. The first letter in the symbol is always capitalized, and the second letter, if there is one, is always written in lowercase. The chart below lists the patterns that the chemical symbols follow, and the Activity will help you investigate two of those patterns further.

Writing the Chemical Symbols	
Pattern of chemical symbols	**Examples**
first letter of the name	S–sulfur
first two letters of the name	Ca–calcium
first letter and third or later letter of the name	Mg–magnesium
letter(s) of a word other than the English name	Pb–lead (from the Latin *plumbum*, meaning "lead")
first letter of root words that stand for the atomic number (used for elements whose official names have not yet been chosen)	Uun–ununnilium (uhn uhn NIL ee uhm) (for atomic number 110)

APPLY

One Set of Symbols

Look at the periodic table shown here. How is it the same as the periodic table you saw earlier? How is it different? Explain why it is important for scientific communication that the chemical symbols used are the same around the world.

元素の周期表

	1 H 1.0079 水素	
1		
2	3 Li 6.941 リチウム	4 Be 9.01218 ベリリウム
3	11 Na 22.98977	12 Mg

SCIENTISTS AT ODDS

When a new element is synthesized, the scientists who created the element submit a proposed name for it. Sometimes two or more scientists claim to have created the new element, and each claimant may submit a proposed name. Names are reviewed and suggested by a committee of the International Union of Pure and Applied Chemistry (IUPAC). But the committee for naming elements is made up of scientists who are competing with each other to create new elements, so the naming process is sometimes difficult. Eventually, the IUPAC designates one name as the official name, and most scientists use it from then on.

Rows Are Called Periods Each horizontal row of elements (from left to right) on the periodic table is called a **period.** For example, the row from lithium (Li) to neon (Ne) is Period 2. A row is called a period because the properties of elements in a row follow a repeating, or periodic, pattern as you move across each period. The physical and chemical properties of elements, such as conductivity and the number of electrons in the outer level of atoms, change gradually from those of a metal to those of a nonmetal in each period, as shown in **Figure 3.**

BRAIN FOOD

To remember that a period goes from left to right across the periodic table, just think of reading a sentence. You read from left to right across the page until you come to a period.

Figure 3 *The elements in a row become less metallic from left to right.*

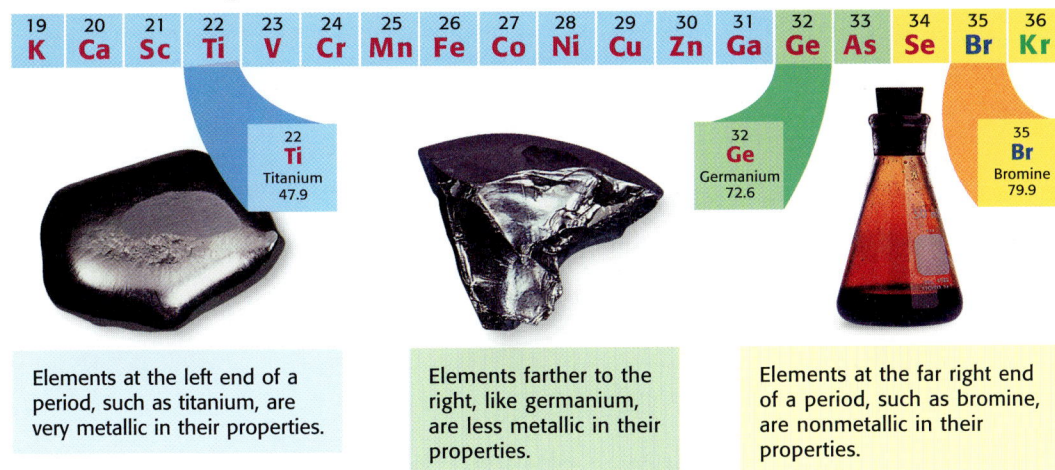

| 19 K | 20 Ca | 21 Sc | 22 Ti | 23 V | 24 Cr | 25 Mn | 26 Fe | 27 Co | 28 Ni | 29 Cu | 30 Zn | 31 Ga | 32 Ge | 33 As | 34 Se | 35 Br | 36 Kr |

22
Ti
Titanium
47.9

32
Ge
Germanium
72.6

35
Br
Bromine
79.9

Elements at the left end of a period, such as titanium, are very metallic in their properties.

Elements farther to the right, like germanium, are less metallic in their properties.

Elements at the far right end of a period, such as bromine, are nonmetallic in their properties.

Columns Are Called Groups Each column of elements (from top to bottom) on the periodic table is called a **group.** Elements in the same group often have similar chemical and physical properties. For this reason, sometimes a group is also called a family. You will learn more about each group in the next section.

SECTION REVIEW

1. Compare a period and a group on the periodic table.
2. How are the elements arranged in the modern periodic table?
3. **Comparing Concepts** Compare metals, nonmetals, and metalloids in terms of their electrical conductivity.

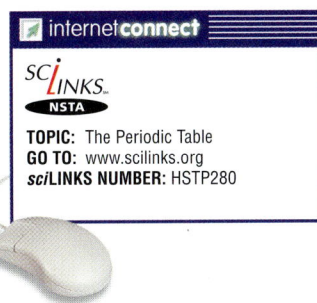

internet connect

SCI LINKS
NSTA

TOPIC: The Periodic Table
GO TO: www.scilinks.org
*sci*LINKS **NUMBER:** HSTP280

111

4) Close

Quiz

1. What does the periodic law state? (The chemical and physical properties of elements are periodic functions of their atomic numbers.)

2. Which elements are in the same group as oxygen? You can use the periodic table to answer this. (sulfur, selenium, tellurium, and polonium)

3. List five elements that have symbols that don't seem to be derived from their English names; for example, Fe is iron. (Others include K—potassium, Na—sodium, W—tungsten, Cu—copper, Ag—silver, Au—gold, and Pb—lead.)

ALTERNATIVE ASSESSMENT

Photocopy the periodic table, and cut the elements apart. Fold each square in half, and put them into a jar or box.

Ask each student to select three squares and then write a short report for each element that includes:

• the full chemical name of each element
• whether each element is a metal, a metalloid, or a nonmetal
• the order of the three by increasing atomic mass
• identification of each element's group and period
• any interesting facts about or important uses of each element

Reinforcement Worksheet "Placing All Your Elements on the Table"

Focus

Grouping the Elements

Students learn how properties of elements are used to group them in the periodic table. Students also study the relationship that elements have to each other and to the overall layout of elements within the table.

🔔 Bellringer

Ask students the following:

How do you know a bird is a bird? a kangaroo is a kangaroo? a shark is a shark? What characteristics of each animal help you to tell them apart? How does this apply to elements?

1 Motivate

DISCUSSION

Pass out several different kinds of cookies. Ask students to brainstorm about the various ingredients. (The goal is to create a long list of all the things from which you can make cookies.) Show students the list of ingredients for the entire universe—the periodic table of the elements. Discuss with students how these eight dozen or so elements combine to make everything we see around us.

Discuss with them the interesting idea that the basic ingredients of the atoms of the elements are simply the proton, neutron, and electron.

Don't forget to eat the cookies!

Terms to Learn

alkali metals
alkaline-earth metals
halogens
noble gases

What You'll Do

- ◆ Explain why elements in a group often have similar properties.
- ◆ Describe the properties of the elements in the groups of the periodic table.

Although the element hydrogen appears above the alkali metals on the periodic table, it is not considered a member of Group 1. It will be described separately at the end of this section.

Directed Reading Worksheet Section 2

Grouping the Elements

You probably know a family with several members that look a lot alike. Or you may have a friend whose little brother or sister acts just like your friend. Members of a family often—but not always—have a similar appearance or behavior. Likewise, the elements in a family or group in the periodic table often—but not always—share similar properties. The properties are similar because the atoms of the elements have the same number of electrons in their outer energy level.

Groups 1 and 2: Very Reactive Metals

The most reactive metals are the elements in Groups 1 and 2. What makes an element reactive? The answer has to do with electrons in the outer energy level of atoms. Atoms will often take, give, or share electrons with other atoms in order to have a complete set of electrons in their outer energy level. Elements whose atoms undergo such processes are *reactive* and combine to form compounds. Elements whose atoms need to take, give, or share only one or two electrons to have a filled outer level tend to be very reactive.

The elements in Groups 1 and 2 are so reactive that they are only found combined with other elements in nature. To study the elements separately, the naturally occurring compounds must first be broken apart through chemical changes.

Group 1: Alkali Metals

| 3 Li Lithium |
| 11 Na Sodium |
| 19 K Potassium |
| 37 Rb Rubidium |
| 55 Cs Cesium |
| 87 Fr Francium |

Group contains: Metals
Electrons in the outer level: 1
Reactivity: Very reactive
Other shared properties: Soft; silver-colored; shiny; low density

Alkali (AL kuh LIE) **metals** are soft enough to be cut with a knife, as shown in **Figure 4**. The densities of the alkali metals are so low that lithium, sodium, and potassium are actually less dense than water.

Figure 4 Metals so soft that they can be cut with a knife? Welcome to the alkali metals.

Alkali metals are the most reactive of the metals. This is because their atoms can easily give away the single electron in their outer level. For example, alkali metals react violently with water, as shown in **Figure 5.** Alkali metals are usually stored in oil to prevent them from reacting with water and oxygen in the atmosphere.

The compounds formed from alkali metals have many uses. Sodium chloride (table salt) can be used to add flavor to your food. Sodium hydroxide can be used to unclog your drains. Potassium bromide is one of several potassium compounds used in photography.

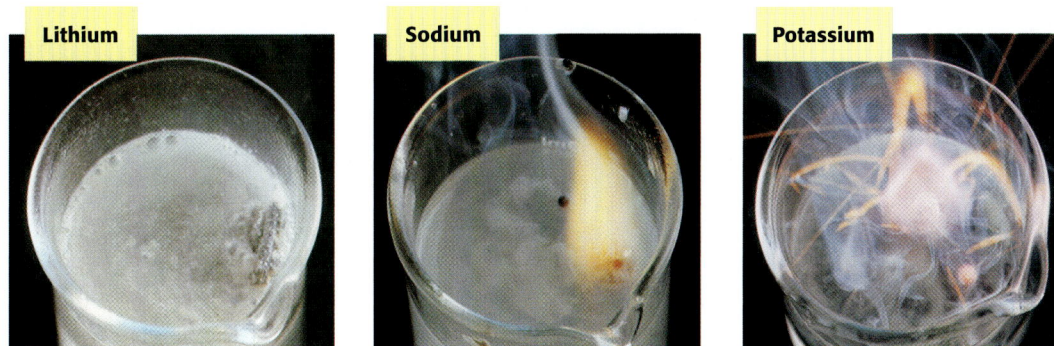

Lithium | Sodium | Potassium

Figure 5 *As alkali metals react with water, they form hydrogen gas.*

Group 2: Alkaline-earth Metals

| 4 Be Beryllium | **Group contains:** Metals |
| 12 Mg Magnesium | **Electrons in the outer level:** 2 **Reactivity:** Very reactive, but less reactive than alkali metals **Other shared properties:** Silver-colored; more dense than alkali metals |

| 20 Ca Calcium |
| 38 Sr Strontium |
| 56 Ba Barium |
| 88 Ra Radium |

Alkaline-earth metals are not as reactive as alkali metals because it is more difficult for atoms to give away two electrons than to give away only one when joining with other atoms.

The alkaline-earth metal magnesium is often mixed with other metals to make low-density materials used in airplanes. Compounds of alkaline-earth metals also have many uses. For example, compounds of calcium are found in cement, plaster, chalk, and even you, as shown in **Figure 6.**

Figure 6 *Smile! Calcium, an alkaline-earth metal, is an important component of a compound that makes your bones and teeth healthy.*

Multicultural CONNECTION

The alkali metals in Group 1, which include potassium, got their group name from Arabic. Hundreds of years ago, to isolate potassium compounds from plant matter, plants were burned, the ashes were dissolved in water, and then the water was boiled off in large pots. The powdery residue left behind, potassium carbonate, was called potash. The Arabic word for potash is *al-qili.*

CONNECT TO LIFE SCIENCE

Calcium is an element that is very important to everyone's health. Calcium is used and stored by the body in the form of calcium compounds. Almost all of the calcium compounds in the body are found in bones. Only 1 percent of calcium in the body is found in compounds in the blood and other tissues. However, this 1 percent is essential for muscle contractions, blood clotting, and nerve functioning. Refer to Teaching Transparency 79, "What's in a Bone?" to help students understand how calcium and other substances make up bones.

Teaching Transparency 79 "What's in a Bone?"

LINK TO LIFE SCIENCE

113

MISCONCEPTION ALERT

Textbooks often show photos of alkali metals reacting explosively with water and producing flames. Students should be aware that the metal itself is not burning. When the metal reacts with water, one of the products of the reaction is hydrogen gas. The energy released during the reaction often ignites the flammable hydrogen, producing flames. Not all alkali metals have the same reactivity with water. **Figure 5** illustrates that as you move down the group (lithium to sodium to potassium), the reactivity of the element with water increases.

REAL-WORLD CONNECTION

Mercury has been the liquid of choice for use in thermometers because it expands and contracts with temperature changes at a nearly constant rate. This means that for every one-degree change in temperature, the volume of mercury changes consistently.

BRAIN FOOD

Titanium is a very light but very strong structural metal. Because of these properties, it is often used in airframes and jet engines. It is also highly resistant to corrosion or erosion, so it can be used in very corrosive places, such as in salt water or in the human body.

Answer to Self-Check

It is easier for atoms of alkali metals to lose one electron than for atoms of alkaline-earth metals to lose two electrons. Therefore, alkali metals are more reactive than alkaline-earth metals.

internet connect

SCI LINKS **NSTA**

TOPIC: Metals
GO TO: www.scilinks.org
*sci*LINKS **NUMBER:** HSTP285

Groups 3–12: Transition Metals

Groups 3–12 do not have individual names. Instead, these groups are described together under the name *transition metals*.

Group contains: Metals
Electrons in the outer level: 1 or 2
Reactivity: Less reactive than alkaline-earth metals
Other shared properties: Shiny; good conductors of thermal energy and electric current; higher densities and melting points (except for mercury) than elements in Groups 1 and 2

21 Sc Scandium	22 Ti Titanium	23 V Vanadium	24 Cr Chromium	25 Mn Manganese	26 Fe Iron	27 Co Cobalt	28 Ni Nickel	29 Cu Copper	30 Zn Zinc
39 Y Yttrium	40 Zr Zirconium	41 Nb Niobium	42 Mo Molybdenum	43 Tc Technetium	44 Ru Ruthenium	45 Rh Rhodium	46 Pd Palladium	47 Ag Silver	48 Cd Cadmium
57 La Lanthanum	72 Hf Hafnium	73 Ta Tantalum	74 W Tungsten	75 Re Rhenium	76 Os Osmium	77 Ir Iridium	78 Pt Platinum	79 Au Gold	80 Hg Mercury
89 Ac Actinium	104 Rf Rutherfordium	105 Db Dubnium	106 Sg Seaborgium	107 Bh Bohrium	108 Hs Hassium	109 Mt Meitnerium	110 Uun Ununnilium	111 Uuu Unununium	112 Uub Ununbium

The atoms of transition metals do not give away their electrons as easily as atoms of the Group 1 and Group 2 metals do, making transition metals less reactive than the alkali metals and the alkaline-earth metals. The properties of the transition metals vary widely, as shown in **Figure 7.**

Figure 7 *Transition metals have a wide range of physical and chemical properties.*

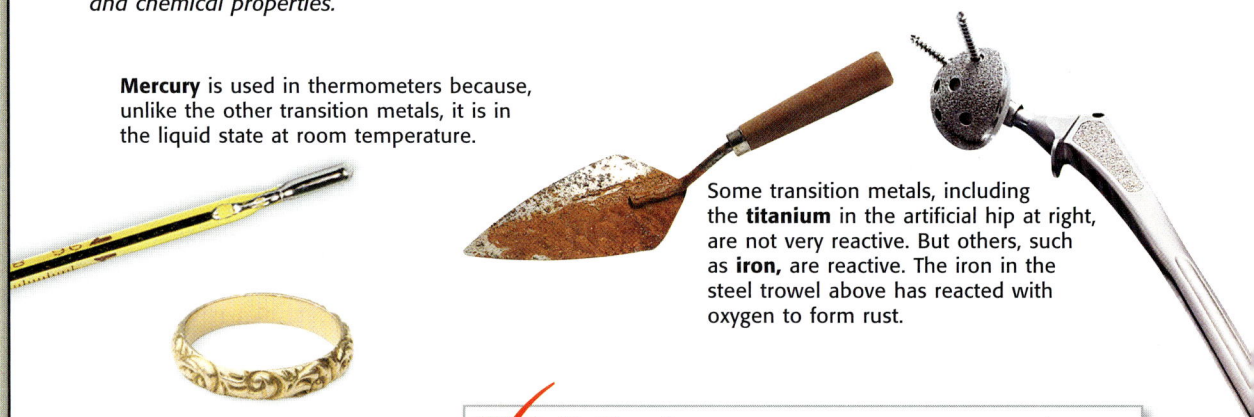

Mercury is used in thermometers because, unlike the other transition metals, it is in the liquid state at room temperature.

Some transition metals, including the **titanium** in the artificial hip at right, are not very reactive. But others, such as **iron,** are reactive. The iron in the steel trowel above has reacted with oxygen to form rust.

Many transition metals are silver-colored—but not all! This **gold** ring proves it!

✔ Self-Check

Why are alkali metals more reactive than alkaline-earth metals? *(See page 168 to check your answer.)*

SCIENCE HUMOR

In a meeting of the transition metals, mercury wished to speak to the entire group. But the group didn't let mercury speak because they didn't like to listen to heavy metal.

57													
La													
Lanthanum													
138.9													

89
Ac
Actinium
(227.0)

Lanthanides and Actinides Some transition metals from Periods 6 and 7 are placed at the bottom of the periodic table to keep the table from being too wide. The properties of the elements in each row tend to be very similar.

Lanthanides	58 **Ce**	59 **Pr**	60 **Nd**	61 **Pm**	62 **Sm**	63 **Eu**	64 **Gd**	65 **Tb**	66 **Dy**	67 **Ho**	68 **Er**	69 **Tm**	70 **Yb**	71 **Lu**
Actinides	90 **Th**	91 **Pa**	92 **U**	93 **Np**	94 **Pu**	95 **Am**	96 **Cm**	97 **Bk**	98 **Cf**	99 **Es**	100 **Fm**	101 **Md**	102 **No**	103 **Lr**

Elements in the first row are called *lanthanides* because they follow the transition metal lanthanum. The lanthanides are shiny, reactive metals. Some of these elements are used to make different types of steel. An important use of a compound of one lanthanide element is shown in **Figure 8.**

Elements in the second row are called *actinides* because they follow the transition metal actinium. All atoms of actinides are radioactive, which means they are unstable. The atoms of a radioactive element can change into atoms of a different element. Elements listed after plutonium, element 94, do not occur in nature but are instead produced in laboratories. You might have one of these elements in your home. Very small amounts of americium (AM uhr ISH ee uhm), element 95, are used in some smoke detectors.

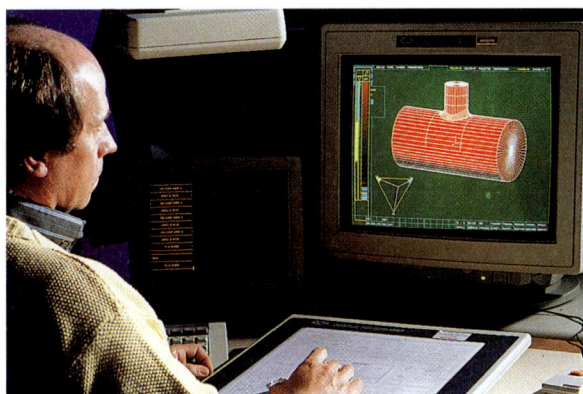

Figure 8 *Seeing red? The color red appears on a computer monitor because of a compound formed from europium that coats the back of the screen.*

SECTION REVIEW

1. What are two properties of the alkali metals?

2. What causes the properties of elements in a group to be similar?

3. **Applying Concepts** Why are neither the alkali metals nor the alkaline-earth metals found uncombined in nature?

internet connect

SC**LINKS**
NSTA

TOPIC: Metals
GO TO: www.scilinks.org
*sci***LINKS NUMBER:** HSTP285

115

Answers to Section Review

1. Answers will vary but could include that they have one electron in their outer level; are very reactive; are soft, silver-colored, and shiny; and have a low density.

2. having the same number of electrons in the outer level of their atoms

3. They are so reactive that they react with water or oxygen in the air.

History Food preservation through canning was invented in 1809 by Frenchman Nicolas-François Appert (c. 1750–1841). Tin-plated cans were first used for canning in 1810 by English inventor Peter Durand. Commercial canning was brought to the United States in 1821 when it was introduced by the William Underwood Company in Boston. In 1874, the canning process was greatly improved when cans were first heated by high-pressure steam. The high pressure in this process kept cans from bursting during heating.

REAL-WORLD CONNECTION

Aluminum alloys are often used in above-deck parts of cruise ships. They are strong, resistant to corrosion, and lightweight.

USING THE FIGURE

Many elements have several forms, called allotropes. For example, oxygen gas and ozone are allotropes of oxygen. Allotropes are usually stable at different temperatures and pressures. For example, diamond, graphite, and buckyballs are allotropes of carbon. Refer to **Figure 10** on page 117.

Groups 13–16: Groups with Metalloids

Moving from Group 13 across to Group 16, the elements shift from metals to nonmetals. Along the way, you find the metalloids. These elements have some properties of metals and some properties of nonmetals.

| 5 B Boron |
| 13 Al Aluminum |
| 31 Ga Gallium |
| 49 In Indium |
| 81 Tl Thallium |

Group 13: Boron Group

Group contains: One metalloid and four metals
Electrons in the outer level: 3
Reactivity: Reactive
Other shared properties: Solid at room temperature

The most common element from Group 13 is aluminum. In fact, aluminum is the most abundant metal in Earth's crust. Until the 1880s, it was considered a precious metal because the process used to produce pure aluminum was very expensive. In fact, aluminum was even more valuable than gold, as shown in **Figure 9.**

Today, the process is not as difficult or expensive. Aluminum is now an important metal used in making lightweight automobile parts and aircraft, as well as foil, cans, and wires.

Figure 9 *During the 1850s and 1860s, Emperor Napoleon III of France used aluminum dinnerware because aluminum was more valuable than gold!*

Environment CONNECTION

Recycling aluminum uses less energy than obtaining aluminum in the first place. Aluminum must be separated from bauxite, a mixture containing naturally occurring compounds of aluminum. Twenty times more electrical energy is required to separate aluminum from bauxite than to recycle used aluminum.

| 6 C Carbon |
| 14 Si Silicon |
| 32 Ge Germanium |
| 50 Sn Tin |
| 82 Pb Lead |
| 114 Uuq Ununquadium |

Group 14: Carbon Group

Group contains: One nonmetal, two metalloids, and two metals
Electrons in the outer level: 4
Reactivity: Varies among the elements
Other shared properties: Solid at room temperature

The metalloids silicon and germanium are used to make computer chips. The metal tin is useful because it is not very reactive. A tin can is really made of steel coated with tin. The tin is less reactive than the steel, and it keeps the steel from rusting.

IS THAT A FACT!

Less than 50 years ago, most scientists believed that silicon had little commercial use. They didn't foresee the invention of the silicon transistor chip, which led the way for the development of computer chips. Now, industrial processes using silicon employ millions of people worldwide. The main silicon product, integrated circuits for computer and game chips, has changed the world.

The nonmetal carbon can be found uncombined in nature, as shown in **Figure 10.** Carbon forms a wide variety of compounds. Some of these compounds, including proteins, fats, and carbohydrates, are essential to life on Earth.

Figure 10 *Diamonds and soot have very different properties, yet both are natural forms of carbon.*

Diamond is the hardest material known. It is used as a jewel and on cutting tools such as saws, drills, and files.

Soot—formed from burning oil, coal, and wood—is used as a pigment in paints and crayons.

Group 15: Nitrogen Group

| 7 N Nitrogen |
| 15 P Phosphorus |
| 33 As Arsenic |
| 51 Sb Antimony |
| 83 Bi Bismuth |

Group contains: Two nonmetals, two metalloids, and one metal
Electrons in the outer level: 5
Reactivity: Varies among the elements
Other shared properties: All but nitrogen are solid at room temperature.

Nitrogen, which is a gas at room temperature, makes up about 80 percent of the air you breathe. Nitrogen removed from air is reacted with hydrogen to make ammonia for fertilizers.

Although nitrogen is unreactive, phosphorus is extremely reactive, as shown in **Figure 11.** In fact, phosphorus is only found combined with other elements in nature.

Figure 11
Simply striking a match on the side of this box causes chemicals on the match to react with phosphorus on the box and begin to burn.

FIRES
32 Strike-on-
SAFETY MAT
Keep Away From Ch
Acme Match Co. Austi

Group 16: Oxygen Group

| 8 O Oxygen |
| 16 S Sulfur |
| 34 Se Selenium |
| 52 Te Tellurium |
| 84 Po Polonium |
| 116 Uuh Ununhexium |

Group contains: Three nonmetals, one metalloid, and one metal
Electrons in the outer level: 6
Reactivity: Reactive
Other shared properties: All but oxygen are solid at room temperature.

Oxygen makes up about 20 percent of air. Oxygen is necessary for substances to burn, such as the chemicals on the match in Figure 11. Sulfur, another common member of Group 16, can be found as a yellow solid in nature. The principal use of sulfur is to make sulfuric acid, the most widely used compound in the chemical industry.

117

3 Extend

RESEARCH

PORTFOLIO
Ask students to find out more about sulfur and sulfuric acid. Sulfuric acid is widely used in the chemical industry. Have students find out how and why it is so widely used. Students may also investigate sulfur, sulfuric acid, smog, and acid rain.

GOING FURTHER

Writing
Sulfur was used by prehistoric people as a pigment for cave drawings. It was also used in Egyptian ceremonies 4,000 years ago and in Chinese fireworks in about 500 B.C. It's even mentioned in Greek mythology. Have students write a report, make a poster, or prepare a presentation on the uses of sulfur before it was recognized as an element in 1777.
Sheltered English

WORKSHEET
Science Skills Worksheet
"Finding Useful Sources"

internet**connect**
SCI LINKS
NSTA
TOPIC: Metalloids
GO TO: www.scilinks.org
*sci***LINKS NUMBER:** HSTP290

Multicultural CONNECTION

About a century ago, Marie Sklodowska Curie (1867–1934) made many contributions to the study of radioactivity and radioactive elements. In 1903, Marie, her husband, Pierre Curie (1859–1906), and French physicist Henri Becquerel (1852–1908) were awarded the Nobel Prize in Physics for their contributions to understanding radioactivity. The Curies discovered the radioactive elements polonium and radium and isolated samples of these elements from tons of ore. For her discoveries of polonium and radium, Marie Curie was awarded the 1911 Nobel Prize in Chemistry.

Astronomers have evidence that the universe began with nothing more than hydrogen and helium. Where did all of the other elements come from?

Stars seem to be the factories that created all of the naturally occurring elements throughout the universe. Students can find information about and photographs of areas in the universe where new stars are born and other areas where old stars have exploded. Have students research how elements may be created or changed in these violent reactions.

MEETING INDIVIDUAL NEEDS

Advanced Learners The word *halogen* comes from the Greek words meaning "salt former." Sodium chloride, table salt, is composed of the halogen chlorine and the alkali metal sodium. Have students research halogens and their uses and then prepare a chart or a poster that shows what they learned.

Interactive Explorations CD-ROM "Element of Surprise"

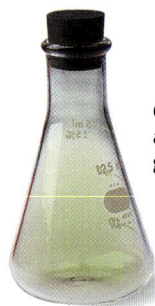

Groups 17 and 18: Nonmetals Only

The elements in Groups 17 and 18 are nonmetals. The elements in Group 17 are the most reactive nonmetals, but the elements in Group 18 are the least reactive nonmetals. In fact, the elements in Group 18 normally won't react at all with other elements.

Chlorine is a yellowish green gas.

Bromine is a dark red liquid.

Iodine is a dark gray solid.

Figure 12 *Physical properties of some halogens at room temperature are shown here.*

| 9 F Fluorine |
| 17 Cl Chlorine |
| 35 Br Bromine |
| 53 I Iodine |
| 85 At Astatine |

Group 17: Halogens

Group contains: Nonmetals
Electrons in the outer level: 7
Reactivity: Very reactive
Other shared properties: Poor conductors of electric current; react violently with alkali metals to form salts; never found uncombined in nature

Halogens are very reactive nonmetals because their atoms need to gain only one electron to have a complete outer level. The atoms of halogens combine readily with other atoms, especially metals, to gain that missing electron.

Although the chemical properties of the halogens are similar, the physical properties are quite different, as shown in **Figure 12.**

Both chlorine and iodine are used as disinfectants. Chlorine is used to treat water, while iodine mixed with alcohol is used in hospitals.

| 2 He Helium |
| 10 Ne Neon |
| 18 Ar Argon |
| 36 Kr Krypton |
| 54 Xe Xenon |
| 86 Rn Radon |
| 118 Uuo Ununoctium |

Group 18: Noble Gases

Group contains: Nonmetals
Electrons in the outer level: 8 (2 for helium)
Reactivity: Unreactive
Other shared properties: Colorless, odorless gases at room temperature

Noble gases are unreactive nonmetals. Because the atoms of the elements in this group have a complete set of electrons in their outer level, they do not need to lose or gain any electrons. Therefore, they do not react with other elements under normal conditions.

All of the noble gases are found in Earth's atmosphere in small amounts. Argon, the most abundant noble gas in the atmosphere, makes up almost 1 percent of the atmosphere.

BRAIN FOOD

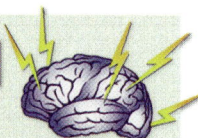

The term *noble gases* describes the nonreactivity of these elements. Just as nobles, such as kings and queens, did not often mix with common people, the noble gases do not normally react with other elements.

118

MISCONCEPTION
/// **ALERT** \\\

Noble gases were originally called *inert gases* because it was thought that they would not react with any elements. However, scientists are able to use high temperatures and pressures to cause some of the elements in Group 18 to react. Thus, the term *inert* is not correct, and the term *noble* is preferred.

The nonreactivity of the noble gases makes them useful. Ordinary light bulbs last longer when filled with argon than they would if filled with a reactive gas. Because argon is unreactive, it does not react with the metal filament in the light bulb even when the filament gets hot. The low density of helium causes blimps and weather balloons to float, and its nonreactivity makes helium safer to use than hydrogen. One popular use of noble gases that does *not* rely on their nonreactivity is shown in **Figure 13.**

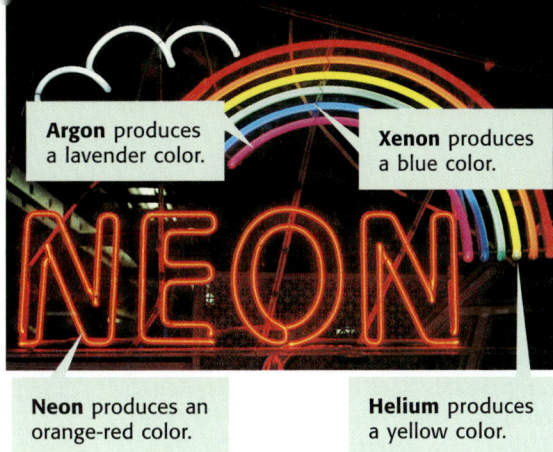

Argon produces a lavender color.

Xenon produces a blue color.

Neon produces an orange-red color.

Helium produces a yellow color.

Figure 13 *Besides neon, other noble gases are often used in "neon" lights.*

Hydrogen Stands Apart

| 1 |
| H |
| Hydrogen |

Electrons in the outer level: 1
Reactivity: Reactive
Other properties: Colorless, odorless gas at room temperature; low density; reacts explosively with oxygen

The properties of hydrogen do not match the properties of any single group, so hydrogen is set apart from the other elements in the table.

Hydrogen is placed above Group 1 in the periodic table because atoms of the alkali metals also have only one electron in their outer level. Atoms of hydrogen, like atoms of alkali metals, can give away one electron when joining with other atoms. However, hydrogen's physical properties are more like the properties of nonmetals than of metals. As you can see, hydrogen really is in a group of its own.

Hydrogen is the most abundant element in the universe. Hydrogen's reactive nature makes it useful as a fuel in rockets, as shown in **Figure 14.**

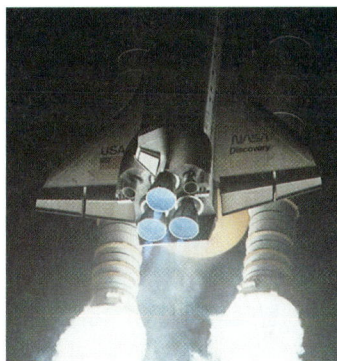

Figure 14 *Hydrogen reacts violently with oxygen. The hot water vapor that forms as a result pushes the space shuttle into orbit.*

SECTION REVIEW

1. In which group are the unreactive nonmetals found?

2. What are two properties of the halogens?

3. **Making Predictions** In the future, a new halogen may be synthesized. Predict its atomic number and properties.

4. **Comparing Concepts** Compare the element hydrogen with the alkali metal sodium.

Create a Periodic Table
Teacher's Notes

Time Required

One or two 45-minute class periods

Lab Ratings

EASY ————————→ HARD

TEACHER PREP
STUDENT SET-UP
CONCEPT LEVEL
CLEAN UP

MATERIALS

The materials listed for this lab are for each group of 2–4 students. For each group of students, assemble a collection of 20 objects (five sets of four objects). You should provide a bag containing 19 of these objects. A recommended collection of objects includes sets of coins (penny, nickel, dime, quarter), sets of buttons that are similar but vary in diameter, and washers that vary in diameter. Other objects, such as nuts, bolts, and paper circles, will work and are easily obtainable. The difference in masses should be large enough for a beam balance to detect. Ideally, each set (one column on the table) should be of the same material and thickness and vary only in diameter.

Preparation Note

You may have students prepare the 20 squares of paper, but the lab will go faster if the squares are prepared ahead of time.

Datasheets for LabBook

Making Models Lab

USING SCIENTIFIC METHODS

Create a Periodic Table

You probably have classification systems for many things in your life, such as your clothes, your books, and your CDs. One of the most important classification systems in science is the periodic table of the elements. In this lab, you will develop your own classification system for a collection of ordinary objects. You will analyze trends in your system and compare your system with the periodic table of the elements.

MATERIALS

- bag of objects
- 20 squares of paper, 3 cm × 3 cm each
- metric balance
- metric ruler
- 2 sheets of graph paper
- computer (optional)

Make Observations

1. Your teacher will give you a bag of objects. It is missing one item. Examine the items carefully. Do you recognize any patterns?

2. Lay out the paper squares on a flat surface so that you have a grid of five rows of four squares each.

3. Analyze the information about the objects to recognize a pattern. Arrange the objects according to the pattern you recognized. You should end up with one blank square for the missing object. In your ScienceLog, describe the basis for your arrangement.

4. Measure the mass (g) and diameter (mm) of each object. Record your results in the appropriate square. Each square except the empty one should have one object placed on it and two measurements written on it.

5. Examine your arrangement again. Does the order in which you arranged your objects still make sense? If necessary, rearrange the squares and their objects to improve your arrangement. Describe the basis for the new arrangement in your ScienceLog.

Form a Hypothesis

6. Based on your observations and your arrangement, form a hypothesis about what you think the identity of the missing object might be. Write your hypothesis in your ScienceLog.

120

Answers

3. Answers will vary. One basis is increasing size from left to right and increasing diameter from top to bottom. This places similar objects in a vertical group or family, like the groups of the periodic table.

6. Accept all reasonable answers.

CLASSROOM TESTED & APPROVED

Norman Holcomb
Marion Elementary School
Maria Stein, Ohio

Test the Hypothesis

7 Working across the rows, number the squares 1 to 20. When you get to the end of a row, continue numbering in the first square of the next row.

8 Copy your grid into your ScienceLog, or create a similar grid using a computer. In each square, be sure to list the type of object and label all measurements with appropriate units.

9 On graph paper or on a computer, construct a graph of your data. Show mass on the *y*-axis and object number on the *x*-axis. Label each axis, and put a title on the graph.

10 Make a second graph showing diameter on the *y*-axis and object number on the *x*-axis. Label each axis, and put a title on the graph.

Analyze the Results

11 Discuss each graph with your classmates. Try to identify any important features of the graph. For example, does the graph form a line or a curve? Is there anything unusual about the graph? What do these features tell you? Write your answers in your ScienceLog.

Draw Conclusions

12 Look back at your hypothesis about the identity of the missing object. Based on your graphs, do you think it is still accurate? Try to improve your description by estimating the mass and diameter of the missing object. Record your estimates in your ScienceLog.

13 How is your arrangement of objects a model of the periodic table of the elements found in this chapter? What are the limitations of your model?

14 How is your experiment similar to the work Mendeleev did with the elements?

121

Science Skills Worksheet
"Grasping Graphing"

Answers

9. Graphs should be similar to sample graph A.

10. Graphs should be similar to sample graph B.

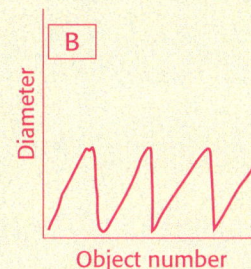

11. Answers will vary. The primary feature is the repeating pattern of increases. This pattern in the first graph indicates the periodic nature of the mass of the items. This pattern in the second graph indicates the periodic nature of the diameter of the items.

12. Answers will vary, depending on the student's original prediction. Accept all reasonable answers. (You may wish to provide the students with the missing object so they can further evaluate their predictions.)

13. Answers may vary. Similarities include repeating patterns (such as mass) across the table. Differences may include no consistent family traits and no chemical properties associated with position in the table.

14. This experiment is similar in that a pattern was identified that helped to identify characteristics of a missing object.

Chapter Highlights

VOCABULARY DEFINITIONS

SECTION 1

periodic having a regular, repeating pattern

periodic law the law that states that the chemical and physical properties of elements are periodic functions of their atomic numbers

period a horizontal row of elements on the periodic table

group a column of elements on the periodic table

Chapter Highlights

SECTION 1

Vocabulary
periodic (p. 104)
periodic law (p. 105)
period (p. 111)
group (p. 111)

Section Notes
- Mendeleev developed the first periodic table. He arranged elements in order of increasing atomic mass. The properties of elements repeated in an orderly pattern, allowing Mendeleev to predict properties for elements that had not yet been discovered.

- Moseley rearranged the elements in order of increasing atomic number.

- The periodic law states that the chemical and physical properties of elements are periodic functions of their atomic numbers.

- Elements in the periodic table are divided into metals, metalloids, and nonmetals.

- Each element has a chemical symbol that is recognized around the world.

- A horizontal row of elements is called a period. The elements gradually change from metallic to nonmetallic from left to right across each period.

- A vertical column of elements is called a group or family. Elements in a group usually have similar properties.

☑ Skills Check

Visual Understanding

PERIODIC TABLE OF THE ELEMENTS Scientists rely on the periodic table as a resource for a large amount of information. Review the periodic table on pages 106–107. Pay close attention to the labels and the key; they will help you understand the information presented in the table.

CLASSES OF ELEMENTS Identifying an element as a metal, nonmetal, or metalloid gives you a better idea of the properties of that element. Review the figures on pages 108–109 to understand how to use the zigzag line on the periodic table to identify the classes of elements and to review the properties of elements in each category.

122

Lab and Activity Highlights

Create a Periodic Table PG 120

Datasheets for LabBook
(blackline masters for these labs)

SECTION 2

Vocabulary

alkali metals *(p. 112)*

alkaline-earth metals *(p. 113)*

halogens *(p. 118)*

noble gases *(p. 118)*

Section Notes

- The alkali metals (Group 1) are the most reactive metals. Atoms of the alkali metals have one electron in their outer level.

- The alkaline-earth metals (Group 2) are less reactive than the alkali metals. Atoms of the alkaline-earth metals have two electrons in their outer level.

- The transition metals (Groups 3–12) include most of the well-known metals as well as the lanthanides and actinides located below the periodic table.

- Groups 13–16 contain the metalloids along with some metals and nonmetals. The atoms of the elements in each of these groups have the same number of electrons in their outer level.

- The halogens (Group 17) are very reactive nonmetals. Atoms of the halogens have seven electrons in their outer level.

- The noble gases (Group 18) are unreactive nonmetals. Atoms of the noble gases have a complete set of electrons in their outer level.

- Hydrogen is set off by itself because its properties do not match the properties of any one group.

SECTION 2

alkali metals the elements in Group 1 of the periodic table; they are the most reactive metals, and their atoms have one electron in their outer level

alkaline-earth metals the elements in Group 2 of the periodic table; they are reactive metals but less reactive than alkali metals; their atoms have two electrons in their outer level

halogens the elements in Group 17 of the periodic table; they are very reactive nonmetals, and their atoms have seven electrons in their outer level

noble gases the unreactive elements in Group 18 of the periodic table; their atoms have eight electrons in their outer level (except for helium, which has two electrons)

internet connect

go.hrw.com

GO TO: go.hrw.com

Visit the **HRW** Web site for a variety of learning tools related to this chapter. Just type in the keyword:

KEYWORD: HSTPRT

SCI LINKSSM

NSTA

GO TO: www.scilinks.org

Visit the **National Science Teachers Association** on-line Web site for Internet resources related to this chapter. Just type in the sciLINKS number for more information about the topic:

TOPIC: The Periodic Table   *sci*LINKS NUMBER: HSTP280
TOPIC: Metals   *sci*LINKS NUMBER: HSTP285
TOPIC: Metalloids   *sci*LINKS NUMBER: HSTP290
TOPIC: Nonmetals   *sci*LINKS NUMBER: HSTP295
TOPIC: Buckminster Fuller and the Buckyball   *sci*LINKS NUMBER: HSTP300

Vocabulary Review Worksheet

Blackline masters of these Chapter Highlights can be found in the **Study Guide.**

123

Lab and Activity Highlights

LabBank

Inquiry Labs, The Chemical Side of Light

Whiz-Bang Demonstrations, Waiter, There's Carbon in My Sugar Bowl!

Long-Term Projects & Research Ideas, It's Element-ary

Interactive Explorations CD-ROM

CD 1, Exploration 5, "Element of Surprise"

1. group
2. period
3. alkali metals
4. noble gases

UNDERSTANDING CONCEPTS

Multiple Choice

5. c
6. b
7. c
8. b
9. c
10. c

Short Answer

11. Mendeleev's periodic table allowed scientists to predict properties of elements that had not yet been found.
12. Moseley arranged elements by increasing atomic number. Mendeleev arranged elements by increasing atomic mass.
13. Both are periodic. The periodic table has repeating properties of elements. The calendar has repeating days and months.
14. Metals are located to the left of the zigzag on the periodic table. Metalloids border the zigzag. Nonmetals are to the right of the zigzag.

Concept Mapping

15.　An answer to this exercise can be found at the front of this book.

Concept Mapping Transparency 12

Chapter Review

USING VOCABULARY

Complete the following sentences by choosing the appropriate term from each pair of terms listed below.

1. Elements in the same vertical column in the periodic table belong to the same __?__. (*group* or *period*)

2. Elements in the same horizontal row in the periodic table belong to the same __?__. (*group* or *period*)

3. The most reactive metals are __?__. (*alkali metals* or *alkaline-earth metals*)

4. Elements that are unreactive are called __?__. (*noble gases* or *halogens*)

UNDERSTANDING CONCEPTS

Multiple Choice

5. An element that is a very reactive gas is most likely a member of the
   a. noble gases.
   b. alkali metals.
   c. halogens.
   d. actinides.

6. Which statement is true?
   a. Alkali metals are generally found in their uncombined form.
   b. Alkali metals are Group 1 elements.
   c. Alkali metals should be stored under water.
   d. Alkali metals are unreactive.

7. Which statement about the periodic table is false?
   a. There are more metals than nonmetals.
   b. The metalloids are located in Groups 13 through 16.
   c. The elements at the far left of the table are nonmetals.
   d. Elements are arranged by increasing atomic number.

8. One property of most nonmetals is that they are
   a. shiny.
   b. poor conductors of electric current.
   c. flattened when hit with a hammer.
   d. solids at room temperature.

9. Which is a true statement about elements?
   a. Every element occurs naturally.
   b. All elements are found in their uncombined form in nature.
   c. Each element has a unique atomic number.
   d. All of the elements exist in approximately equal quantities.

10. Which is NOT found on the periodic table?
    a. the atomic number of each element
    b. the symbol of each element
    c. the density of each element
    d. the atomic mass of each element

Short Answer

11. Why was Mendeleev's periodic table useful?

12. How is Moseley's basis for arranging the elements different from Mendeleev's?

13. How is the periodic table like a calendar?

14. Describe the location of metals, metalloids, and nonmetals on the periodic table.

Concept Mapping

15. Use the following terms to create a concept map: periodic table, elements, groups, periods, metals, nonmetals, metalloids.

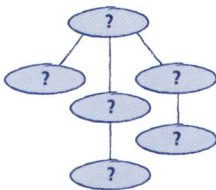

CRITICAL THINKING AND PROBLEM SOLVING

16. When an element with 115 protons in its nucleus is synthesized, will it be a metal, a nonmetal, or a metalloid? Explain.

17. Look at Mendeleev's periodic table in Figure 2. Why was Mendeleev not able to make any predictions about the noble gas elements?

18. Your classmate offers to give you a piece of sodium he found while hiking. What is your response? Explain.

19. Determine the identity of each element described below:
    a. This metal is very reactive, has properties similar to magnesium, and is in the same period as bromine.
    b. This nonmetal is in the same group as lead.
    c. This metal is the most reactive metal in its period and cannot be found uncombined in nature. Each atom of the element contains 19 protons.

MATH IN SCIENCE

20. The chart below shows the percentages of elements in the Earth's crust.

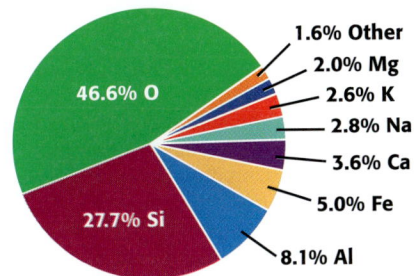

- 46.6% O
- 27.7% Si
- 8.1% Al
- 5.0% Fe
- 3.6% Ca
- 2.8% Na
- 2.6% K
- 2.0% Mg
- 1.6% Other

Excluding the "Other" category, what percentage of the Earth's crust is
    a. alkali metals?
    b. alkaline-earth metals?

INTERPRETING GRAPHICS

21. Study the diagram below to determine the pattern of the images. Predict the missing image, and draw it. Identify which properties are periodic and which properties are shared within a group.

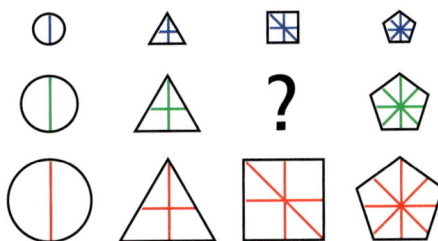

Reading Check-up

Take a minute to review your answers to the Pre-Reading Questions found at the bottom of page 102. Have your answers changed? If necessary, revise your answers based on what you have learned since you began this chapter.

125

CRITICAL THINKING AND PROBLEM SOLVING

16. Metal; it will be located below the metal bismuth to the left of the zigzag.

17. Mendeleev could only make predictions about elements where there were clear gaps in his table. Because no noble gases were known at the time, there were no obvious gaps in the table and no way he could have known a whole column was missing.

18. I would tell my classmate that he didn't find sodium. Sodium is very reactive and cannot be found uncombined in nature. It would react with oxygen and water in the air and form a compound.

19. a. calcium
    b. carbon
    c. potassium

MATH IN SCIENCE

20. a. 5.4 percent (sodium and potassium)
    b. 5.6 percent (magnesium and calcium)

INTERPRETING GRAPHICS

21.

Periodic properties are the order of the shapes and the number of lines inside the shape. The properties shared in a group are the shape and the color of the lines inside the shape.

Blackline masters of this Chapter Review can be found in the **Study Guide.**

Background

The colors in a fireworks display depend on the wavelengths of the light emitted by different chemicals as they burn. Light with the shortest wavelength appears violet in color. Light with the longest wavelength appears red. Refer to the chart at the bottom of the page for the colors produced by various elements.

When the fuse in the fireworks is lit, the gunpowder ignites and produces gases that propel the fireworks into the air. Charcoal gives the fireworks a sparkling, flaming tail.

Energy is necessary to start the reaction in a fireworks display. If the fireworks are not packed correctly, the thermal reaction fails, resulting in what is called a dud.

Teaching Strategy

Show students that each element produces a certain color. Obtain samples of calcium chloride, strontium chloride, and sodium chloride. To prepare 0.5 M solutions, dissolve the following quantities in separate containers with enough water to make 100 mL of each solution: 5.5 g of $CaCl_2$, 8.83 g of $SrCl_2 \cdot H_2O$, and 2.9 g of NaCl. Dip a different wooden splint in each solution, and use tongs to insert each splint into the flame of a portable burner to burn the chemical from the splint. Try not to ignite the splints. The splints can be dipped into the solutions again if necessary.

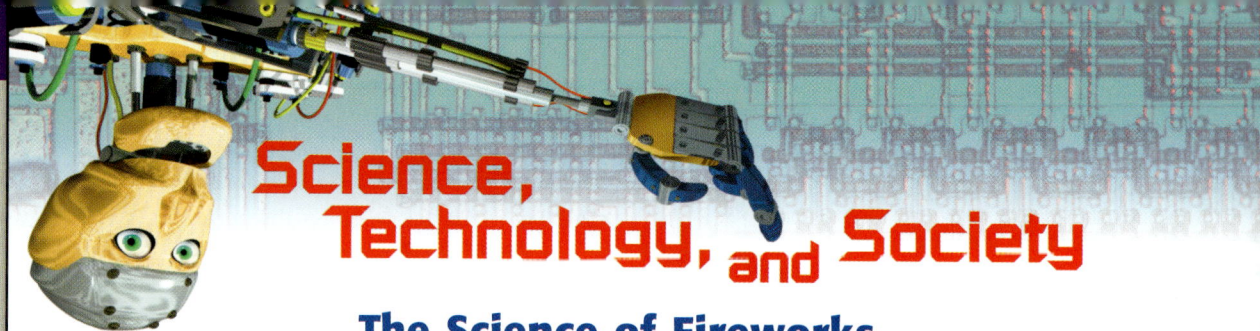

Science, Technology, and Society

The Science of Fireworks

What do the space shuttle and the Fourth of July have in common? The same scientific principles that help scientists launch a space shuttle also help pyrotechnicians create spectacular fireworks shows. The word *pyrotechnics* comes from the Greek words for "fire art." Explosive and dazzling, a fireworks display is both a science and an art.

An Ancient History

More than 1,000 years ago, Chinese civilizations made black powder, the original gunpowder used in pyrotechnics. They used the powder to set off firecrackers and primitive missiles. Black powder is still used today to launch fireworks into the air and to give fireworks an explosive charge. Even the ingredients—saltpeter (potassium nitrate), charcoal, and sulfur—haven't changed since ancient times.

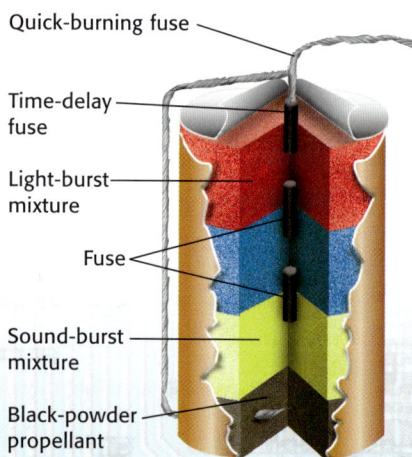

Quick-burning fuse

Time-delay fuse

Light-burst mixture

Fuse

Sound-burst mixture

Black-powder propellant

▲ *Cutaway view of a typical firework. Each shell creates a different type of display.*

126

Snap, Crackle, Pop!

The shells of fireworks contain the ingredients that create the explosions. Inside the shells, black powder and other chemicals are packed in layers. When ignited, one layer may cause a bright burst of light while a second layer produces a loud booming sound. The shell's shape affects the shape of the explosion. Cylindrical shells produce a trail of lights that looks like an umbrella. Round shells produce a star-burst pattern of lights.

The color and sound of fireworks depend on the chemicals used. To create colors, chemicals like strontium (for red), magnesium (for white), and copper (for blue) can be mixed with the gunpowder.

Explosion in the Sky

Fireworks are launched from metal, plastic, or cardboard tubes. Black powder at the bottom of the shell explodes and shoots the shell into the sky. A fuse begins to burn when the shell is launched. Seconds later, when the explosive chemicals are high in the air, the burning fuse lights another charge of black powder. This ignites the rest of the ingredients in the shell, causing an explosion that lights up the sky!

Bang for Your Buck

▶ The fireworks used during New Year's Eve and Fourth of July celebrations can cost anywhere from $200 to $2,000 apiece. Count the number of explosions at the next fireworks show you see. If each of the fireworks cost just $200 to produce, how much would the fireworks for the entire show cost?

Answer to Bang for Your Buck

Students should be aware that a fireworks display is costly. If a fireworks display consists of 50 explosions at $200 per explosion, the display would cost $10,000.

Element	Color
sodium	yellow
barium	green
nickel	green
copper	blue
strontium	crimson
lithium	bright red
calcium	dark red
magnesium	white

WEIRD SCIENCE

BUCKYBALLS

Researchers are scrambling for the ball—the buckyball, that is. This special form of carbon has 60 carbon atoms linked together in a shape much like a soccer ball. Scientists are having a field day trying to find new uses for this unusual molecule.

Potassium atom trapped inside buckyball

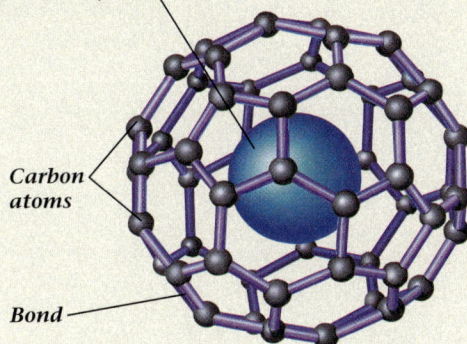

Carbon atoms

Bond

▲ *The buckyball, short for buckminster-fullerene, was named after architect Buckminster Fuller.*

The Starting Lineup

Named for architect Buckminster Fuller, bucky-balls resemble the geodesic domes that are characteristic of the architect's work. Excitement over buckyballs began in 1985 when scientists projected light from a laser onto a piece of graphite. In the soot that remained, researchers found a completely new kind of molecule! Buckyballs are also found in the soot from a candle flame. Some scientists claim to have detected buckyballs in outer space. In fact, one

hypothesis suggests that buckyballs might be at the center of the condensing clouds of gas, dust, and debris that form galaxies.

The Game Plan

Ever since buckyballs were discovered, chemists have been busy trying to identify the molecules' properties. One interesting property is that substances can be trapped inside a buckyball. A buckyball can act like a cage that surrounds smaller substances, such as individual atoms. Buckyballs also appear to be both slippery and strong. They can be opened to insert materials, and they can even link together in tubes.

How can buckyballs be used? They may have a variety of uses, from carrying messages through atom-sized wires in computer chips to delivering medicines right where the body needs them. Making tough plastics and cutting tools are uses that are also under investigation. With so many possibilities, scientists expect to get a kick out of bucky-balls for some time!

The Kickoff

▶ A soccer ball is a great model for a buckyball. On the model, the places where three seams meet correspond to the carbon atoms on a buckyball. What represents the bonds between carbon atoms? Does your soccer-ball model have space for all 60 carbon atoms? You'll have to count and see for yourself.

127

WEIRD SCIENCE
Buckyballs

Background

The interlocking hexagons and pentagons of Buckminster Fuller's geodesic domes provide great stability because they distribute stress evenly. The bucky-ball (C_{60}) is one member of a large family of carbon "cages" called fullerenes. Fullerenes with fewer than 60 carbon atoms are called buckybabies. Buckytubes have more than 60 carbon atoms and are shaped like cylinders of spiraling honeycombs.

Scientists know that forming buckyballs requires high temperatures, whether in the lab or in nature. Researchers looking for buckyballs in nature concentrate on intensely heated sites, such as asteroid craters and areas of lightning strikes. Buckyballs are sometimes formed in the crevices between rocks at these sites. Buckyball molecules are found in greatest abundance in soot.

Answers to The Kickoff

The seams represent the bonds between atoms. A standard soccer ball has 60 places where three seams meet. In other words, the soccer-ball model has 60 places for exactly 60 carbon atoms.

internet**connect**

SCILINKS
NSTA

TOPIC: Buckminster Fuller and the Buckyball
GO TO: www.scilinks.org
*sci*LINKS NUMBER: HSTP300

SAFETY FIRST!

Exploring, inventing, and investigating are essential to the study of science. However, these activities can also be dangerous. To make sure that your experiments and explorations are safe, you must be aware of a variety of safety guidelines.

You have probably heard of the saying, "It is better to be safe than sorry." This is particularly true in a science classroom where experiments and explorations are being performed. Being uninformed and careless can result in serious injuries. Don't take chances with your own safety or with anyone else's.

Following are important guidelines for staying safe in the science classroom. Your teacher may also have safety guidelines and tips that are specific to your classroom and laboratory. Take the time to be safe.

Safety Rules!

Start Out Right

Always get your teacher's permission before attempting any laboratory exploration. Read the procedures carefully, and pay particular attention to safety information and caution statements. If you are unsure about what a safety symbol means, look it up or ask your teacher. You cannot be too careful when it comes to safety. If an accident does occur, inform your teacher immediately, regardless of how minor you think the accident is.

If you are instructed to note the odor of a substance, wave the fumes toward your nose with your hand. Never put your nose close to the source.

Safety Symbols

All of the experiments and investigations in this book and their related worksheets include important safety symbols to alert you to particular safety concerns. Become familiar with these symbols so that when you see them, you will know what they mean and what to do. It is important that you read this entire safety section to learn about specific dangers in the laboratory.

Eye protection

Clothing protection

Hand safety

Heating safety

Electric safety

Chemical safety

Animal safety

Sharp object

Plant safety

Eye Safety

Wear safety goggles when working around chemicals, acids, bases, or any type of flame or heating device. Wear safety goggles any time there is even the slightest chance that harm could come to your eyes. If any substance gets into your eyes, notify your teacher immediately, and flush your eyes with running water for at least 15 minutes. Treat any unknown chemical as if it were a dangerous chemical. Never look directly into the sun. Doing so could cause permanent blindness.

Avoid wearing contact lenses in a laboratory situation. Even if you are wearing safety goggles, chemicals can get between the contact lenses and your eyes. If your doctor requires that you wear contact lenses instead of glasses, wear eye-cup safety goggles in the lab.

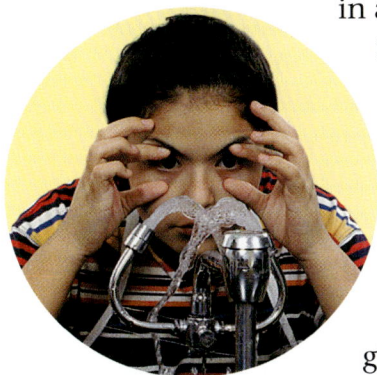

Safety Equipment

Know the locations of the nearest fire alarms and any other safety equipment, such as fire blankets and eyewash fountains, as identified by your teacher, and know the procedures for using them.

Be extra careful when using any glassware. When adding a heavy object to a graduated cylinder, tilt the cylinder so the object slides slowly to the bottom.

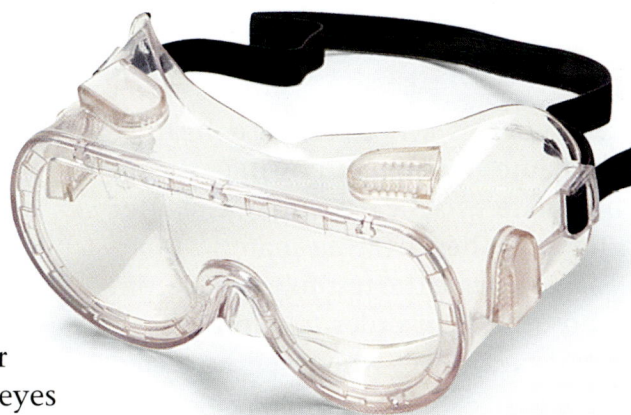

Neatness

Keep your work area free of all unnecessary books and papers. Tie back long hair, and secure loose sleeves or other loose articles of clothing, such as ties and bows. Remove dangling jewelry. Don't wear open-toed shoes or sandals in the laboratory. Never eat, drink, or apply cosmetics in a laboratory setting. Food, drink, and cosmetics can easily become contaminated with dangerous materials.

Certain hair products (such as aerosol hair spray) are flammable and should not be worn while working near an open flame. Avoid wearing hair spray or hair gel on lab days.

Sharp/Pointed Objects

Use knives and other sharp instruments with extreme care. Never cut objects while holding them in your hands. Place objects on a suitable work surface for cutting.

Heat

Wear safety goggles when using a heating device or a flame. Whenever possible, use an electric hot plate as a heat source instead of an open flame. When heating materials in a test tube, always angle the test tube away from yourself and others. In order to avoid burns, wear heat-resistant gloves whenever instructed to do so.

Chemicals

Wear safety goggles when handling any potentially dangerous chemicals, acids, or bases. If a chemical is unknown, handle it as you would a dangerous chemical. Wear an apron and safety gloves when working with acids or bases or whenever you are told to do so. If a spill gets on your skin or clothing, rinse it off immediately with water for at least 5 minutes while calling to your teacher.

Never mix chemicals unless your teacher tells you to do so. Never taste, touch, or smell chemicals unless you are specifically directed to do so. Before working with a flammable liquid or gas, check for the presence of any source of flame, spark, or heat.

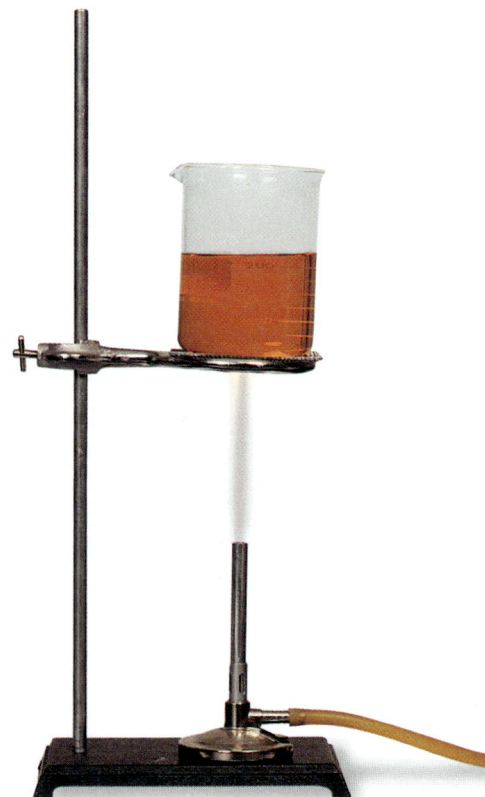

Electricity

Be careful with electrical cords. When using a microscope with a lamp, do not place the cord where it could trip someone. Do not let cords hang over a table edge in a way that could cause equipment to fall if the cord is accidentally pulled. Do not use equipment with damaged cords. Be sure your hands are dry and that the electrical equipment is in the "off" position before plugging it in. Turn off and unplug electrical equipment when you are finished.

Animal Safety

Always obtain your teacher's permission before bringing any animal into the school building. Handle animals only as your teacher directs. Always treat animals carefully and with respect. Wash your hands thoroughly after handling any animal.

Plant Safety

Do not eat any part of a plant or plant seed used in the laboratory. Wash hands thoroughly after handling any part of a plant. When in nature, do not pick any wild plants unless your teacher instructs you to do so.

Glassware

Examine all glassware before use. Be sure that glassware is clean and free of chips and cracks. Report damaged glassware to your teacher. Glass containers used for heating should be made of heat-resistant glass.

Measuring Liquid Volume
Teacher's Notes

Time Required

One 45-minute class period

Lab Ratings

TEACHER PREP	🜂🜂
STUDENT SET-UP	🜂
CONCEPT LEVEL	🜂🜂
CLEAN UP	🜂🜂

MATERIALS

The materials listed for this lab are for each group of 2–4 students. Each lab group should have one 30 mL beaker each of red, yellow, and blue water. The large test tubes must have a capacity of at least 14 mL.

Safety Caution

Remind students to review all safety cautions and icons before beginning this lab activity.

Preparation Notes

On the board, set up a chart with one row for each group and six columns, one for each test tube.

Answers

11. Groups may have different results because of measuring inaccuracy. Have students discuss how these inaccuracies might have resulted. Remind students that the purpose of this lab is to practice a skill, not necessarily to get the same results.

12. The graduated cylinder should not be filled to the top because the scale does not go all the way to the top, so accurate measurements would not be possible.

Measuring Liquid Volume

SKILL BUILDER

In this lab you will use a graduated cylinder to measure and transfer precise amounts of liquids. Remember, in order to accurately measure liquids in a graduated cylinder, you should read the level at the bottom of the meniscus, the curved surface of the liquid.

Procedure

1. Using the masking tape and marker, label the test tubes A, B, C, D, E, and F. Place them in the test-tube rack. Be careful not to confuse the test tubes.

2. Using the 10 mL graduated cylinder and the funnel, pour 14 mL of the red liquid into test tube A. (To do this, first pour 10 mL of the liquid into the test tube and then add 4 mL of liquid.)

3. Rinse the graduated cylinder and funnel between uses.

4. Measure 13 mL of the yellow liquid, and pour it into test tube C. Then measure 13 mL of the blue liquid, and pour it into test tube E.

5. Transfer 4 mL of liquid from test tube C into test tube D. Transfer 7 mL of liquid from test tube E into test tube D.

6. Measure 4 mL of blue liquid from the beaker, and pour it into test tube F. Measure 7 mL of red liquid from the beaker, and pour it into test tube F.

7. Transfer 8 mL of liquid from test tube A into test tube B. Transfer 3 mL of liquid from test tube C into test tube B.

Collect Data

8. Make a data table in your ScienceLog, and record the color of the liquid in each test tube.

9. Use the graduated cylinder to measure the volume of liquid in each test tube, and record the volumes in your data table.

10. Record your color observations in a table of class data prepared by your teacher. Copy the completed table into your ScienceLog.

Analysis

11. Did all of the groups report the same colors? Explain why the colors were the same or different.

12. Why should you not fill the graduated cylinder to the very top?

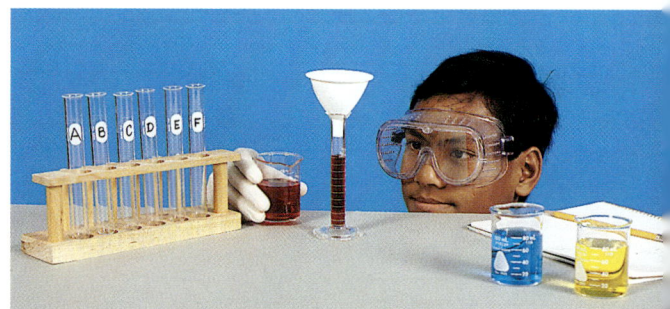

Materials

- masking tape
- marker
- 6 large test tubes
- test-tube rack
- 10 mL graduated cylinder
- 3 beakers filled with colored liquid
- small funnel

Datasheets for LabBook

Science Skills Worksheet "Measuring"

Norman Holcomb
Marion Elementary School
Maria Stein, Ohio

Coin Operated

All pennies are exactly the same, right? Probably not! After all, each penny was made in a certain year at a specific mint, and each has traveled a unique path to reach your classroom. But all pennies *are* similar. In this lab you will investigate differences and similarities among a group of pennies.

Procedure

1. Write the numbers 1 through 10 on a page in your ScienceLog, and place a penny next to each number.

2. Use the metric balance to find the mass of each penny to the nearest 0.1 g. Record each measurement in your ScienceLog.

3. On a table that your teacher will provide, make a mark in the correct column of the table for each penny you measured.

4. Separate your pennies into piles based on the class data. Place each pile on its own sheet of paper.

5. Measure and record the mass of each pile. Write the mass on the paper you are using to identify the pile.

6. Fill a graduated cylinder about halfway with water. Carefully measure the volume, and record it.

7. Carefully place the pennies from one pile in the graduated cylinder. Measure and record the new volume.

8. Carefully remove the pennies from the graduated cylinder, and dry them off.

9. Repeat steps 6 through 8 for each pile of pennies.

Analyze the Results

10. Determine the volume of the displaced water by subtracting the initial volume from the final volume. This amount is equal to the volume of the pennies. Record the volume of each pile of pennies.

11. Calculate the density of each pile. To do this, divide the total mass of the pennies by the volume of the pennies. Record the density in your ScienceLog.

Draw Conclusions

12. How is it possible for the pennies to have different densities?

13. What clues might allow you to separate the pennies into the same groups without experimentation? Explain.

Materials

- 10 pennies
- metric balance
- few sheets of paper
- 100 mL graduated cylinder
- water
- paper towels

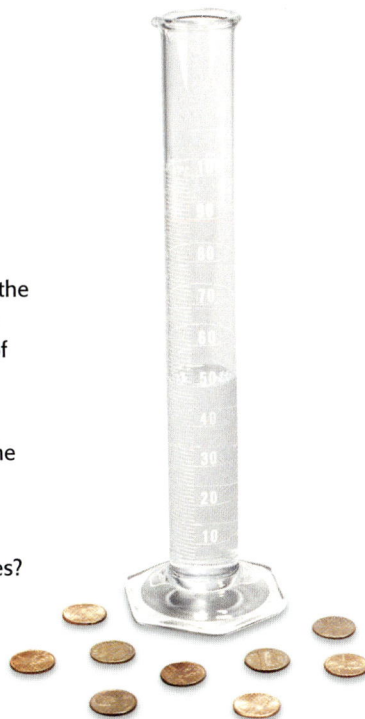

Coin Operated
Teacher's Notes

Time Required

One 45-minute class period

Lab Ratings

EASY ——————→ HARD

TEACHER PREP ▲▲
STUDENT SET-UP ▲
CONCEPT LEVEL ▲▲
CLEAN UP ▲

MATERIALS

The materials listed for this lab are for each group of 2–4 students. Be sure each group has both pre-1982 and post-1982 pennies. (1982 was the year the U.S. Department of the Treasury changed the penny's composition from copper to zinc with copper coating.)

Safety Caution

Remind students to review all safety cautions and icons before beginning this lab activity.

Procedure Note

3. Create the table (step 3) with one row for each group and 11 columns in 0.1 g increments. A twelfth column should be added for density. Record the number of pennies in the appropriate columns.

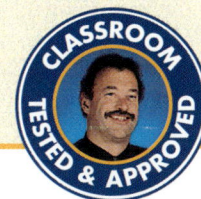

Answers

4. Students should have two piles of pennies based on the mass of the pennies.

11. The density of the copper pennies is approximately 8.85 g/cm^3; of the zinc-and-copper pennies, 7.14 g/cm^3.

12. The pennies can have different densities if they are made with different materials.

13. Pennies can be sorted by their dates. (Encourage students to examine the pennies closely for clues and to compare their piles with the piles of other students.)

Datasheets for LabBook

CLASSROOM TESTED & APPROVED

Paul Boyle
Perry Heights Middle School
Evansville, Indiana

Volumania!
Teacher's Notes

Time Required

One 45-minute class period

Lab Ratings

EASY ———————————————→ HARD

TEACHER PREP
STUDENT SET-UP
CONCEPT LEVEL
CLEAN UP

Part A: Finding the Volume of Small Objects

MATERIALS

The materials listed are for each group of 2–3 students. The objects used must fit in the graduated cylinders but be large enough to make a measurable change in volume. Rock or mineral samples, hardware (such as bolts or screws), and fishing weights work well.

Safety Caution

To avoid breaking glass graduated cylinders, caution students to tilt the graduated cylinder so objects can slide in gently. Remind students to read the volume when the meniscus is at eye level. Caution students to wear goggles during this lab.

Answers

6. The units of milliliters should be changed to cubic centimeters because you are measuring the volume of a solid object.

7. No; sometimes a heavier object will have a smaller volume than a lighter object because the matter is more tightly packed.

Volumania!

You have learned how to measure the volume of a solid object that has square or rectangular sides. But there are lots of objects in the world that have irregular shapes. In this lab activity, you'll learn some ways to find the volume of objects that have irregular shapes.

Part A: Finding the Volume of Small Objects

Procedure

1. Fill a graduated cylinder half full with water. Read the volume of the water, and record it in your ScienceLog. Be sure to look at the surface of the water at eye level and to read the volume at the bottom of the meniscus, as shown below.

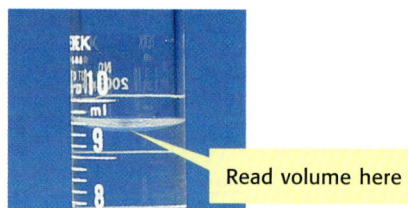

Read volume here

2. Carefully slide one of the objects into the tilted graduated cylinder, as shown below.

3. Read the new volume, and record it in your ScienceLog.

4. Subtract the old volume from the new volume. The resulting amount is equal to the volume of the solid object.

5. Use the same method to find the volume of the other objects. Record your results in your ScienceLog.

Analysis

6. What changes do you have to make to the volumes you determine in order to express them correctly?

7. Do the heaviest objects always have the largest volumes? Why or why not?

Materials

Part A
- graduated cylinder
- water
- various small objects supplied by your teacher

Part B
- bottom half of a 2 L plastic bottle or similar container
- water
- aluminum pie pan
- paper towels
- funnel
- graduated cylinder

Datasheets for LabBook

CLASSROOM TESTED & APPROVED

Alyson Mike
Radley Middle School
East Helena, Montana

Part B: Finding the Volume of Your Hand

Procedure

8. Completely fill the container with water. Put the container in the center of the pie pan. Be sure not to spill any of the water into the pie pan.

9. Make a fist, and put your hand into the container up to your wrist.

10. Remove your hand, and let the excess water drip into the container, not the pie pan. Dry your hand with a paper towel.

11. Use the funnel to pour the overflow water into the graduated cylinder. Measure the volume. This is the volume of your hand. Record the volume in your ScienceLog. (Remember to use the correct unit of volume for a solid object.)

12. Repeat this procedure with your other hand.

Analysis

13. Was the volume the same for both of your hands? If not, were you surprised? What might account for a person's hands having different volumes?

14. Would it have made a difference if you had placed your open hand into the container instead of your fist? Explain your reasoning.

15. Compare the volume of your right hand with the volume of your classmates' right hands. Create a class graph of right-hand volumes. What is the average right-hand volume for your class?

Going Further

■ Design an experiment to determine the volume of a person's body. In your plans, be sure to include the materials needed for the experiment and the procedures that must be followed. Include a sketch that shows how your materials and methods would be used in this experiment.

■ Using an encyclopedia, the Internet, or other reference materials, find out how the volumes of very large samples of matter—such as an entire planet—are determined.

Part B: Finding the Volume of Your Hand

MATERIALS

Plastic containers from whipped toppings and the like can also be used. Containers should be deep enough so students' fists can be submerged. Remind students that their container must be completely filled with water so that it overflows as their hand enters. The pie pan should be dry at the start. When students remove their hand from the container, they should allow the water cupped in their hand to drip back into the container and not into the pie pan. This water was not displaced and should not be measured.

Answers

13. Answers will vary. Often the preferred hand will be slightly larger due to greater muscle development.

14. It would not make a difference; a hand's volume remains the same regardless of its shape.

15. Answers will vary by class. Check for correct graphing technique and interpretation.

Going Further

• Designs should center on finding the volume of the body through water displacement. The equipment designed should be large enough to perform the experiment and allow for overflow.

• Accept all reasonable answers and findings.

Science Skills Worksheet
"Finding Useful Sources"

Science Skills Worksheet
"Researching on the Web"

Determining Density
Teacher's Notes

Time Required

One 45-minute class period

Lab Ratings

EASY ————————————→ HARD

TEACHER PREP	🧪
STUDENT SET-UP	🧪
CONCEPT LEVEL	🧪🧪
CLEAN UP	🧪

Safety Caution

Caution students to tilt the graduated cylinder so marbles can slide in gently.

Answers

8. The total mass increases. The volume of the marbles increases. The density of the marbles remains the same.

9. The graph is a straight line (see graph below).

10. No; the density is independent of the amount of substance. The density is the same for one marble as it is for several marbles.

Going Further

To calculate the slope of the graph, pick two points on the line. The slope is the difference between the *y*-values of the points divided by the difference between the *x*-values of the points. The slope of the graph should be equal to the density of the marbles because the graph shows mass versus volume, which means the slope is mass divided by volume—in other words, density.

📝 **Datasheets for LabBook**

SKILL BUILDER

Determining Density

The density of an object is its mass divided by its volume. But how does the density of a small amount of a substance relate to the density of a larger amount of the same substance? In this lab, you will calculate the density of one marble and of a group of marbles. Then you will confirm the relationship between the mass and volume of a substance.

Materials

- 100 mL graduated cylinder
- water
- paper towels
- 8 to 10 glass marbles
- metric balance
- graph paper

Collect Data

1. Copy the table below in your ScienceLog. Include one row for each marble.

Mass of marble, g	Total mass of marbles, g	Total volume, mL	Volume of marbles, mL (total volume minus 50.0 mL)	Density of marbles, g/mL (total mass of marbles divided by volume of marbles)
DO NOT WRITE IN BOOK			DO NOT WRITE IN BOOK	

2. Fill the graduated cylinder with 50.0 mL of water. If you put in too much water, twist one of the paper towels and use its end to absorb excess water.

3. Measure the mass of a marble as accurately as you can (to at least one-tenth of a gram). Record the marble's mass in the table.

4. Carefully drop the marble in the tilted cylinder, and measure the total volume. Record the volume in the third column.

5. Measure and record the mass of another marble. Add the masses of the marbles together, and record this value in the second column of the table.

6. Carefully drop the second marble in the graduated cylinder. Complete the row of information in the table.

7. Repeat steps 5 and 6, adding one marble at a time. Stop when you run out of marbles, the water no longer completely covers the marbles, or the graduated cylinder is full.

Analyze the Results

8. Examine the data in your table. As the number of marbles increases, what happens to the total mass of the marbles? What happens to the volume of the marbles? What happens to the density of the marbles?

9. Graph the total mass of the marbles (*y*-axis) versus the volume of the marbles (*x*-axis). Is the graph a straight line or a curved line?

Draw Conclusions

10. Does the density of a substance depend on the amount of substance present? Explain how your results support your answer.

Going Further

Calculate the slope of the graph. How does the slope compare with the values in the column titled "Density of marbles"? Explain.

136

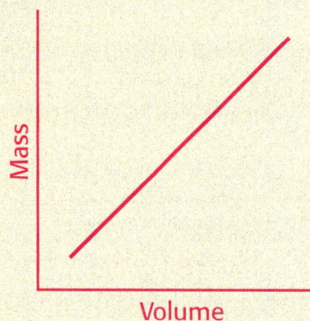

Mass / Volume

CLASSROOM TESTED & APPROVED

Alyson Mike
Radley Middle School
East Helena, Montana

Using Scientific Methods

Layering Liquids

You have learned that liquids form layers according to their densities. In this lab, you'll discover whether it matters in which order you add the liquids.

Make a Prediction

1. Does the order in which you add liquids of different densities to a container affect the order of the layers formed by those liquids?

Conduct an Experiment

2. Using the graduated cylinders, add 10 mL of each liquid to the clear container. Remember to read the volume at the bottom of the meniscus, as shown below. In your ScienceLog, record the order in which you added the liquids.

3. Observe the liquids in the container. In your ScienceLog, sketch what you see. Be sure to label the layers and the colors.

4. Add 10 mL more of liquid C. Observe what happens, and write your observations in your ScienceLog.

5. Add 20 mL more of liquid A. Observe what happens, and write your observations in your ScienceLog.

Analyze Your Results

6. Which of the liquids has the greatest density? Which has the least density? How can you tell?

7. Did the layers change position when you added more of liquid C? Explain your answer.

8. Did the layers change position when you added more of liquid A? Explain your answer.

Communicate Your Results

9. Find out in what order your classmates added the liquids to the container. Compare your results with those of a classmate who added the liquids in a different order. Were your results different? In your ScienceLog, explain why or why not.

Draw Conclusions

10. Based on your results, evaluate your prediction from step 1.

Materials

- liquid A
- liquid B
- liquid C
- beaker or other small, clear container
- 10 mL graduated cylinders (3)
- 3 funnels

LabBook

Layering Liquids
Teacher's Notes

Time Required

One 45-minute class period

Lab Ratings

EASY ————————→ HARD

TEACHER PREP   🧪🧪🧪
STUDENT SET-UP   🧪
CONCEPT LEVEL   🧪🧪
CLEAN UP   🧪🧪

Preparation Notes

Liquid A is red-colored water, liquid B is vegetable oil, and liquid C is dark corn syrup.

Disposal Information

To keep the oil out of the drains, have students empty their containers into several disposable containers. These can be capped, refrigerated, and thrown in the trash. It might be interesting to let these waste bottles stand overnight to see if the layers are visible the following day.

Datasheets for LabBook

Science Skills Worksheet
"Working with Hypotheses"

CLASSROOM TESTED & APPROVED

Alyson Mike
Radley Middle School
East Helena, Montana

Answers

1. Accept all reasonable predictions.

6. Liquid C has the greatest density. Liquid B has the least density. The liquids form layers with the least dense on top and the most dense on bottom.

7. The position of the layers did not change. Adding more of liquid C does not change its density, so its position stays the same.

8. The position of the layers did not change.

Adding more of liquid A does not change its density, so its position stays the same.

9. All results should be identical. Liquid B is the top layer, liquid A is the middle layer, and liquid C is the bottom layer.

10. Answers will vary, depending on the original prediction. The order in which the liquids are added does not affect the order of the layers formed.

Full of Hot Air!
Teacher's Notes

Time Required

One 45-minute class period

Lab Ratings

EASY ——————→ HARD

TEACHER PREP
STUDENT SET-UP
CONCEPT LEVEL
CLEAN UP

Safety Caution

Remind students to review all safety cautions and icons before beginning this lab activity. Keep all power cords away from the beakers and pans of hot water. Be careful—hot plates may stay hot for a long time. Students should wear heat-resistant gloves when handling the hot beaker.

Answers

1. Accept all reasonable hypotheses.

9. When the balloon cooled, it contracted. When heated, it expanded. These observations confirm Charles's law.

10. Answers will vary, depending on the original hypothesis. Sample supported hypothesis: Increasing temperature increases the volume of a balloon, while decreasing temperature decreases the volume of a balloon.

11. As the temperature increased, volume increased and mass remained constant. Therefore the density decreased. Conversely, density increases when temperature decreases.

Using Scientific Methods

DISCOVERY LAB

Full of Hot Air!

Why do hot-air balloons float gracefully above Earth, while balloons you blow up fall to the ground? The answer has to do with the density of the air inside the balloon. Density is mass per unit volume, and volume is affected by changes in temperature. In this experiment, you will investigate the relationship between the temperature of a gas and its volume. Then you will be able to determine how the temperature of a gas affects its density.

Form a Hypothesis

1. How does an increase or decrease in temperature affect the volume of a balloon? Write your hypothesis in your ScienceLog.

Test the Hypothesis

2. Fill an aluminum pan with water about 4 to 5 cm deep. Put the pan on the hot plate, and turn the hot plate on.

3. While the water is heating, fill the other pan 4 to 5 cm deep with ice water.

4. Blow up a balloon inside the 500 mL beaker, as shown. The balloon should fill the beaker but should not extend outside the beaker. Tie the balloon at its opening.

5. Place the beaker and balloon in the ice water. Observe what happens. Record your observations in your ScienceLog.

6. Remove the balloon and beaker from the ice water. Observe the balloon for several minutes. Record any changes.

Materials

- 2 aluminum pans
- water
- metric ruler
- hot plate
- ice water
- balloon
- 250 mL beaker
- heat-resistant gloves

7. Put on heat-resistant gloves. When the hot water begins to boil, put the beaker and balloon in the hot water. Observe the balloon for several minutes, and record your observations.

8. Turn off the hot plate. When the water has cooled, carefully pour it into a sink.

Analyze the Results

9. Summarize your observations of the balloon. Relate your observations to Charles's law.

10. Was your hypothesis for step 1 supported? If not, revise your hypothesis.

Draw Conclusions

11. Based on your observations, how is the density of a gas affected by an increase or decrease in temperature?

12. Explain in terms of density and Charles's law why heating the air allows a hot-air balloon to float.

---

12. When you heat the air, it expands and becomes less dense than the surrounding air. The balloon begins to float.

Datasheets for LabBook

CLASSROOM TESTED & APPROVED

Sharon L. Woolf
Langston Hughes Middle School
Reston, Virginia

Can Crusher

Condensation can occur when gas particles come near the surface of a liquid. The gas particles slow down because they are attracted to the liquid. This reduction in speed causes the gas particles to condense into a liquid. In this lab, you'll see that particles that have condensed into a liquid don't take up as much space and therefore don't exert as much pressure as they did in the gaseous state.

Conduct an Experiment

1. Place just enough water in an aluminum can to slightly cover the bottom.

2. Put on heat-resistant gloves. Place the aluminum can on a hot plate turned to the highest temperature setting.

3. Heat the can until the water is boiling. Steam should be rising vigorously from the top of the can.

4. Using tongs, quickly pick up the can and place the top 2 cm of the can upside down in the 1 L beaker filled with room-temperature water.

5. Describe your observations in your ScienceLog.

Analyze the Results

6. The can was crushed because the atmospheric pressure outside the can became greater than the pressure inside the can. Explain what happened inside the can to cause this.

Draw Conclusions

7. Inside every popcorn kernel is a small amount of water. When you make popcorn, the water inside the kernels is heated until it becomes steam. Explain how the popping of the kernels is the opposite of what you saw in this lab. Be sure to address the effects of pressure in your explanation.

Materials

- water
- 2 empty aluminum cans
- heat-resistant gloves
- hot plate
- tongs
- 1 L beaker

Going Further
Try the experiment again, but use ice water instead of room-temperature water. Explain your results in terms of the effects of temperature.

Going Further
The can was crushed more quickly because the ice water made the steam condense more quickly. So the pressure inside the can decreased further.

Can Crusher
Teacher's Notes

Time Required
One 45-minute class period

Lab Ratings

EASY ——————→ HARD

TEACHER PREP
STUDENT SET-UP
CONCEPT LEVEL
CLEAN UP

Safety Caution
Remind students to review all safety cautions and icons before beginning this lab activity. To prevent spills, caution students to keep all power cords away from beakers and pans of hot water. Heat-resistant gloves may not be necessary if tongs are properly used.

Answers

6. The steam inside the can cooled and condensed. The volume of water (condensed steam) is smaller than the volume of the steam, so the pressure inside the can was reduced.

7. When the water inside the kernel becomes steam, it expands about 100 times. The pressure inside the kernel increases. The pressure outside is unchanged, so the pressure inside forces the kernel to "explode."

📄 **Datasheets for LabBook**

A Sugar Cube Race!
Teacher's Notes

Time Required

One 45-minute class period

Lab Ratings

EASY			HARD

TEACHER PREP

STUDENT SET-UP

CONCEPT LEVEL

CLEAN UP

Preparation Notes

Materials listed are for each student. Remind students not to eat the sugar cube. Have hot water or hot plates and heat-resistant gloves ready for students who want to test temperature. Have paper towels on hand for students to wrap their cube in as they crush it. Caution students to wear goggles.

Answers

1. Accept all reasonable predictions. Variables include water temperature, surface area of cube, motion of water due to stirring, and time. Sample predictions: Increasing the temperature of the water will make the sugar cube dissolve faster.

5. Answers will vary, depending on the original prediction.

6. Observing the sugar cube dissolving on its own provides a control so that you can measure the effect of the variable.

7. Changing two variables that each increase the dissolving rate should increase the rate of dissolving even more, but it would be difficult to determine which variable had the greater effect.

A Sugar Cube Race!

If you drop a sugar cube into a glass of water, how long will it take to dissolve? Will it take 5 minutes, 10 minutes, or longer? What can you do to speed up the rate at which it dissolves? Should you change something about the water, the sugar cube, or the process? In other words, what variable should you change? Before reading further, make a list of variables that could be changed in this situation. Record your list in your ScienceLog.

Materials

- water
- graduated cylinder
- 2 sugar cubes
- 2 beakers or other clear containers
- clock or stopwatch
- other materials approved by your teacher

Make a Prediction

1. Choose one variable to test. In your ScienceLog, record your choice, and predict how changing your variable will affect the rate of dissolving.

Conduct an Experiment

2. Pour 150 mL of water into one of the beakers. Add one sugar cube, and use the stopwatch to measure how long it takes for the sugar cube to dissolve. You must not disturb the sugar cube in any way! Record this time in your ScienceLog.

3. Tell your teacher how you wish to test the variable. Do not proceed without his or her approval. You may need additional equipment.

4. Prepare your materials to test the variable you have picked. When you are ready, start your procedure for speeding up the dissolving of the sugar cube. Use the stopwatch to measure the time. Record this time in your ScienceLog.

Analyze the Results

5. Compare your results with the results obtained in step 2. Was your prediction correct? Why or why not?

Draw Conclusions

6. Why was it necessary to observe the sugar cube dissolving on its own before you tested the variable?

7. Do you think that changing more than one variable would speed up the rate of dissolving even more? Explain your reasoning.

Communicate Results

8. Discuss your results with a group that tested a different variable. Which variable had a greater effect on the rate of dissolving?

8. Accept all reasonable answers based on class data.

Datasheets for LabBook

Kenneth J. Horn
Fallston Middle School
Fallston, Maryland

Making Butter

A colloid is an interesting substance. It has properties of both solutions and suspensions. Colloidal particles are not heavy enough to settle out, so they remain evenly dispersed throughout the mixture. In this activity, you will make butter—a very familiar colloid—and observe the characteristics that classify butter as a colloid.

SKILL BUILDER

Materials

- marble
- small, clear container with lid
- heavy cream
- clock or stopwatch

Procedure

1. Place a marble inside the container, and fill the container with heavy cream. Put the lid tightly on the container.

2. Take turns shaking the container vigorously and constantly for 10 minutes. Record the time when you begin shaking in your ScienceLog. Every minute, stop shaking the container and hold it up to the light. Record your observations.

3. Continue shaking the container, taking turns if necessary. When you see, hear, or feel any changes inside the container, note the time and change in your ScienceLog.

4. After 10 minutes of shaking, you should have a lump of "butter" surrounded by liquid inside the container. Describe both the butter and the liquid in detail in your ScienceLog.

5. Let the container sit for about 10 minutes. Observe the butter and liquid again, and record your observations in your ScienceLog.

Analysis

6. When you noticed the change in the container, what did you think was happening at that point?

7. Based on your observations, explain why butter is classified as a colloid.

8. What kind of mixture is the liquid that is left behind? Explain.

141

CLASSROOM TESTED & APPROVED

Kenneth J. Horn
Fallston Middle School
Fallston, Maryland

Datasheets for LabBook

LabBook

Making Butter
Teacher's Notes

Time Required
One 45-minute class period

Lab Ratings

EASY ————————→ HARD

TEACHER PREP ▲▲
STUDENT SET-UP ▲
CONCEPT LEVEL ▲▲
CLEAN UP ▲▲

MATERIALS

Materials listed are for each pair of students. If using glass containers, students should shake the container vigorously but not violently because it might break. Be sure each lid fits tightly. A small or medium-sized ball bearing may be substituted for the marble. For best results, the cream should be room temperature, not cold.

Safety Caution

Caution students to wear safety goggles while performing this activity.

Answers

6. Answers may vary. Students should mention that the suspended materials were starting to settle out.

7. The butter appears to have characteristics of both a solution and a suspension.

8. The liquid left behind appears to be a suspension.

Unpolluting Water
Teacher's Notes

Time Required

One or two 45-minute class periods

Lab Ratings

EASY ———————————————→ HARD

TEACHER PREP △△△
STUDENT SET-UP △△△
CONCEPT LEVEL △△△
CLEAN UP △△

MATERIALS

Materials listed are for each group of 2–3 students. Use large filter paper for part D, or place filter paper in a funnel.

Special notes on supplies:

1. Sand must be thoroughly washed to eliminate as much dust as possible. Put the sand in a bowl and run water into it while stirring until the water runs clear. The finer the sand, the better the filtering action.

2. Use activated charcoal, available from pet-supply stores. This charcoal can be washed by quickly running water through the charcoal in a sieve or colander. Do not allow the charcoal to remain in water too long or it will lose its adsorbing power.

Safety Caution

Remind students to review all safety cautions and icons before beginning this lab activity. Make sure all spills are cleaned up immediately.

Unpolluting Water

In many cities, the water supply comes from a river, lake, or reservoir. This water may include several mixtures, including suspensions (with suspended dirt, oil, or living organisms) and solutions (with dissolved chemicals). To make the water safe to drink, your city's water supplier must remove impurities. In this lab, you will model the procedures used in real water-treatment plants.

Part A: Untreated Water

Procedure

1. Measure 100 mL of "polluted" water into a graduated cylinder. Be sure to shake the bottle of water before you pour so your sample will include all the impurities.

2. Pour the contents of the graduated cylinder into one of the beakers.

3. Copy the table below into your ScienceLog, and record your observations of the water in the "Before treatment" row.

Materials

- "polluted" water
- graduated cylinder
- 250 mL beakers (4)
- 2 plastic spoons
- small nail
- 8 oz plastic-foam cup (2)
- scissors
- 2 pieces of filter paper
- washed fine sand
- metric ruler
- washed activated charcoal
- rubber band

Observations						
	Color	Clearness	Odor	Any layers?	Any solids?	Water volume
Before treatment						
After oil separation						
After sand filtration						
After charcoal						

DO NOT WRITE IN BOOK

Part B: Settling In

If a suspension is left standing, the suspended particles will settle to the top or bottom. You should see a layer of oil at the top.

Procedure

4. Separate the oil by carefully pouring the oil into another beaker. You can use a plastic spoon to get the last bit of oil from the water. Record your observations.

142

Preparation Notes

Make "polluted water" as follows: Put the following into a half-gallon milk jug:

   1 cup cooking oil

   3/4 to 1 cup of dirt

   1 or 2 drops of food coloring (yellow or red works best)

Fill the jug with water, put the cap on, and shake the jug well. It is important that students shake the mixture well before pouring their 100 mL sample. Students can estimate the water volume after parts B–D using the approximate volume markings on the side of the beaker. In part D, have students use a clean graduated cylinder to measure the volume of treated water.

Part C: Filtration

Cloudy water can be a sign of small particles still in suspension. These particles can usually be removed by filtering. Water-treatment plants use sand and gravel as filters.

Procedure

5. Make a filter as follows:
   a. Use the nail to poke 5 to 10 small holes in the bottom of one of the cups.
   b. Cut a circle of filter paper to fit inside the bottom of the cup. (This will keep the sand in the cup.)
   c. Fill the cup to 2 cm below the rim with wet sand. Pack the sand tightly.
   d. Set the cup inside an empty beaker.

6. Pour the polluted water on top of the sand, and let it filter through. Do not pour any of the settled mud onto the sand. (Dispose of the mud as instructed by your teacher.) In your table, record your observations of the water collected in the beaker.

Part D: Separating Solutions

Something that has been dissolved in a solvent cannot be separated using filters. Water-treatment plants use activated charcoal to absorb many dissolved chemicals.

Procedure

7. Place activated charcoal about 3 cm deep in the unused cup. Pour the water collected from the sand filtration into the cup, and stir for a minute with a spoon.

8. Place a piece of filter paper over the top of the cup, and fasten it in place with a rubber band. With the paper securely in place, pour the water through the filter paper and back into a clean beaker. Record your observations in your table.

Analysis (Parts A–D)

9. Is your unpolluted water safe to drink? Why or why not?

10. When you treat a sample of water, do you get out exactly the same amount of water that you put in? Explain your answer.

11. Some groups may still have cloudy water when they finish. Explain a possible cause for this.

143

Datasheets for LabBook

CLASSROOM TESTED & APPROVED

Joseph Price
H. M. Browne Junior High
Washington, D.C.

Answers

9. Students will have different opinions, depending on their results. Many will likely say that unpolluted water is safe to drink because it goes through so many filtering processes, but some may say it is unsafe because their samples still look cloudy after the experiment. (Students should be discouraged from tasting the water. This activity does not include treatment with chlorine that takes place at most water treatment plants to kill bacteria.)

10. No; some of the water is lost in the treatment processes.

11. Accept all reasonable answers. Students may note that dust came from the charcoal and sand or that bacteria in the water caused it to appear cloudy.

Disposal Information

1. Solid charcoal should be dried and buried in a landfill that is approved for chemical disposal. You may want to consider drying and reusing the charcoal, although it will eventually lose its adsorbing power.

2. Pour cooking oil into disposable containers, refrigerate (if possible) until the oil congeals, and put in the trash.

3. The sand can be reused if it is washed after this activity.

4. Spoon or pour the mud into disposable containers and put them in the trash.

Concept Mapping: A Way to Bring Ideas Together

What Is a Concept Map?

Have you ever tried to tell someone about a book or a chapter you've just read and found that you can remember only a few isolated words and ideas? Or maybe you've memorized facts for a test and then weeks later discovered you're not even sure what topics those facts covered.

In both cases, you may have understood the ideas or concepts by themselves but not in relation to one another. If you could somehow link the ideas together, you would probably understand them better and remember them longer. This is something a concept map can help you do. A concept map is a way to see how ideas or concepts fit together. It can help you see the "big picture."

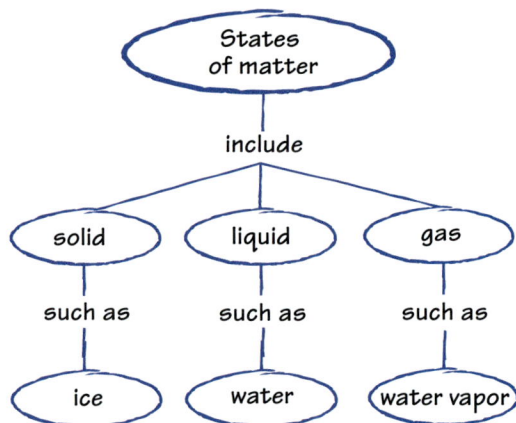

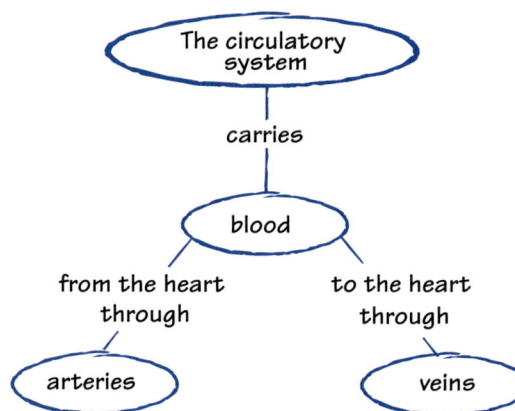

How to Make a Concept Map

1 Make a list of the main ideas or concepts.

It might help to write each concept on its own slip of paper. This will make it easier to rearrange the concepts as many times as necessary to make sense of how the concepts are connected. After you've made a few concept maps this way, you can go directly from writing your list to actually making the map.

2 Arrange the concepts in order from the most general to the most specific.

Put the most general concept at the top and circle it. Ask yourself, "How does this concept relate to the remaining concepts?" As you see the relationships, arrange the concepts in order from general to specific.

3 Connect the related concepts with lines.

4 On each line, write an action word or short phrase that shows how the concepts are related.

Look at the concept maps on this page, and then see if you can make one for the following terms:

plants, water, photosynthesis, carbon dioxide, sun's energy

One possible answer is provided at right, but don't look at it until you try the concept map yourself.

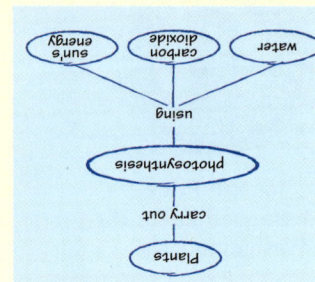

SI Measurement

The International System of Units, or SI, is the standard system of measurement used by many scientists. Using the same standards of measurement makes it easier for scientists to communicate with one another.

SI works by combining prefixes and base units. Each base unit can be used with different prefixes to define smaller and larger quantities. The table below lists common SI prefixes.

SI Prefixes			
Prefix	**Abbreviation**	**Factor**	**Example**
kilo-	k	1,000	kilogram, 1 kg = 1,000 g
hecto-	h	100	hectoliter, 1 hL = 100 L
deka-	da	10	dekameter, 1 dam = 10 m
		1	meter, liter
deci-	d	0.1	decigram, 1 dg = 0.1 g
centi-	c	0.01	centimeter, 1 cm = 0.01 m
milli-	m	0.001	milliliter, 1 mL = 0.001 L
micro-	μ	0.000 001	micrometer, 1 μm = 0.000 001 m

SI Conversion Table		
SI units	**From SI to English**	**From English to SI**
Length		
kilometer (km) = 1,000 m	1 km = 0.621 mi	1 mi = 1.609 km
meter (m) = 100 cm	1 m = 3.281 ft	1 ft = 0.305 m
centimeter (cm) = 0.01 m	1 cm = 0.394 in.	1 in. = 2.540 cm
millimeter (mm) = 0.001 m	1 mm = 0.039 in.	
micrometer (μm) = 0.000 001 m		
nanometer (nm) = 0.000 000 001 m		
Area		
square kilometer (km^2) = 100 hectares	1 km^2 = 0.386 mi^2	1 mi^2 = 2.590 km^2
hectare (ha) = 10,000 m^2	1 ha = 2.471 acres	1 acre = 0.405 ha
square meter (m^2) = 10,000 cm^2	1 m^2 = 10.765 ft^2	1 ft^2 = 0.093 m^2
square centimeter (cm^2) = 100 mm^2	1 cm^2 = 0.155 $in.^2$	1 $in.^2$ = 6.452 cm^2
Volume		
liter (L) = 1,000 mL = 1 dm^3	1 L = 1.057 fl qt	1 fl qt = 0.946 L
milliliter (mL) = 0.001 L = 1 cm^3	1 mL = 0.034 fl oz	1 fl oz = 29.575 mL
microliter (μL) = 0.000 001 L		
Mass		
kilogram (kg) = 1,000 g	1 kg = 2.205 lb	1 lb = 0.454 kg
gram (g) = 1,000 mg	1 g = 0.035 oz	1 oz = 28.349 g
milligram (mg) = 0.001 g		
microgram (μg) = 0.000 001 g		

Temperature Scales

Temperature can be expressed using three different scales: Fahrenheit, Celsius, and Kelvin. The SI unit for temperature is the kelvin (K).

Although 0 K is much colder than 0°C, a change of 1 K is equal to a change of 1°C.

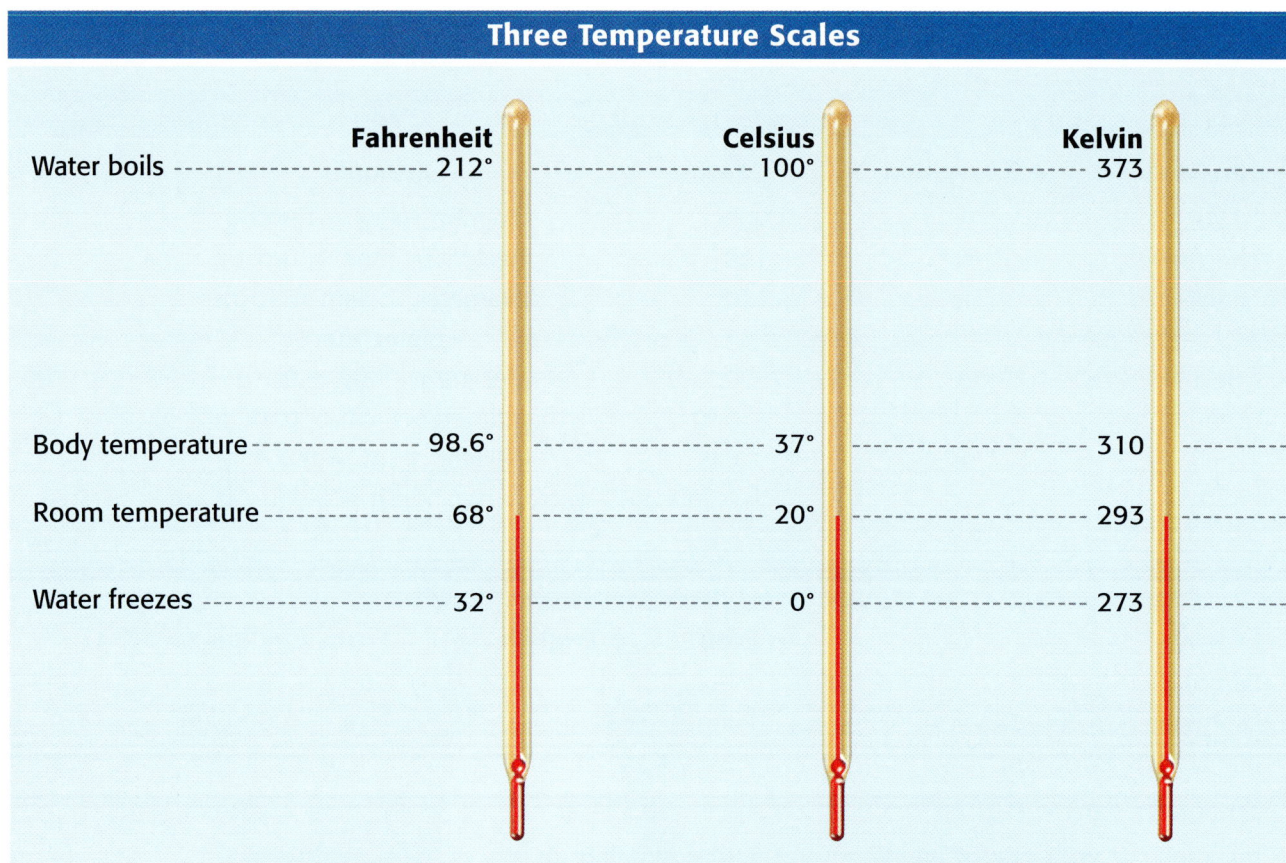

Three Temperature Scales

	Fahrenheit	Celsius	Kelvin
Water boils	212°	100°	373
Body temperature	98.6°	37°	310
Room temperature	68°	20°	293
Water freezes	32°	0°	273

Temperature Conversions Table

To convert	Use this equation:	Example
Celsius to Fahrenheit °C ⟶ °F	$°F = \left(\dfrac{9}{5} \times °C\right) + 32$	Convert 45°C to °F. $°F = \left(\dfrac{9}{5} \times 45°C\right) + 32 = 113°F$
Fahrenheit to Celsius °F ⟶ °C	$°C = \dfrac{5}{9} \times (°F - 32)$	Convert 68°F to °C. $°C = \dfrac{5}{9} \times (68°F - 32) = 20°C$
Celsius to Kelvin °C ⟶ K	$K = °C + 273$	Convert 45°C to K. $K = 45°C + 273 = 318\ K$
Kelvin to Celsius K ⟶ °C	$°C = K - 273$	Convert 32 K to °C. $°C = 32\ K - 273 = -241°C$

Measuring Skills

Using a Graduated Cylinder

When using a graduated cylinder to measure volume, keep the following procedures in mind:

1. Make sure the cylinder is on a flat, level surface.

2. Move your head so that your eye is level with the surface of the liquid.

3. Read the mark closest to the liquid level. On glass graduated cylinders, read the mark closest to the center of the curve in the liquid's surface.

Using a Meterstick or Metric Ruler

When using a meterstick or metric ruler to measure length, keep the following procedures in mind:

1. Place the ruler firmly against the object you are measuring.

2. Align one edge of the object exactly with the zero end of the ruler.

3. Look at the other edge of the object to see which of the marks on the ruler is closest to that edge. **Note:** Each small slash between the centimeters represents a millimeter, which is one-tenth of a centimeter.

Using a Triple-Beam Balance

When using a triple-beam balance to measure mass, keep the following procedures in mind:

1. Make sure the balance is on a level surface.

2. Place all of the countermasses at zero. Adjust the balancing knob until the pointer rests at zero.

3. Place the object you wish to measure on the pan. **Caution:** Do not place hot objects or chemicals directly on the balance pan.

4. Move the largest countermass along the beam to the right until it is at the last notch that does not tip the balance. Follow the same procedure with the next-largest countermass. Then move the smallest countermass until the pointer rests at zero.

5. Add the readings from the three beams together to determine the mass of the object.

6. When determining the mass of crystals or powders, use a piece of filter paper. First find the mass of the paper. Then add the crystals or powder to the paper and re-measure. The actual mass of the crystals or powder is the total mass minus the mass of the paper. When finding the mass of liquids, first find the mass of the empty container. Then find the mass of the liquid and container together. The mass of the liquid is the total mass minus the mass of the container.

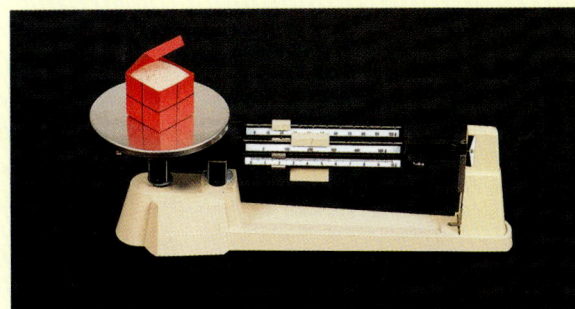

Scientific Method

The series of steps that scientists use to answer questions and solve problems is often called the **scientific method.** The scientific method is not a rigid procedure. Scientists may use all of the steps or just some of the steps of the scientific method. They may even repeat some of the steps. The goal of the scientific method is to come up with reliable answers and solutions.

Six Steps of the Scientific Method

1 **Ask a Question** Good questions come from careful **observations.** You make observations by using your senses to gather information. Sometimes you may use instruments, such as microscopes and telescopes, to extend the range of your senses. As you observe the natural world, you will discover that you have many more questions than answers. These questions drive the scientific method.

Questions beginning with *what, why, how,* and *when* are very important in focusing an investigation, and they often lead to a hypothesis. (You will learn what a hypothesis is in the next step.) Here is an example of a question that could lead to further investigation.

Question: How does acid rain affect plant growth?

Ask a Question

2 **Form a Hypothesis** After you come up with a question, you need to turn the question into a **hypothesis.** A hypothesis is a clear statement of what you expect the answer to your question to be. Your hypothesis will represent your best "educated guess" based on your observations and what you already know. A good hypothesis is testable. If observations and information cannot be gathered or if an experiment cannot be designed to test your hypothesis, it is untestable, and the investigation can go no further.

Here is a hypothesis that could be formed from the question, "How does acid rain affect plant growth?"

Hypothesis: Acid rain causes plants to grow more slowly.

Form a Hypothesis

Notice that the hypothesis provides some specifics that lead to methods of testing. The hypothesis can also lead to predictions. A **prediction** is what you think will be the outcome of your experiment or data collection. Predictions are usually stated in an "if . . . then" format. For example, **if** meat is kept at room temperature, **then** it will spoil faster than meat kept in the refrigerator. More than one prediction can be made for a single hypothesis. Here is a sample prediction for the hypothesis that acid rain causes plants to grow more slowly.

Prediction: If a plant is watered with only acid rain (which has a pH of 4), then the plant will grow at half its normal rate.

3 **Test the Hypothesis** After you have formed a hypothesis and made a prediction, you should test your hypothesis. There are different ways to do this. Perhaps the most familiar way is to conduct a **controlled experiment.** A controlled experiment tests only one factor at a time. A controlled experiment has a **control group** and one or more **experimental groups.** All the factors for the control and experimental groups are the same except for one factor, which is called the **variable.** By changing only one factor, you can see the results of just that one change.

Sometimes, the nature of an investigation makes a controlled experiment impossible. For example, dinosaurs have been extinct for millions of years, and the Earth's core is surrounded by thousands of meters of rock. It would be difficult, if not impossible, to conduct controlled experiments on such things. Under such circumstances, a hypothesis may be tested by making detailed observations. Taking measurements is one way of making observations.

Test the Hypothesis

4 **Analyze the Results** After you have completed your experiments, made your observations, and collected your data, you must analyze all the information you have gathered. Tables and graphs are often used in this step to organize the data.

Analyze the Results

5 **Draw Conclusions** Based on the analysis of your data, you should conclude whether or not your results support your hypothesis. If your hypothesis is supported, you (or others) might want to repeat the observations or experiments to verify your results. If your hypothesis is not supported by the data, you may have to check your procedure for errors. You may even have to reject your hypothesis and make a new one. If you cannot draw a conclusion from your results, you may have to try the investigation again or carry out further observations or experiments.

Draw Conclusions

Do they support your hypothesis?

No

Yes

6 **Communicate Results** After any scientific investigation, you should report your results. By doing a written or oral report, you let others know what you have learned. They may want to repeat your investigation to see if they get the same results. Your report may even lead to another question, which in turn may lead to another investigation.

Communicate Results

Scientific Method in Action

The scientific method is not a "straight line" of steps. It contains loops in which several steps may be repeated over and over again, while others may not be necessary. For example, sometimes scientists will find that testing one hypothesis raises new questions and new hypotheses to be tested. And sometimes, testing the hypothesis leads directly to a conclusion. Furthermore, the steps in the scientific method are not always used in the same order. Follow the steps in the diagram below, and see how many different directions the scientific method can take you.

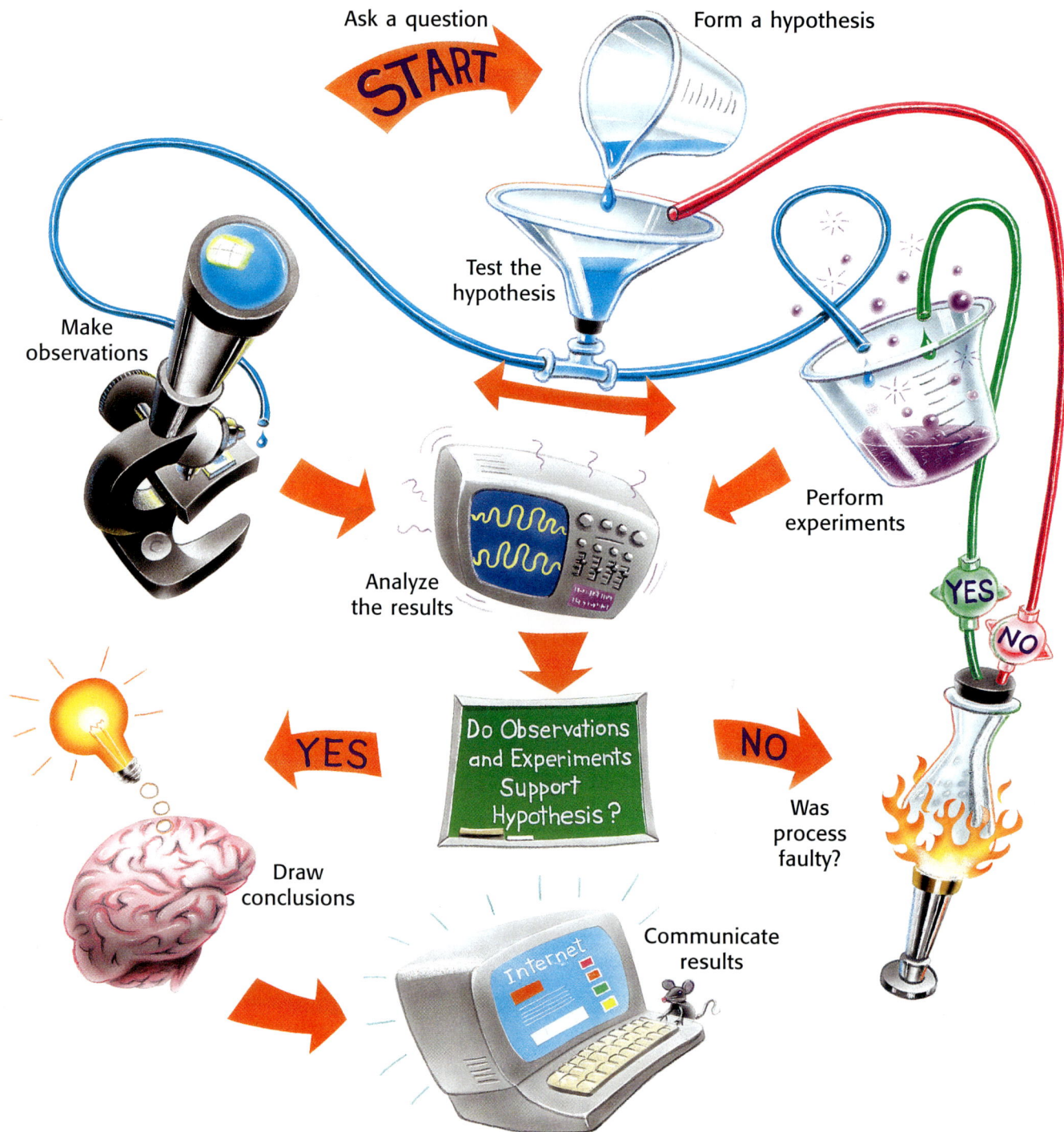

Ask a question

START

Form a hypothesis

Test the hypothesis

Make observations

Perform experiments

Analyze the results

YES

NO

Do Observations and Experiments Support Hypothesis?

YES

NO

Was process faulty?

Draw conclusions

Communicate results

Internet

Making Charts and Graphs

Circle Graphs

A circle graph, or pie chart, shows how each group of data relates to all of the data. Each part of the circle represents a category of the data. The entire circle represents all of the data. For example, a biologist studying a hardwood forest in Wisconsin found that there were five different types of trees. The data table at right summarizes the biologist's findings.

Wisconsin Hardwood Trees	
Type of tree	**Number found**
Oak	600
Maple	750
Beech	300
Birch	1,200
Hickory	150
Total	3,000

How to Make a Circle Graph

1 In order to make a circle graph of this data, first find the percentage of each type of tree. To do this, divide the number of individual trees by the total number of trees and multiply by 100.

$$\frac{600 \text{ oak}}{3{,}000 \text{ trees}} \times 100 = 20\%$$

$$\frac{750 \text{ maple}}{3{,}000 \text{ trees}} \times 100 = 25\%$$

$$\frac{300 \text{ beech}}{3{,}000 \text{ trees}} \times 100 = 10\%$$

$$\frac{1{,}200 \text{ birch}}{3{,}000 \text{ trees}} \times 100 = 40\%$$

$$\frac{150 \text{ hickory}}{3{,}000 \text{ trees}} \times 100 = 5\%$$

2 Now determine the size of the pie shapes that make up the chart. Do this by multiplying each percentage by 360°. Remember that a circle contains 360°.

$20\% \times 360° = 72°$ $25\% \times 360° = 90°$
$10\% \times 360° = 36°$ $40\% \times 360° = 144°$
$5\% \times 360° = 18°$

3 Then check that the sum of the percentages is 100 and the sum of the degrees is 360.

$20\% + 25\% + 10\% + 40\% + 5\% = 100\%$
$72° + 90° + 36° + 144° + 18° = 360°$

4 Use a compass to draw a circle and mark its center.

5 Then use a protractor to draw angles of 72°, 90°, 36°, 144°, and 18° in the circle.

6 Finally, label each part of the graph, and choose an appropriate title.

A Community of Wisconsin Hardwood Trees

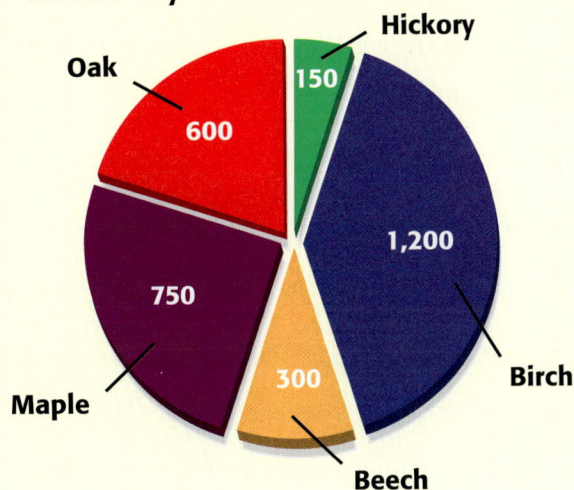

Line Graphs

Line graphs are most often used to demonstrate continuous change. For example, Mr. Smith's science class analyzed the population records for their hometown, Appleton, between 1900 and 2000. Examine the data at left.

Because the year and the population change, they are the *variables*. The population is determined by, or dependent on, the year. Therefore, the population is called the **dependent variable**, and the year is called the **independent variable.** Each set of data is called a **data pair.** To prepare a line graph, data pairs must first be organized in a table like the one at left.

Population of Appleton, 1900–2000

Year	Population
1900	1,800
1920	2,500
1940	3,200
1960	3,900
1980	4,600
2000	5,300

How to Make a Line Graph

1 Place the independent variable along the horizontal (x) axis. Place the dependent variable along the vertical (y) axis.

2 Label the x-axis "Year" and the y-axis "Population." Look at your largest and smallest values for the population. Determine a scale for the y-axis that will provide enough space to show these values. You must use the same scale for the entire length of the axis. Find an appropriate scale for the x-axis too.

3 Choose reasonable starting points for each axis.

4 Plot the data pairs as accurately as possible.

5 Choose a title that accurately represents the data.

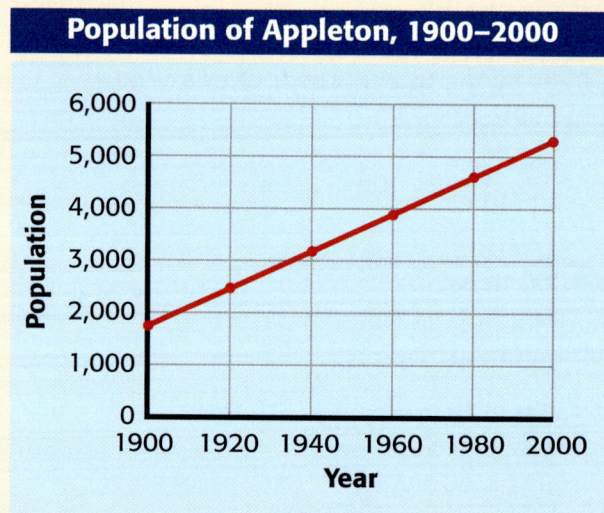

Population of Appleton, 1900–2000

How to Determine Slope

Slope is the ratio of the change in the y-axis to the change in the x-axis, or "rise over run."

1 Choose two points on the line graph. For example, the population of Appleton in 2000 was 5,300 people. Therefore, you can define point a as (2000, 5,300). In 1900, the population was 1,800 people. Define point b as (1900, 1,800).

2 Find the change in the y-axis.
(y at point a) − (y at point b)
5,300 people − 1,800 people = 3,500 people

3 Find the change in the x-axis.
(x at point a) − (x at point b)
2000 − 1900 = 100 years

4 Calculate the slope of the graph by dividing the change in y by the change in x.

$$slope = \frac{change\ in\ y}{change\ in\ x}$$

$$slope = \frac{3,500\ people}{100\ years}$$

slope = 35 people per year

In this example, the population in Appleton increased by a fixed amount each year. The graph of this data is a straight line. Therefore, the relationship is **linear.** When the graph of a set of data is not a straight line, the relationship is **nonlinear.**

Using Algebra to Determine Slope

The equation in step 4 may also be arranged to be:

$$y = kx$$

where *y* represents the change in the *y*-axis, *k* represents the slope, and *x* represents the change in the *x*-axis.

$$slope = \frac{change\ in\ y}{change\ in\ x}$$

$$k = \frac{y}{x}$$

$$k \times x = \frac{y \times x}{x}$$

$$kx = y$$

Bar Graphs

Bar graphs are used to demonstrate change that is not continuous. These graphs can be used to indicate trends when the data are taken over a long period of time. A meteorologist gathered the precipitation records at right for Hartford, Connecticut, for April 1–15, 1996, and used a bar graph to represent the data.

Precipitation in Hartford, Connecticut April 1–15, 1996

Date	Precipitation (cm)	Date	Precipitation (cm)
April 1	0.5	April 9	0.25
April 2	1.25	April 10	0.0
April 3	0.0	April 11	1.0
April 4	0.0	April 12	0.0
April 5	0.0	April 13	0.25
April 6	0.0	April 14	0.0
April 7	0.0	April 15	6.50
April 8	1.75		

How to Make a Bar Graph

1 Use an appropriate scale and a reasonable starting point for each axis.

2 Label the axes, and plot the data.

3 Choose a title that accurately represents the data.

Precipitation in Hartford, Connecticut, April 1–15, 1996

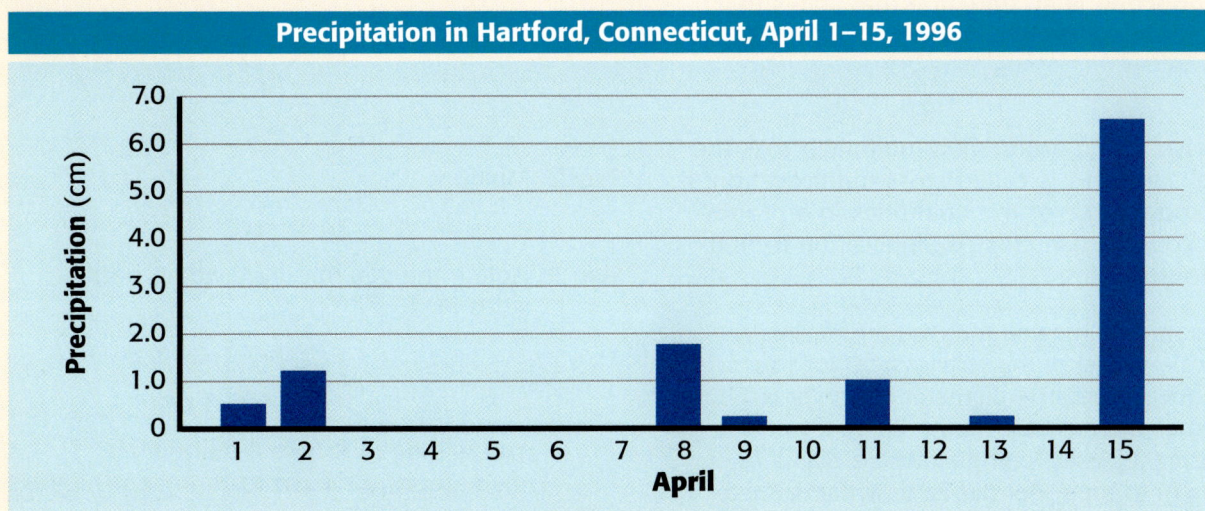

Math Refresher

Science requires an understanding of many math concepts. The following pages will help you review some important math skills.

Averages

An **average**, or **mean**, simplifies a list of numbers into a single number that *approximates* their value.

> **Example:** Find the average of the following set of numbers: 5, 4, 7, and 8.

Step 1: Find the sum.

$$5 + 4 + 7 + 8 = 24$$

Step 2: Divide the sum by the amount of numbers in your set. Because there are four numbers in this example, divide the sum by 4.

$$\frac{24}{4} = 6$$

The average, or mean, is **6.**

Ratios

A **ratio** is a comparison between numbers, and it is usually written as a fraction.

> **Example:** Find the ratio of thermometers to students if you have 36 thermometers and 48 students in your class.

Step 1: Make the ratio.

$$\frac{36 \text{ thermometers}}{48 \text{ students}}$$

Step 2: Reduce the fraction to its simplest form.

$$\frac{36}{48} = \frac{36 \div 12}{48 \div 12} = \frac{3}{4}$$

The ratio of thermometers to students is **3 to 4,** or $\frac{3}{4}$. The ratio may also be written in the form 3:4.

Proportions

A **proportion** is an equation that states that two ratios are equal.

$$\frac{3}{1} = \frac{12}{4}$$

To solve a proportion, first multiply across the equal sign. This is called cross-multiplication. If you know three of the quantities in a proportion, you can use cross-multiplication to find the fourth.

> **Example:** Imagine that you are making a scale model of the solar system for your science project. The diameter of Jupiter is 11.2 times the diameter of the Earth. If you are using a plastic-foam ball with a diameter of 2 cm to represent the Earth, what diameter does the ball representing Jupiter need to be?
>
> $$\frac{11.2}{1} = \frac{x}{2 \text{ cm}}$$

Step 1: Cross-multiply.

$$\frac{11.2}{1} \diagdown \diagup \frac{x}{2}$$

$$11.2 \times 2 = x \times 1$$

Step 2: Multiply.

$$22.4 = x \times 1$$

Step 3: Isolate the variable by dividing both sides by 1.

$$x = \frac{22.4}{1}$$
$$x = 22.4 \text{ cm}$$

You will need to use a ball with a diameter of **22.4 cm** to represent Jupiter.

Percentages

A **percentage** is a ratio of a given number to 100.

> **Example:** What is 85 percent of 40?

Step 1: Rewrite the percentage by moving the decimal point two places to the left.

$$.85$$

Step 2: Multiply the decimal by the number you are calculating the percentage of.

$$0.85 \times 40 = 34$$

85 percent of 40 is **34.**

Decimals

To **add** or **subtract decimals,** line up the digits vertically so that the decimal points line up. Then add or subtract the columns from right to left, carrying or borrowing numbers as necessary.

> **Example:** Add the following numbers: 3.1415 and 2.96.

Step 1: Line up the digits vertically so that the decimal points line up.

$$
\begin{array}{r}
3.1415 \\
+\ 2.96 \\
\hline
\end{array}
$$

Step 2: Add the columns from right to left, carrying when necessary.

$$
\begin{array}{r}
1\ 1 \\
3.1415 \\
+\ 2.96 \\
\hline
6.1015
\end{array}
$$

The sum is **6.1015.**

Fractions

Numbers tell you how many; **fractions** tell you *how much of a whole.*

> **Example:** Your class has 24 plants. Your teacher instructs you to put 5 in a shady spot. What fraction does this represent?

Step 1: Write a fraction with the total number of parts in the whole as the denominator.

$$\frac{?}{24}$$

Step 2: Write the number of parts of the whole being represented as the numerator.

$$\frac{5}{24}$$

$\frac{5}{24}$ of the plants will be in the shade.

Reducing Fractions

It is usually best to express a fraction in simplest form. This is called *reducing* a fraction.

> **Example:** Reduce the fraction $\frac{30}{45}$ to its simplest form.

Step 1: Find the largest whole number that will divide evenly into both the numerator and denominator. This number is called the greatest common factor (GCF).

factors of the numerator 30: 1, 2, 3, 5, 6, 10, **15,** 30

factors of the denominator 45: 1, 3, 5, 9, **15,** 45

Step 2: Divide both the numerator and the denominator by the GCF, which in this case is 15.

$$\frac{30}{45} = \frac{30 \div 15}{45 \div 15} = \frac{2}{3}$$

$\frac{30}{45}$ reduced to its simplest form is $\frac{2}{3}$.

Adding and Subtracting Fractions

To **add** or **subtract fractions** that have the **same denominator,** simply add or subtract the numerators.

Examples:

$$\frac{3}{5} + \frac{1}{5} = ? \quad \text{and} \quad \frac{3}{4} - \frac{1}{4} = ?$$

Step 1: Add or subtract the numerators.

$$\frac{3}{5} + \frac{1}{5} = \frac{4}{} \quad \text{and} \quad \frac{3}{4} - \frac{1}{4} = \frac{2}{}$$

Step 2: Write the sum or difference over the denominator.

$$\frac{3}{5} + \frac{1}{5} = \frac{4}{5} \quad \text{and} \quad \frac{3}{4} - \frac{1}{4} = \frac{2}{4}$$

Step 3: If necessary, reduce the fraction to its simplest form.

$$\frac{4}{5} \text{ cannot be reduced, and } \frac{2}{4} = \frac{1}{2}.$$

To **add** or **subtract fractions** that have **different denominators,** first find the least common denominator (LCD).

Examples:

$$\frac{1}{2} + \frac{1}{6} = ? \quad \text{and} \quad \frac{3}{4} - \frac{2}{3} = ?$$

Step 1: Write the equivalent fractions with a common denominator.

$$\frac{3}{6} + \frac{1}{6} = ? \quad \text{and} \quad \frac{9}{12} - \frac{8}{12} = ?$$

Step 2: Add or subtract.

$$\frac{3}{6} + \frac{1}{6} = \frac{4}{6} \quad \text{and} \quad \frac{9}{12} - \frac{8}{12} = \frac{1}{12}$$

Step 3: If necessary, reduce the fraction to its simplest form.

$$\frac{4}{6} = \frac{2}{3}, \text{ and } \frac{1}{12} \text{ cannot be reduced.}$$

Multiplying Fractions

To **multiply fractions,** multiply the numerators and the denominators together, and then reduce the fraction to its simplest form.

Example:

$$\frac{5}{9} \times \frac{7}{10} = ?$$

Step 1: Multiply the numerators and denominators.

$$\frac{5}{9} \times \frac{7}{10} = \frac{5 \times 7}{9 \times 10} = \frac{35}{90}$$

Step 2: Reduce.

$$\frac{35}{90} = \frac{35 \div 5}{90 \div 5} = \frac{7}{18}$$

Dividing Fractions

To **divide fractions,** first rewrite the divisor (the number you divide *by*) upside down. This is called the reciprocal of the divisor. Then you can multiply and reduce if necessary.

Example:

$$\frac{5}{8} \div \frac{3}{2} = ?$$

Step 1: Rewrite the divisor as its reciprocal.

$$\frac{3}{2} \rightarrow \frac{2}{3}$$

Step 2: Multiply.

$$\frac{5}{8} \times \frac{2}{3} = \frac{5 \times 2}{8 \times 3} = \frac{10}{24}$$

Step 3: Reduce.

$$\frac{10}{24} = \frac{10 \div 2}{24 \div 2} = \frac{5}{12}$$

Scientific Notation

Scientific notation is a short way of representing very large and very small numbers without writing all of the place-holding zeros.

Example: Write 653,000,000 in scientific notation.

Step 1: Write the number without the place-holding zeros.

653

Step 2: Place the decimal point after the first digit.

6.53

Step 3: Find the exponent by counting the number of places that you moved the decimal point.

6̫53000000

The decimal point was moved eight places to the left. Therefore, the exponent of 10 is positive 8. Remember, if the decimal point had moved to the right, the exponent would be negative.

Step 4: Write the number in scientific notation.

6.53×10^8

Area

Area is the number of square units needed to cover the surface of an object.

Formulas:
Area of a square = side × side
Area of a rectangle = length × width
Area of a triangle = $\frac{1}{2}$ × base × height

Examples: Find the areas.

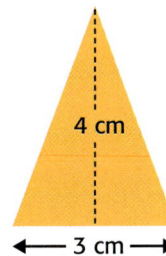

Triangle
Area = $\frac{1}{2}$ × base × height
Area = $\frac{1}{2}$ × 3 cm × 4 cm
Area = **6 cm²**

4 cm

3 cm

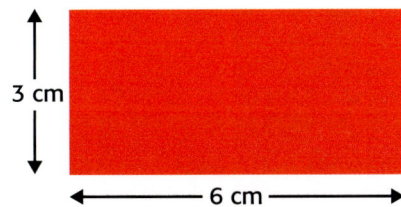

Rectangle
Area = length × width
Area = 6 cm × 3 cm
Area = **18 cm²**

3 cm

6 cm

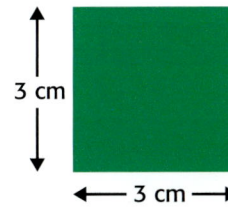

Square
Area = side × side
Area = 3 cm × 3 cm
Area = **9 cm²**

3 cm

3 cm

Volume

Volume is the amount of space something occupies.

Formulas:
Volume of a cube =
side × side × side

Volume of a prism =
area of base × height

Examples:
Find the volume
of the solids.

Cube
Volume = side × side × side
Volume = 4 cm × 4 cm × 4 cm
Volume = **64 cm³**

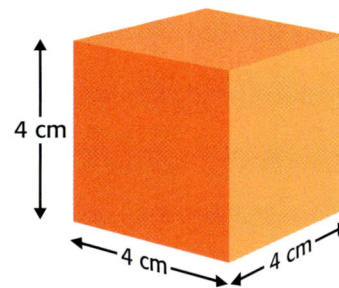

4 cm

4 cm

4 cm

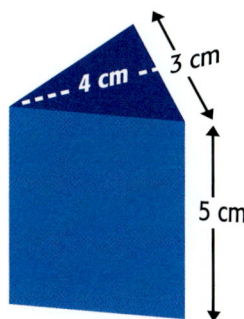

3 cm

4 cm

5 cm

Prism
Volume = area of base × height
Volume = (area of triangle) × height
Volume = $\left(\frac{1}{2} \times 3 \text{ cm} \times 4 \text{ cm} \right) \times 5 \text{ cm}$
Volume = 6 cm² × 5 cm
Volume = **30 cm³**

Periodic Table of the Elements

Each square on the table includes an element's name, chemical symbol, atomic number, and atomic mass.

Atomic number —————— 6

Chemical symbol —————— C

Element name —————— Carbon

Atomic mass —————— 12.0

The background color indicates the type of element. Carbon is a nonmetal.

The color of the chemical symbol indicates the physical state at room temperature. Carbon is a solid.

Background

Metals

Metalloids

Nonmetals

Chemical Symbol

Solid

Liquid

Gas

Period 1

1
H
Hydrogen
1.0

	Group 1	Group 2
Period 2	3 **Li** Lithium 6.9	4 **Be** Beryllium 9.0
Period 3	11 **Na** Sodium 23.0	12 **Mg** Magnesium 24.3

	Group 1	Group 2	Group 3	Group 4	Group 5	Group 6	Group 7	Group 8	Group 9
Period 4	19 **K** Potassium 39.1	20 **Ca** Calcium 40.1	21 **Sc** Scandium 45.0	22 **Ti** Titanium 47.9	23 **V** Vanadium 50.9	24 **Cr** Chromium 52.0	25 **Mn** Manganese 54.9	26 **Fe** Iron 55.8	27 **Co** Cobalt 58.9
Period 5	37 **Rb** Rubidium 85.5	38 **Sr** Strontium 87.6	39 **Y** Yttrium 88.9	40 **Zr** Zirconium 91.2	41 **Nb** Niobium 92.9	42 **Mo** Molybdenum 95.9	43 **Tc** Technetium (97.9)	44 **Ru** Ruthenium 101.1	45 **Rh** Rhodium 102.9
Period 6	55 **Cs** Cesium 132.9	56 **Ba** Barium 137.3	57 **La** Lanthanum 138.9	72 **Hf** Hafnium 178.5	73 **Ta** Tantalum 180.9	74 **W** Tungsten 183.8	75 **Re** Rhenium 186.2	76 **Os** Osmium 190.2	77 **Ir** Iridium 192.2
Period 7	87 **Fr** Francium (223.0)	88 **Ra** Radium (226.0)	89 **Ac** Actinium (227.0)	104 **Rf** Rutherfordium (261.1)	105 **Db** Dubnium (262.1)	106 **Sg** Seaborgium (263.1)	107 **Bh** Bohrium (262.1)	108 **Hs** Hassium (265)	109 **Mt** Meitnerium (266)

A row of elements is called a period.

A column of elements is called a group or family.

Lanthanides	58 **Ce** Cerium 140.1	59 **Pr** Praseodymium 140.9	60 **Nd** Neodymium 144.2	61 **Pm** Promethium (144.9)	62 **Sm** Samarium 150.4
Actinides	90 **Th** Thorium 232.0	91 **Pa** Protactinium 231.0	92 **U** Uranium 238.0	93 **Np** Neptunium (237.0)	94 **Pu** Plutonium 244.1

These elements are placed below the table to allow the table to be narrower.

This zigzag line reminds you where the metals, nonmetals, and metalloids are.

	Group 13	Group 14	Group 15	Group 16	Group 17	Group 18
						2 **He** Helium 4.0
	5 **B** Boron 10.8	6 **C** Carbon 12.0	7 **N** Nitrogen 14.0	8 **O** Oxygen 16.0	9 **F** Fluorine 19.0	10 **Ne** Neon 20.2

Group 10	Group 11	Group 12	13 **Al** Aluminum 27.0	14 **Si** Silicon 28.1	15 **P** Phosphorus 31.0	16 **S** Sulfur 32.1	17 **Cl** Chlorine 35.5	18 **Ar** Argon 39.9
28 **Ni** Nickel 58.7	29 **Cu** Copper 63.5	30 **Zn** Zinc 65.4	31 **Ga** Gallium 69.7	32 **Ge** Germanium 72.6	33 **As** Arsenic 74.9	34 **Se** Selenium 79.0	35 **Br** Bromine 79.9	36 **Kr** Krypton 83.8
46 **Pd** Palladium 106.4	47 **Ag** Silver 107.9	48 **Cd** Cadmium 112.4	49 **In** Indium 114.8	50 **Sn** Tin 118.7	51 **Sb** Antimony 121.8	52 **Te** Tellurium 127.6	53 **I** Iodine 126.9	54 **Xe** Xenon 131.3
78 **Pt** Platinum 195.1	79 **Au** Gold 197.0	80 **Hg** Mercury 200.6	81 **Tl** Thallium 204.4	82 **Pb** Lead 207.2	83 **Bi** Bismuth 209.0	84 **Po** Polonium (209.0)	85 **At** Astatine (210.0)	86 **Rn** Radon (222.0)
110 **Uun*** Ununnilium (271)	111 **Uuu*** Unununium (272)	112 **Uub*** Ununbium (277)		114 **Uuq*** Ununquadium (285)		116 **Uuh*** Ununhexium (289)		118 **Uuo*** Ununoctium (293)

A number in parenthesis is the mass number of the most stable form of that element.

63 **Eu** Europium 152.0	64 **Gd** Gadolinium 157.3	65 **Tb** Terbium 158.9	66 **Dy** Dysprosium 162.5	67 **Ho** Holmium 164.9	68 **Er** Erbium 167.3	69 **Tm** Thulium 168.9	70 **Yb** Ytterbium 173.0	71 **Lu** Lutetium 175.0
95 **Am** Americium (243.1)	96 **Cm** Curium (247.1)	97 **Bk** Berkelium (247.1)	98 **Cf** Californium (251.1)	99 **Es** Einsteinium (252.1)	100 **Fm** Fermium (257.1)	101 **Md** Mendelevium (258.1)	102 **No** Nobelium (259.1)	103 **Lr** Lawrencium (262.1)

*The official names and symbols for the elements greater than 109 will eventually be approved by a committee of scientists.

Glossary

A

alkaline-earth metals the elements in Group 2 of the periodic table; they are reactive metals but are less reactive than alkali metals; their atoms have two electrons in their outer level (113)

alloys solid solutions of metals or nonmetals dissolved in metals (65)

atom the smallest particle into which an element can be divided and still be the same substance (80)

atomic mass the weighted average of the masses of all the naturally occurring isotopes of an element (92)

atomic mass unit (amu) the SI unit used to express the masses of particles in atoms (88)

atomic number the number of protons in the nucleus of an atom (90)

B

boiling vaporization that occurs throughout a liquid (40)

boiling point the temperature at which a liquid boils and becomes a gas (40)

Boyle's law the law that states that for a fixed amount of gas at a constant temperature, the volume of a gas increases as its pressure decreases (35)

C

change of state the conversion of a substance from one physical form to another (38)

characteristic property a property of a substance that is always the same whether the sample observed is large or small (16)

Charles's law the law that states that for a fixed amount of gas at a constant pressure, the volume of a gas increases as its temperature increases (36)

chemical change a change that occurs when one or more substances are changed into entirely new substances with different properties; cannot be reversed using physical means (17)

chemical property a property of matter that describes a substance based on its ability to change into a new substance with different properties (15)

colloid (KAWL oyd) a mixture in which the particles are dispersed throughout but are not heavy enough to settle out (69)

compound a pure substance composed of two or more elements that are chemically combined (58)

concentration a measure of the amount of solute dissolved in a solvent (66)

condensation the change of state from a gas to a liquid (41)

condensation point the temperature at which a gas becomes a liquid (41)

D

density the amount of matter in a given space; mass per unit volume (12)

ductility (duhk TIL uh tee) the ability of a substance to be drawn or pulled into a wire (12)

E

electron clouds the regions inside an atom where electrons are likely to be found (86)

electrons the negatively charged particles found in all atoms; electrons are involved in the formation of chemical bonds (83)

element a pure substance that cannot be separated or broken down into simpler substances by physical or chemical means (54)

endothermic the term used to describe a physical or a chemical change in which energy is absorbed (39)

evaporation (ee VAP uh RAY shuhn) vaporization that occurs at the surface of a liquid below its boiling point (40)

exothermic the term used to describe a physical or a chemical change in which energy is released or removed (39)

F

freezing the change of state from a liquid to a solid (39)

freezing point the temperature at which a liquid changes into a solid (39)

G

gas the state in which matter changes in both shape and volume (33)

gravity a force of attraction between objects that is due to their masses (7)

group a column of elements on the periodic table (111)

H

halogens the elements in Group 17 of the periodic table; they are very reactive nonmetals, and their atoms have seven electrons in their outer level (118)

heterogeneous (HET uhr OH JEE nee uhs) **mixture** a combination of substances in which different components are easily observed (68)

homogeneous (HOH moh JEE nee uhs) **mixture** a combination of substances in which the appearance and properties are the same throughout (64)

hypothesis a possible explanation or answer to a question (148)

I

inertia the tendency of all objects to resist any change in motion (10)

isotopes atoms that have the same number of protons but have different numbers of neutrons (90)

L

liquid the state in which matter takes the shape of its container and has a definite volume (32)

M

malleability (MAL ee uh BIL uh tee) the ability of a substance to be pounded into thin sheets (12)

mass the amount of matter that something is made of (6)

mass number the sum of the protons and neutrons in an atom (91)

matter anything that has volume and mass (4)

melting the change of state from a solid to a liquid (39)

melting point the temperature at which a substance changes from a solid to a liquid (39)

meniscus (muh NIS kuhs) the curve at a liquid's surface by which you measure the volume of the liquid (5)

metalloids elements that have properties of both metals and nonmetals; sometimes referred to as semiconductors (57)

metals elements that are shiny and are good conductors of thermal energy and electric current; most metals are malleable and ductile (57)

mixture a combination of two or more substances that are not chemically combined (62)

model a representation of an object or system (83)

N

neutrons the particles of the nucleus that have no charge (88)

newton (N) the SI unit of force (9)

noble gases the unreactive elements in Group 18 of the periodic table; their atoms have eight electrons in their outer level (except for helium, which has two electrons) (118)

nonmetals elements that are dull and are poor conductors of thermal energy and electric current (57)

nucleus (NOO klee uhs) the tiny, extremely dense, positively charged region in the center of an atom; made up of protons and neutrons (85)

P

period a horizontal row of elements on the periodic table (111)

periodic having a regular, repeating pattern (104)

periodic law the law that states that the chemical and physical properties of elements are periodic functions of their atomic numbers (105)

physical change a change that affects one or more physical properties of a substance; many physical changes are easy to undo (16)

physical property a property of matter that can be observed or measured without changing the identity of the matter (11)

plasma the state of matter that does not have a definite shape or volume and whose particles have broken apart; plasma is composed of electrons and positively charged ions (37)

protons the positively charged particles of the nucleus; the number of protons in a nucleus is the atomic number that determines the identity of an element (88)

S

saturated solution a solution that contains all the solute it can hold at a given temperature (66)

scientific method a series of steps that scientists use to answer questions and solve problems (148)

solid the state in which matter has a definite shape and volume (31)

solubility (SAHL yoo BIL uh tee) the ability to dissolve in another substance; more specifically, the amount of solute needed to make a saturated solution using a given amount of solvent at a certain temperature (12, 66)

solute the substance that is dissolved to form a solution (64)

solution a mixture that appears to be a single substance but is composed of particles of two or more substances that are distributed evenly amongst each other (64)

solvent the substance in which a solute is dissolved to form a solution (64)

states of matter the physical forms in which a substance can exist; states include solid, liquid, gas, and plasma (30)

sublimation (SUHB luh MAY shuhn) the change of state from a solid directly into a gas (42)

surface tension the force acting on the particles at the surface of a liquid that causes the liquid to form spherical drops (33)

suspension a mixture in which particles of a material are dispersed throughout a liquid or gas but are large enough that they settle out (68)

T

theory a unifying explanation for a broad range of hypotheses and observations that have been supported by testing (80)

V

vaporization the change of state from a liquid to a gas; includes boiling and evaporation (40)

viscosity (vis KAHS uh tee) a liquid's resistance to flow (33)

volume the amount of space that something occupies or the amount of space that something contains (4)

W

weight a measure of the gravitational force exerted on an object, usually by the Earth (8)

Index

A **boldface** number refers to an illustration on that page.

A

acid precipitation, 19
air, 64, **68**
airplanes, 32
alcohols, 16
Alexander the Great, 81
alkali metals, **112,** 112–113
alkaline-earth metals, 113
alloys, 65
aluminum, 60–61, 87, **87,** 116
amber, 83
ammonia, 61
amorphous solids, 31, **31**
area
    calculation of, 157
argon, 91, 118, **119**
Aristotle, 81
atmospheric pressure, 41
atomic mass, 92, 104–105
atomic mass unit (amu), 88
atomic nucleus, 85, 88, **88**
atomic number, 90, 105
atomic theory, 80–86, **86**
atoms, 93
    defined, 80
    forces in, 93
    size of, 87, **88**
    structure of, **88,** 88–89
auroras, **37**
automobiles. *See* cars
averages, 154

B

baking soda, 16
bar graphs, 153, **153**
bleach, 16, **16**
blood, 68
Bohr, Niels, 86
boiling, 40, **40**
boiling point, 40–41, 43
    pressure and, 41
boron, **91,** 109
Boron group, 116
Boyle, Robert, 35
Boyle's law, 35, **35**
brake fluid, 32
bromine, **111, 118**
buckyballs, 127, **127**

C

calcium, 113
    in bones, 8
carbon
    bonding, 127
    compounds of, 117
    isotopes, 92
carbonated drinks, 60, **60**
carbon dioxide
    composition of, 58
    sublimation, 42, **42**
Carbon group, 116–117
carbon monoxide, 58
carbonic acid, **60**
careers,
    experimental physicist, 101
cars
    brake fluid, 32
    hydraulic brakes, 32
    maintenance of, 15, **15**
cathode-ray tubes (CRT), 82, **82**
Celsius scale, 146
centrifuges, 63
cerium sulfate, **67**
changes of state, 38–43
    boiling, 40
    chart, 42
    condensation, 41
    defined, 38, **38**
    endothermic, 39
    evaporation, 40
    exothermic, 39
    freezing, 39, **39**
    graph of, 43
    melting, 39, **39**
    sublimation, 42
    temperature and, 43
    vaporization, 40, **40**
characteristic properties, 16, 39–40, 55
charge. *See* electric charge
Charles's law, 36
chemical changes, 17–19
    clues to, **18,** 18–19
chemical properties, 15–17
chemical symbols, 110
chlorine
    properties of, **59,** 118
circle graph, **151**
cobalt, **55**
colloids, 69, **69,** 73
color, 11
comets, 100
compounds
    breakdown of, 60, **60**
    defined, 58
    formation of, 58, **58**
    in nature, 61, **61**
    properties of, 59

concentration, 66
concept mapping
    defined, 144
condensation, **38,** 41–42
condensation point, 41
conductivity
    electrical, 108–109
    thermal, 12, 108–109
control group, 149
controlled experiment, 149
conversion tables
    SI, 145
    temperature, 146
copper, 92, **92, 108,** 126
    density of, 13
copper carbonate, **18**
CRT (cathode-ray tube), 82, **82**
crystalline solids, 31, **31**

D

Dalton, John, **81,** 81–83
dark matter, 26
decimals, 155
Democritus, 80–81
density
    calculation of, 13
    defined, 12–13, 23
    identifying substances using, 13
    liquid layers and, 14, **14**
    of pennies, 13
diamond, **117**
dimension, 6
dissolution, 64, 67
distillation, 63
dry ice, 42, **42.** *See also* carbon dioxide
ductility, 12, 57, 108

E

effervescent tablets, **18**
electrical energy
    electric charge and, 82–83
electric charge, 82
electrolysis, 60
electromagnetic force, 93
electron clouds, 86, **86**
electrons
    in atomic models, 83, **83, 85,** 85–86, **86**
    discovery of, 83
    overview, 83, **88,** 89

Index

Credits

Abbreviations used: (t) top, (c) center, (b) bottom, (l) left, (r) right, (bkgd) background

ILLUSTRATIONS

All illustrations, unless noted below, by Holt, Rinehart and Winston.

Table of Contents Page iv(tl), Kristy Sprott; iv(br), Dan Stuckenschneider/Uhl Studios, Inc.

Scope and Sequence: T11, Paul DiMare, T13, Dan Stuckenschneider/Uhl Studios, Inc.

Chapter One Page 6(t), 7, Stephen Durke/Washington Artists; 10(l), Gary Locke/Suzanne Craig; 11, Blake Thornton/Rita Marie; 18, 19, 23, Marty Roper/Planet Rep; 25(lc); Terry Kovalcik; 27(tc), Daniels & Daniels.

Chapter Two Page 30(t), Mark Heine; 30(b), 31, 32, 33, 34, 35, 36(cl,cr), Stephen Durke/Washington Artists; 36(bl), Preface, Inc.; 38, David Schleinkofer/Mendola Ltd.; 40(t), Marty Roper/Planet Rep; 40(b), Mark Heine; 43, David Schleinkofer/Mendola Ltd. and Preface, Inc.; 46(t), Stephen Durke/Washington Artists; 46(b), Preface, Inc.; 47, Marty Roper/Planet Rep; 49(cr), Preface, Inc.

Chapter Three Page 54, Marty Roper/Planet Rep; 56(b), Preface, Inc.; 60, Blake Thornton/Rita Marie; 67, Preface, Inc.

Chapter Four Page 81(c), Preface, Inc.; 82, Mark Heine; 83, Stephen Durke/Washington Artists; 84(c), Mark Heine; 84(b), Preface, Inc.; 85(t), Stephen Durke/Washington Artists; 85(br), Preface, Inc.; 86, 88, 89(t,b), Stephen Durke/Washington Artists; 89(cr), Terry Kovalcik; 90, 91(b), 93, 94(br), Stephen Durke/Washington Artists; 97, Terry Kovalcik; 98, Mark Heine; 99(r), Stephen Durke/Washington Artists.

Chapter Five Page 104, Michael Jaroszko/American Artists; 106, 107, Kristy Sprott; 108(tr), 109, Stephen Durke/Washington Artists; 111, 112(bc), 113(bl), 114(t), 115(t), 116(tc, b), Preface, Inc.; 116(l), Gary Locke/Suzanne Craig; 117, 118, 119(lc), Preface, Inc.; 123, Gary Locke/Suzanne Craig; 125(tr), Preface, Inc.; 125(bl), Keith Locke/Suzanne Craig; 125(br), Annie Bissett; 126-127(l), Dan Stuckenschneider/Uhl Studios Inc.

LabBook Page 140, Blake Thornton/Rita Marie.

Appendix Page 146(t), Terry Guyer; 150(b), Mark Mille/Sharon Langley Artist Rep.; 151, 152, 153, Preface, Inc.; 158, 159, Kristy Sprott.

PHOTOGRAPHY

Front Cover John Higginson/Stone

Table of Contents iv(bl), Victoria Smith/HRW Photo; v(l), Runk/Schoenberg/Grant Heilman Photography, Inc., Richard Megna/Fundemental Photographs; v(r), Joseph Drivas/The Image Bank; vi(tl), Dr. Harold E. Edgerton/©The Harold E. Edgerton 1992 Trust/courtesy Palm Press, Inc.; vi(cl), Stuart Westmoreland/Stone; vi(br), Scott Van Osdol/HRW Photo; vii(tr), Brett H. Froomer/The Image Bank; vii(cr), Kennan Ward Photography; vii(br), Scott Van Osdol/HRW Photo

Scope and Sequence: T8(l), Lee F. Snyder/Photo Researchers, Inc.; T8(r), Stephen Dalton/Photo Researchers, Inc.; T10, E. R. Degginger/Color-Pic, Inc.; T12(l), Rob Matheson/The Stock Market

Master Materials List: T25(cl, br), Image ©2001 PhotoDisc; T26(tr), Image ©2001 PhotoDisc; T26(c), John Langford/HRW Photo; T27(bl, cr), Image ©2001 PhotoDisc

Feature Borders Unless otherwise noted below, all images ©2001 PhotoDisc/HRW: "Across the Sciences" Pages 26, 100, all images by HRW; "Careers" 101, sand bkgd and saturn, Corbis Images, DNA, Morgan Cain & Associates, scuba gear, ©1997 Radlund & Associates for Artville; "Eureka" 51, ©2001 PhotoDisc/HRW; "Health Watch" 27, dumbbell, Sam Dudgeon/HRW Photo, aloe vera and EKG, Victoria Smith/ HRW Photo, basketball, ©1997 Radlund & Associates for Artville, shoes and Bubbles, Greg Geisler; "Science Fiction" 77, saucers, Ian Christopher/Greg Geisler, book, HRW, bkgd, Stock Illustration Source; "Science, Technology, and Society" 50, 76, 126, robot, Greg Geisler; "Weird Science" 127, mite, David Burder/Stone, atom balls, J/B Woolsey Associates, walking stick and turtle, EclectiCollection.

Chapter One pp. 2-3 Ken Reid/FPG International; 3 HRW Photo; 4(b), NASA, Media Services Corp.; 8(all), John Morrison/Morrison Photography; 9(t), Michelle Bridwell/HRW Photo; 11-12(all), John Morrison/Morrison Photography; 13, Neal Nishler/Stone; 14(all), John Morrison/Morrison Photography; 15, Rob Boudreau/Stone; 16(t), Brett H. Froomer/The Image Bank; 16(cl,cr), John Morrison/Morrison Photography; 17, Lance Schriner/HRW Photo; 18(tl,tr), John Morrison/Morrison Photography; 18(bl), Joseph Drivas/The Image Bank; 18(br), SuperStock; 22(r), Michelle Bridwell/HRW Photo; 25, Lance Schriner/HRW Photo; 26, David Malin/©Anglo-Australian Observatory/Royal Observatory, Edinburgh.

Chapter Two pp. 28-29 Phil Degginger/Stone; 29 HRW Photo; 31(t), Gilbert J. Charbonneau; 33(t), Dr. Harold E. Edgerton/©The Harold E. Edgerton 1992 Trust/courtesy Palm Press, Inc.; 37(b), Pekka Parviainen/Science Photo Library/ Photo Researchers, Inc.; 38, Union Pacific Museum Collection; 39(t), Richard Megna/Fundamental Photographs; 44 Victoria Smith/HRW Photo; 45 Victoria Smith/HRW Photo; 48(t), Myrleen Ferguson/PhotoEdit; 49, Charles D. Winters/ Photo Researchers, Inc.; 50(l), Kennan Ward Photography; 50(r), Dr. Jean-Claude Diels/University of New Mexico; 51, Union Pacific Museum Collection.

Chapter Three pp. 52-53 Scott Van Osdol/HRW Photo; 53 HRW Photo; 54(l), Jonathan Blair/Woodfin Camp & Associates; 54(r), David R. Frazier Photolibrary; 55(b), Russ Lappa/Photo Researchers, Inc.; 55(t,c), Charles D. Winters/Photo Researchers, Inc.; 56(t,l), Walter Chandoha; 56(r), Zack Burris; 56(b), Yann Arthus-Bertrand/Corbis; 57(tc), HRW Victoria Smith/HRW Photo Victoria Smith; 57(cl), Dr. E.R. Degginger/Color-Pic, Inc.; 57(cr), Runk/Schoenberg/Grant Heilman Photography Inc.; 57(br), Joyce Photographics/Photo Researchers, Inc.; 57(l), Russ Lappa/Photo Researchers, Inc.; 57(bc), Charles D. Winters/Photo Researchers, Inc.; 58(c), Runk/Schoenberger/Grant Heilman; 59(l), Runk/ Shoenberger/Grant Heilman Photography; 59(c), Richard Megna/ Fundamental Photographs; 60, Runk/Schoenberg/Grant Heilman Photography Inc.; 60, Richard Megna/Fundamental Photographs; 61(t), Kapriellan/Photo Researchers, Inc.; 62(l), Victoria Smith/HRW Photo; 63(t), Charles D. Winters/ Timeframe Photography; 63(c), Charles Winters/Photo Researchers, Inc.; 63(cr), Klaus Guldbrandsen/Science Photo Library/Photo Researchers, Inc.; 65(b), Richard Haynes/HRW Photo; 65(c), Image ©2001 PhotoDisc, Inc.; 66(c), Dr. E.R. Degginger/Color-Pic, Inc.; 68(c), Michelle Bridwell/HRW Photo; 69(t), Lance Schriner/HRW Photo; 69(b), Dr. E.R. Degginger/Color-Pic, Inc.; 72(b), Victoria Smith/HRW Photo; 72(t), David R. Frazier Photolibrary; 74(t), Yann Arthus-Bertrand/Corbis; 75(t), Richard Megna/Fundamental Photographs; 76(l), Anthony Bannister/Photo Researchers, Inc.; 76(r), Richard Steedman/The Stock Market.

Chapter Four pp. 78-79 P. Loiez Cern/Science Photo Library/Photo Researchers, Inc.; 79 HRW Photo; 80(t), Dr. Mitsuo Ohtsuki/Photo Researchers, Inc.; 80(b), Nawrocki Stock Photography; 81, Corbis-Bettmann; 84, Stephen Maclone; 85(l), John Zoiner; 85(r), Mavourina Hay/HRW Photo; 87(t), Lawrence Berkeley National Lab; 88, Richard Megna/Fundamental Photographs; 91, Charles D. Winters/Timeframe Photography Inc./Photo Researchers, Inc.; 92, Superstock; 96, Nawrocki Stock Photography; 99, Fermilab National Laboratory; 100, NASA Ames ; 101(t), Stephen Maclone; 101(b), Fermi National Lab/ Corbis.

Chapter Five pp. 102-103 Jeff Goldberg/Esto; 103 HRW Photo; 109(tr), Richard Megna/Fundamental Photographs; 109(bl), Russ Lappa/Photo Researchers, Inc.; 109(br), Lester V. Bergman/Corbis-Bettman; 111(l,c), Dr. E.R. Degginger/Color-Pic, Inc.; 111(r), Richard Megna/Fundamental Photography; 112, Charles D. Winters/ Photo Researchers, Inc.; 113(l,c,r), Richard Megna/Fundamental Photographs; 114(tr), Petersen/Custom Medical Stock Photo; 114(tc), Victoria Smith/HRW Photo; 115, David Parker/ Science Photo Library/Photo Researchers, Inc.; 117(t), Phillip Hayson/Photo Researchers; 118(t,c), Richard Megna/ Fundamental Photographs; 118(b), Dr. E.R. Degginger/Color-Pic, Inc.; 119(t), Michael Dalton/Fundamental Photographers; 119(b)NASA; 120 John Langford/ HRW Photo; 121 Victoria Smith/HRW Photo; 122(b), 124, Richard Megna/ Fundamental Photographs; 127, Image copyright 2001 PhotoDisc, Inc.

LabBook "LabBook Header": "L," Corbis Images, "a," Letraset Phototone, "b" and "B," HRW, "o" and "k," images ©2001 PhotoDisc/HRW; 129(c), Michelle Bridwell/ HRW Photo; 129(br), Image copyright © 2001 PhotoDisc, Inc.; 130(cl), Victoria Smith/HRW Photo; 130(bl), Stephanie Morris/HRW Photo; 131(tr), Jana Birchum/ HRW Photo; 131(b), Peter Van Steen/HRW Photo; 135, NASA, 142, Gareth Trevor/ Stone.

Appendix Page 147(t), Peter Van Steen/HRW Photo.

Sam Dudgeon/HRW Photo Pages viii-1, 5(bl); 21; 31(bl); 34(r); 36; 48(b); 57(tl,tr,c); 59(r); 62(t); 63(cl); 65(b); 66(bl); 68(t,b); 71; 73; 74(b); 75(br); 82-83; 94; 105; 108; 109(c,tl); 110; 113(b); 114(tl,b); 117(tl, br); 122(t); 128; 129(bc); 130(br,t); 131(tl); 132; 134; 136-137; 138-139; 141; 143; 147.

John Langford/HRW Photo Pages 4(t); 5(t); 6-7; 9(b); 10; 22(l); 24(all); 34(l); 63(bl,bc,br); 67(all); 129(tl); 135(t).

Scott Van Osdol/HRW Photo Pages 31(br); 32(all); 33(b); 37(t); 39(b); 41-42; 133.

Self-Check Answers

Chapter 1—The Properties of Matter

Page 9: approximately 30 N

Chapter 2—States of Matter

Page 34: The pressure would increase.

Page 40: endothermic

Chapter 3—Elements, Compounds, and Mixtures

Page 59: No, the properties of pure water are the same no matter what its source is.

Page 65: Copper and silver are solutes. Gold is the solvent.

Chapter 4—Introduction to Atoms

Page 85: The particles Thomson discovered had negative charges. Because an atom has no charge, it must contain positively charged particles to cancel the negative charges.

Chapter 5—The Periodic Table

Page 114: It is easier for atoms of alkali metals to lose one electron than for atoms of alkaline-earth metals to lose two electrons. Therefore, alkali metals are more reactive than alkaline-earth metals.